Heinemann
Physics
for *CSEC*®

Norman Lambert ■ **Natasha Lewis dos Santos**
Tricia A. Samuel

CSEC® is a registered trade mark of the **Caribbean
Examinations Council (*CXC*). Heinemann Physics for
CSEC®** is an independent publication and has not been
authorised, sponsored, or otherwise approved by *CXC*.

Orders: please contact Hachette UK Distribution, Hely Hutchinson Centre, Milton Road,
Didcot, Oxfordshire, OX11 7HH. Telephone: +44 (0)1235 827827. Email education@hachette.co.uk
Lines are open from 9 a.m. to 5 p.m., Monday to Friday.
You can also order through our website: www.hoddereducation.com

Heinemann is the registered trademark of
Pearson Education Limited

© Norman Lambert, Natasha Lewis Dos Santos, Tricia A. Samuel 2000
First published by Heinemann Educational Publishers in 2000

Published from 2015 by Hodder Education, An Hachette UK Company, Carmelite
House, 50 Victoria Embankment, London EC4Y 0DZ

British Library Cataloguing in Publication Data
A catalogue record for this book is available from the British Library.

Design by   Jackie Hill
Cover by   Gabrielle Kern
Cover photographs by Piero Guerrini (pilots), PhotoDisc (windsurfer and wind turbines),
Powerstock/Zefa (prism).
Typesetting and illustrations by David Gregson Associates, Beccles, Suffolk

ISBN: 978 0 435975 33 3

Printed and bound by CPI Group (UK) Ltd, Croydon, CR0 4YY

2022
IMP 7

Acknowledgements

The publishers would like to thank the following people and organizations for supplying
photographs for this book:

Advertising Archives 4.9; B & C Alexander 9.1; BNFL Magnox Generation 28.12; Corbis/
Carl Purcell 5.38; Corbis/Danny Lehman 20.11; Corbis/George Hall 9.21; Deutsches Museum 10.1;
Douglas Hall 4.22; EFDA-JET 29.6; Peter Gould 1.4,1.5, 2.8, 2.9, 4.25,15.13,16.5,18.10;
Fisher Co 2.7; Microscopix Photo Library 4.8; Milepost 921/2 11.2; Ontario Science Centre/Craig
Hyde Parker 20.10; PhotoDisc 5.8, 7. 7; PowerStock/Craig J McCormick 9.8; Science Museum 1.1;
Science Photo Library/Adam Hart-Davis 12.4; Science Photo Library/Conor Caffrey 8.2;
Science Photo Library/NASA 5.29; Science Photo Library/Prof. Harold Edgerton 6.13;
Science Photo Library/US Dept of Energy/RNFI/Craig Miller Productions 8.9; Science Photo
Library/US Library of Congress/E Willard Spurr 29.1; Science Photo Library 9.18, 28.11;
Sporting Pictures (UK) Ltd 6.10; Telegraph Colour Library 4.18; Thermotex 10.6 (and a special
thanks to BWIA West Indies Airways for allowing us to photograph their pilots in Trinidad).

Contents

Measurements and units

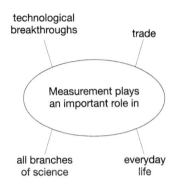

technological
breakthroughs   trade

Measurement plays
an important role in

all branches   everyday
of science   life

Figure 1.1 *The water clock was
an early instrument for measuring
time.*

There are two parts to physics: observations about the world
and the theories to explain them. To be able to test theories we
need accurate measurement. The methods of measurement
explained here are used throughout physics.

Measurements, accurately taken and appropriately recorded,
help us to make sense of natural phenomena and to understand
the physical world better.

1.1 Metrology

Metrology – the science of measurement – has changed over
time. At one time there were so many standards and units in
use that the communication and interpretation of experimental
results caused much confusion and difficulty.

1.2 The International System (SI)
Fundamental quantities and their base units

Measurements involve comparing unknown quantities with
known standard units.

Since 1960 the system of units used in the scientific community
is the International System (SI). The SI is based on the seven
fundamental quantities listed in Table 1.1. A **fundamental
quantity** is a quantity from which others may be derived, e.g.
length and time are fundamental quantities from which
quantities such as velocity are derived.

Table 1.1	SI units			
Fundamental quantity			**Base SI unit**	
Name	Symbol		Name	Symbol
mass	m		kilogram	kg
length	l		metre	m
time	t		second	s
current	I		ampere	A
temperature*	T		kelvin	K
amount of substance	n		mole	mol
luminous intensity	I_ν		candela	cd

*Although temperature (T) [unit: degree Celsius (°C)] is non-SI, it is widely
used in everyday situations.

Table 1.2	Common prefixes used with SI					
Multiple	**Prefix**	**Symbol**	**Sub-multiple**	**Prefix**	**Symbol**
10^3 (1000)	kilo	k	10^{-3} (1/1000)	milli	m
10^6 (1 000 000)	mega	M	10^{-6} ( )	micro	μ
10^9 ( )	giga	G	10^{-9} ( )	nano	n
10^{12} ( )	tera	T	10^{-12} ( )	pico	p

Thinking it through

Copy Table 1.2 and complete it by filling in the missing information.

A **unit** is a specified measure of a physical quantity.

When recording measurements you should give both the numerical value and the unit. For example:

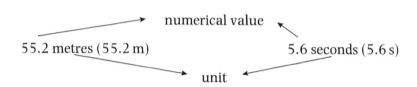

55.2 metres (55.2 m)   numerical value   5.6 seconds (5.6 s)

unit

The SI is based on the decimal system of numbers. Greek prefixes may be used to indicate fractions (sub-multiples) and multiples of 10. Table 1.2 gives some of these prefixes and their values. Note that capital letters are used for most multiples and lower case letters are used for sub-multiples.

Thinking it through

What problems are likely to arise if measured values of physical quantities are expressed without units?

Derived quantities and their units

Fundamental quantities and their base units are combined to produce derived quantities and their units.

For example:

1 The derived quantity velocity is obtained by combining the SI quantities displacement (length) and time. Velocity is defined as displacement per unit time. It has the unit: metres per second (m/s or $m\,s^{-1}$).

Note

A few physical quantities have *no* units. These are quantities which are the ratios of like quantities. Some examples are refractive index, relative atomic mass and relative density.

2 Charge is the product of current (in amperes) and time (in seconds). Charge has the unit: ampere.second (A s) or coulomb (C).

3 Density is mass (kg) per unit volume (m^3). Density has the unit kg/m^3 or $kg\,m^{-3}$.

Other examples of derived quantities and their units are given in Table 1.3 opposite.

Table 1.3	Some derived quantities and units		
Derived quantity (and symbol)	**Defining equation**	**Derived unit (SI)**	**Derived unit in use**
force (F)	force = mass × acceleration	$kg \times m\,s^{-2}$ $[kg\,m\,s^{-2}]$	newton (N) $1\,N = 1\,kg\,m\,s^{-2}$
work done (W)	work = force × displacement	$kg\,m\,s^{-2} \times m$ $[kg\,m^2\,s^{-2}]$	joule (J) $1\,J = 1\,N\,m$
potential difference (V)	p.d. = energy transferred or work done per unit charge	$kg\,m\,s^{-2} \times A^{-1} \times s^{-1}$	volt (V) $1\,V = 1\,J\,C^{-1}$

1.3 Scales

Explain what is meant by the term a 'non-linear scale.'

A scale is a set of marks (graduations) at intervals on a measuring instrument. The smaller the value of the subdivisions on a scale, the greater the precision of the instrument (see p. 4). Scales are linear if the marks are evenly spaced and non-linear if the marks are not evenly spaced. On a linear scale, such as in Figure 1.2, equal changes in the value of the physical quantity being measured give rise to equal changes on the scale of the measuring instrument. Figure 1.3 shows a non-linear scale.

Figure 1.2 *A linear scale.*

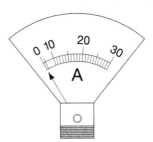

Figure 1.3 *A non-linear scale.*

Calibrating a scale

Calibration is the process of setting or checking the graduations on the scale of a measuring instrument by comparing it with a standard or some universally accepted reference.

Things to do

You have a measuring cylinder and a supply of tap water. Describe how you would calibrate a large boiling tube to measure volumes of liquids to the nearest $2\,cm^3$.

How to calibrate a scale

1 Choose two known values at opposite ends of the range of the instrument (for example, the boiling point and freezing point of water on a thermometer).
2 Mark the points on the scale.
3 Divide the scale into equally spaced intervals, using graduation marks.

Figure 1.4 *An analogue instrument.*

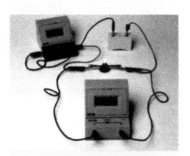

Figure 1.5 *Digital instrument.*

Analogue and digital scales

Signals may be varied continuously or simply by switching on and off.

An analogue device is one which has an output signal that varies continuously with the input. An analogue scale is one in which the pointer or indicator moves continuously over the calibrated scale (see Figure 1.4).

When using an analogue scale:

1 Ensure that you read the scale accurately.

2 Use the correct multiplication factors.

3 Take care when estimating readings which fall between two adjacent marks.

The reading displayed or the information sent by a digital system does not vary continuously. It jumps from one number to the next. When using a digital scale:

1 Note that the last digit usually fluctuates. This means that there is an uncertainty in this digit.

2 Ensure that the instrument is set to the appropriate range for the experiment being undertaken.

3 A convenient way of getting to the appropriate range is to start with the highest range and work down to the most suitable.

1.4 Accuracy, precision, sensitivity and range

Progress in science depends heavily on our ability to measure physical quantities with greater accuracy and precision. Often, we use the words accuracy and precision as if they mean the same thing, but they have different meanings.

Accuracy means how close the experimental value is to the true value. Accuracy is improved if errors due to the measuring instruments and experimental procedures are identified and their effects reduced.

Precision refers to how small an uncertainty the measuring instrument will give. For example, a thermometer marked at every degree will give a more precise reading than one marked at every five degrees. For example, 10.5° is not a useful reading if there has been inaccuracy in the reading and the true value is 15°.

Sensitivity measures the response of an instrument to the smallest change in input. The greater the response of an

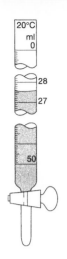

Figure 1.6 *This burette reading is between 27.5 cm³ and 27.6 cm³, say 27.55 cm³. There is an uncertainty in the last digit of the reading 27.55 cm³.*

instrument to a small change in, for example, voltage or current, the greater its sensitivity. In general, sensitive readings are more precise.

Range is the size of the interval between the maximum and the minimum quantities that a measuring instrument can measure. Some instruments have more than one range.

You should select measuring instruments for particular experiments or tasks only after carefully considering their accuracy, precision, sensitivity and range.

1.5 The numerical part of a physical measurement

Numbers may be exact or inexact. Exact numbers are determined by counting or by definition. For example, if Vibert has ten coins, then ten here is an exact number.

On the other hand, numbers obtained from measurements are inexact (uncertain). These uncertainties may be due to:

- the limitations of the measuring instrument;
- variations in how the measurements are made;
- the conditions of the experiment.

Uncertainties in measurements

All measurements are inaccurate to some degree. For the burette of Figure 1.6, on which the smallest scale divisions are 0.1 cm³, an estimate is made when reading a volume of 27.55 cm³, say.

The smallest scale divisions on a metre rule are 1 mm (0.1 cm) apart. To measure a length that lies between 30.3 cm and 30.4 cm requires an estimation.

The recorded quantity 27.6 mm suggests that the actual reading lies somewhere between 27.5 mm and 27.7 mm but that there is some doubt as to whether it is 27.60 mm.

Using standard form

It is awkward to work with numbers such as 0.000 057 6 and 710 000 000. They can be more elegantly written in a simplified form using exponents or powers of 10, as 5.76×10^{-5} and 7.1×10^8, respectively. (The exponents -5 and 8 are not always the number of noughts but the number of times the decimal point is moved: 10^{-5} indicates dividing by 10, 5 times, 10^8 indicates multiplying by 10, 8 times.) The use of standard form makes it easier to compare similar quantities.

Note

In standard form (scientific notation) numbers are expressed in the form $B \times 10^n$ (where $1 \leqslant B < 10$) and n is a whole number or integer.

When recording measurements in standard form, write the numerical part of the measurement in such a way that there is only one figure before the decimal point, i.e. the measurement is stated as a number between 1 and 10 multiplied by a power of 10. Remember that all quantities should be in the appropriate SI unit, whether basic or derived.

Note

When multiplying numbers in standard form, e.g. $5.76 \times 10^{-5} \times 7.1 \times 10^8$, you can rearrange them so that the prefactors 5.76 and 7.1 can be multiplied separately. To multiply powers of 10, add the exponents. Thus the final sum is

$5.76 \times 7.1 \times 10^{-5} \times 10^8$

$= 40.896 \times 10^{(-5+8)}$

$= 40.9 \times 10^3$

$= 4.09 \times 10^4$

Problems

1 Write the following numbers in standard form:

  a 15 789   b 149.5   c 0.0059   d 0.0702

2 Use standard form in each of the following operations:

  a 24 623 ÷ 324   b 147 × 25 × 0.123

Note

It is important that the number of significant figures in the final answer of any arithmetic operation reflects the uncertainty in the original measurements.

Using significant figures

Because of uncertainty, the number used to express a given measurement should have a limited number of digits or significant figures. When using significant figures the following rules of thumb apply:

- Zeros used to position the decimal point are not significant. For example, the number 0.0056 has only two significant figures since the zeros are used to position the decimal point.
- When multiplying or dividing, give the answer to the same number of significant figures as the measured quantity with the *least* number of significant figures. This is because the least precise measurement determines the precision of the final answer.

Example

Length of a plot $= 52.2$ m, width of plot $= 12.3$ m, calculated area $= 642.06$ m^2. Answer should be stated as 642 m^2, i.e. to three significant figures.

- In addition or subtraction, the uncertainty in the result is the same as that in the quantity which is least precise.

Note

Although the *actual* value of a physical quantity cannot be found in practice, we assume that such a value exists.

$40.55\,\text{N} - 26.2\,\text{N} = 14.4\,\text{N}$ and $31.3\,\text{s} + 21.57\,\text{s} = 52.9\,\text{s}$.

The results of the addition and the subtraction are both given to three significant figures since the less precise quantity in each case has only three significant figures.

1 Evaluate the following, expressing your answer to the correct number of significant figures:

   a 2079.65×1.13 **b** $13\,579 \div 2468$

   c $4596.35 + 205.7$ **d** $468.20 - 1.3756$

2 Express each of the above answers in standard form.

> **Note**
>
> Random errors may be reduced by controlling environmental fluctuations. Random errors reduce precision.

> **Note**
>
> Systematic errors tend to be in one direction – either too high or too low.

> **Note**
>
> Every measuring instrument is imperfect. It is the responsibility of the experimenter to reduce these imperfections.

1.6 Errors and how to treat them

The error of observation is the difference between the observed or measured value and the true value. Errors in measurement may be of two types – random and systematic.

Random errors

Random errors are caused by experimental factors (variables). Let us consider the following situation. The 'true' value for the time for 20 swings of a pendulum is 30.6 seconds, but in repeated determinations of this interval an experimenter obtained the results: 30.4 s, 30.5 s, 30.7 s, 30.5 s, 30.8 s, 30.6 s and 30.7 s. All but one of these measurements are in error. Observe that (1) the measurements fluctuate about the true value and (2), in this special case, the mean or average of the experimental values is the true value.

Random errors mean that readings cannot be repeated exactly. Random errors reduce precision. Their effects may be reduced by taking several observations of the same event and finding the mean.

Systematic errors

Systematic errors cause readings to be *consistently* too high or too low when compared with the true value. They are not reduced or eliminated by repeating the measurement. Systematic errors affect the closeness of the result of a measurement to the true value. They therefore reduce accuracy.

The ammeters in Figure 1.7 show the position of the pointer when no current is flowing. These meters have a **zero error**. There will be a systematic error in every measurement made with these meters.

Sources of errors

Errors arise from:

- lack of precision or a lack of uniformity in the measuring instrument;
- other features of the measuring instrument. For example,

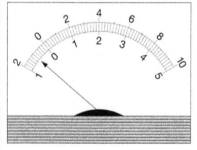

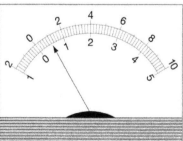

Figure 1.7 *Each of these ammeters has a zero error – it shows a reading when it should show zero.*

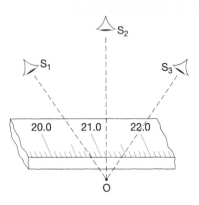

Figure 1.8 *Parallax.*

there is a systematic error in all measurements made with vernier callipers that have a zero error;

- experimental conditions. For example, there may be fluctuations in temperature or pressure and draughts;
- the experimenter. For example, each experimenter has a reaction time which affects her or his determination of time intervals.

Parallax errors

Parallax is the apparent motion of objects in front of the eye when the eye position changes. It is a common source of error in scale reading.

Consider the transparent scale placed over point O in Figure 1.8. For student S_2 viewing at right angles to the scale the scale reads 21.1 cm. Student S_1 reads the value as 20.8 cm while student S_3 reads it as 21.4 cm.

Reducing experimental errors

Some uncertainties/errors in measurement can be minimized, for example, by:

- taking several observations of the same event and finding the mean;
- measuring a large number of events and calculating the value for one. The time for one swing of a pendulum may be obtained by timing 20 swings, say, and dividing the time by 20;
- using an instrument of appropriate range and sensitivity. For example, if voltmeters of ranges 0–3 V and 0–12 V are available, use the 0–3 V instrument to measure a potential difference of the order of 1.5 V;
- taking readings in such a manner as to reduce parallax errors.

Improving your computational skills

It is sometimes necessary to convert sub-multiples (e.g. mm) and multiples (e.g. km) to base SI units (e.g. m).

A **conversion factor** is the factor by which a quantity expressed in one set of units must be multiplied in order to express that quantity in different units. For example, when converting m to mm the conversion factor is 10^3.

What is the conversion factor for converting $km\,h^{-1}$ to $m\,s^{-1}$?

$$1\,m = 1000\,mm = 10^3\,mm$$

$$1\,m^2 = 1 \times 10^6\,mm^2 \quad (10^3\,mm \times 10^3\,mm)$$

$$1\,m^3 = 1 \times 10^9\,mm^3 \quad (10^3 \times 10^3 \times 10^3)\,mm^3$$

So:

to convert m^2 to mm^2 $\times 10^6$

to convert mm^2 to m^2 $\div 10^6$ i.e. $\times 10^{-6}$

to convert m^3 to mm^3 $\times 10^9$

to convert mm^3 to m^3 $\div 10^9$ i.e. $\times 10^{-9}$

When converting from a small unit (e.g. mm) to a larger one (i.e. m), *divide* by the conversion factor (e.g. 10^3). This is because there are many mm in 1 m: you expect the number of m to be smaller than the number of mm. To convert from m to mm, *multiply* by 10^3.

> **Note**
>
> If the units do not combine algebraically in a conversion, then the conversion has not been carried out correctly.

Problems

1 Show that

     a $2000\,mm^2 = 0.002\,m^2 = 2 \times 10^{-3}\,m^2$
     b $400\,km^3 = 4 \times 10^{11}\,m^3$
     c $7.92\,g\,cm^{-3} = 7920\,kg\,m^{-3}$
     d $450\,nm = 4.5 \times 10^{-7}\,m$
     e $36\,km\,h^{-1} = 10\,m\,s^{-1}$

2 Which is the smaller member of each of the following pairs?

     a   50 m or 0.35 km   b   500 mg or 0.050 g
     c   2500 nm or 25 m

Checklist

After studying Chapter 1 you should be able to:

- discuss the importance/relevance of measurements
- recall fundamental quantities, their units and symbols
- recall and use common prefixes such as milli-, kilo-
- recall and use common derived quantities and their units
- define scales, differentiating between linear and non-linear scales
- outline steps in calibration methods
- compare and contrast analogue and digital scales
- express numbers in standard form (scientific notation)

- perform simple operations involving standard form and significant figures
- define and use the terms: accuracy, precision, sensitivity and range
- distinguish between random and systematic errors, citing examples
- list sources of errors associated with given experiments
- discuss ways of reducing sources of errors
- perform simple conversions

Questions

1 A Grenadian quoted his body mass as 160 lb, a visiting Englishman quoted his as 12 stone while a Frenchman quoted his as 68 kg. Use this information to argue the case for the use of common units. Who had the greatest body mass? (1 stone = 14 lb = 6.35 kg)

2 **a** Distinguish between a fundamental quantity and a derived quantity.
b What is meant by the term 'a base unit'?
c Which of the following is an SI 'derived unit': the candela, the joule, the volt, the second, the coulomb?

3 Distinguish between the following pairs:
a accuracy and precision
b random error and systematic error
c sensitivity and range
d analogue scale and digital scale.

4 Using suitable examples, explain what is meant by each of the following:
a standard form
b significant figures.

5 Red light has a wavelength of 700 nm. What is the wavelength of red light in metres?

6 A newly discovered virus was found to have a length of 1.7×10^{-8} m. Convert this length to
a nm   **b** mm   **c** km.
In each case state the conversion factor. Explain what is meant by the term 'conversion factor'.

7 Galileo invented what was probably the first thermometer. His device was called the thermoscope.

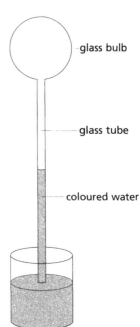

glass bulb

glass tube

coloured water

Galileo's thermoscope.

a Explain how you think the thermoscope worked.
b Suggest two reasons why Galileo's thermoscope was not very accurate.
c Describe and explain what steps could have been taken to make the thermoscope more reliable for reading temperatures.
d Newton was the first to calibrate a thermometer.
(i) Explain what is meant by the term 'calibration'.

(ii) Newton, it is generally agreed, was one of the greatest scientists. Using any suitable resource materials, including the Internet where possible, make a list of Newton's contributions to science.

8 In rough or approximate calculations one may round off numbers to the nearest power of ten. Such rounded-off numbers are called 'orders of magnitude'. By consulting appropriate resource materials, obtain values for the following:

 a the radius of the Earth, the mass of the moon, the mass of a virus, the height of the Blue Mountains and the charge on an electron.

 b Express the values in (**a**)(i) as orders of magnitude and (ii) to one significant figure.

9 The standard metre could be defined as the distance travelled by light in the time $1/299\,792\,458$ of a second. Use this data to determine a value for the speed of light.

10 Clock A is consistently 5% fast compared with the standard caesium clock. Clock B fluctuates by 2%. Which of Clocks A or B would be more suitable as a secondary standard? Justify your answer. Why is it convenient to have secondary standards?

11 a Find out the age of each student in your form. Express these ages to 3 significant figures.

 b Calculate the average age of members of your form.

 c Do the individual ages vary randomly or systematically when compared with the average of the form?

 d How do the other ages vary (randomly or systematically) when compared with (i) the age of the youngest member of the form, (ii) the age of the oldest member of the form?

12 The treads on a new motor car tyre are 1.5 cm thick. Motorists estimate that they replace tyres every 50 000 km. Determine a value for the reduction in the thickness of the treads of tyres for every kilometre travelled, expressing your answer in standard form and to 2 significant figures.

Using common laboratory equipment

In the last few hundred years, people have invented measuring instruments of unbelievable accuracy and precision. We use these instruments to help us understand the world better.

The use of precise and accurate measuring instruments has led to:

- the 'uncovering' of some of the secrets of the physical world;
- dramatic technological breakthroughs.

These discoveries and breakthroughs have produced, in turn, even more sophisticated instruments which have given rise to the next generation of discoveries and inventions . . . and the cycle continues.

In this chapter, however, our discussions will be confined to those measuring instruments which are used regularly in the school laboratory.

2.1 Measuring length

You use a ruler or measuring tape to measure length. However, you can only read the scale on a ruler accurately to the nearest millimetre. To measure length more precisely, you need a more sophisticated instrument.

Reading a vernier

A **vernier** is a short auxiliary (secondary) scale which is placed beside the main scale on a measuring instrument. The vernier enables subdivisions of the main scale to be read accurately. Figure 2.1 shows a main scale and a vernier.

The vernier works as follows. Each division on the vernier scale measures 0.9 of the smallest division on the main scale. That is, the difference in length between a division of main scale (1 mm) and a division on the vernier scale (0.9 mm) is 0.1 mm.

Things to do

What does this instrument read? Check your answer with your teacher.

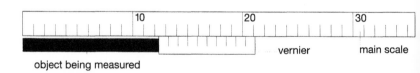

Figure 2.1 *A main scale and a vernier.*

Note

Vernier callipers are a versatile measuring instrument. They normally read to within 0.1 mm.

Vernier callipers should not be used to measure compressible materials. Why?

The vernier scale gives one extra decimal place over and above that which the main scale can measure. A vernier improves the precision of a measuring instrument. If an ordinary scale measures to the nearest millimetre then the addition of a vernier enables readings to be taken to the nearest 0.1 mm.

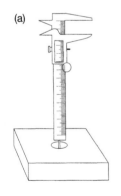

Figure 2.2 *Using vernier callipers to measure the diameter of a rod.*

Callipers

Callipers are used for measuring the diameter of tubes and other 'convex' bodies. They are also used for measuring the internal diameters of cavities. When callipers are fitted with a vernier they are called vernier callipers.

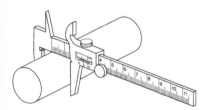

(a)

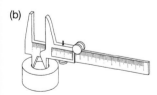

(b)

Figure 2.3 *Using vernier callipers to measure (a) the depth and (b) the internal diameter of a cavity.*

Things to do

1

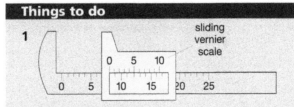

What is the reading on this instrument? Check your answer with your teacher.

2 This is an enlarged picture of the reading of the diameter of a cylindrical rod as measured by vernier callipers. What is the diameter of the rod?

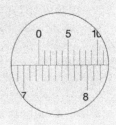

3 Practise using vernier callipers.

Note

The micrometer is used for measuring small lengths accurately.

It is important to make allowance for zero errors, if any, when using the micrometer screw gauge.

The micrometer screw gauge

The micrometer screw gauge is a G-shaped device. It measures diameters or thicknesses accurately. The device is adjusted by an accurately calibrated screw. It measures to within 0.01 mm.

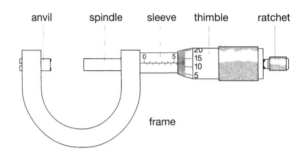

anvil   spindle   sleeve   thimble   ratchet

frame

Figure 2.4 *The micrometer screw gauge.*

How to use a micrometer screw gauge

1 The object to be measured is held snugly between the anvil and the spindle.
2 The horizontal scale (on the sleeve) is marked in millimetres. (Note that the divisions above the horizontal line on the sleeve are in whole millimetres while those below the horizontal line mark the half (0.5) millimetres.) The edge of the thimble marks the reading on the sleeve scale: in Figure 2.5 it is 2.5 mm to the nearest 0.5 mm.
3 The pitch of the thread is 0.5 mm, which means that each rotation of the screw opens the micrometer by 0.5 mm. There are 50 divisions on the thimble scale. This means that each division on the thimble is equal to $0.5 \div 50 = 0.01$ mm. Ask your teacher for help if you are having difficulty with reading the micrometer.

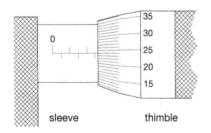

sleeve   thimble

Figure 2.5
Sleeve reading = 2.5 mm;
thimble reading = 0.24 mm;
total reading = 2.74 mm.

Things to do

1

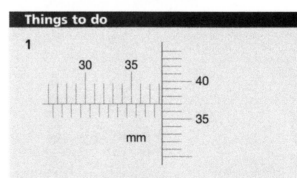

30   35

40

35

mm

What is the reading on this micrometer screw gauge?
2 This reading represents the diameter of a sphere. What is the volume of the sphere? (The volume of a sphere $= \frac{4}{3}\pi r^3$.)

Thinking it through

Which measuring instrument is likely to give a more precise reading – the vernier callipers of Figure 2.2 or the micrometer screw gauge of Figure 2.4?

Figure 2.6 *The weight of an object is measured by a spring balance or other force measurer. This spring balance is graduated in newtons (N).*

> **Note**
>
> The weight of an object varies from planet to planet and with latitude on a given planet.

> **Note**
>
> Strictly speaking, only beam balances measure mass, by comparing the unknown mass with a standard mass. Other types of balance are really measuring weight, but are calibrated in units of mass.

2.2 Measuring weight and mass

In everyday life we tend to talk about weight and mass as if they were the same thing. Scientifically, however, weight and mass are different.

The **weight** of an object is the gravitational force acting on it. Weight, then, is the pull of gravity on an object by the Earth or some other large body such as a planet.

The SI unit of weight is the **newton** (**N**). Weight, like all other forces, is a vector quantity. The weight of an object depends on:

- the mass of the object

    weight $\propto$ mass

- where the object is located. For example, a given object weighs about 0.5% less at the poles than at the equator and about six times less on the Moon than on the Earth.

Differences in the weight of a particular object at different places result from differences in the pull of gravity. Weight (W) is the product of mass (m in kg) and gravity (g in $N\,kg^{-1}$).

$$W(N) = m(kg) \times g(N\,kg^{-1})$$

The **mass** of an object is the amount of matter in it. The SI unit of mass is the kilogram. The mass of an object is a measure of its resistance to acceleration (see Chapter 6). The mass of an object remains constant, no matter where it is measured.

The first known instrument for measuring mass was invented in Egypt some nine thousand years ago. This instrument was called a **balance** (from the Latin *bi lanx* meaning two pans).

Mass may be measured on a lever balance (**Figure 2.7**), a beam balance (**Figure 2.8**) or a modern top-pan balance (**Figure 2.9**).

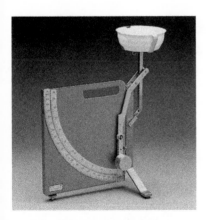

Figure 2.7 *A lever balance is a direct reading instrument.*

Figure 2.8 *A double-pan beam balance. The unknown mass is placed on one pan, standard masses on the other.*

Figure 2.9 *The top-pan balance has largely replaced the lever arm balance and the double-pan balance.*

2.3 Measuring time

What is time? Nobody really knows. But through the ages we have devised ingenious methods for measuring time. Some of these are shown in Figure 2.10.

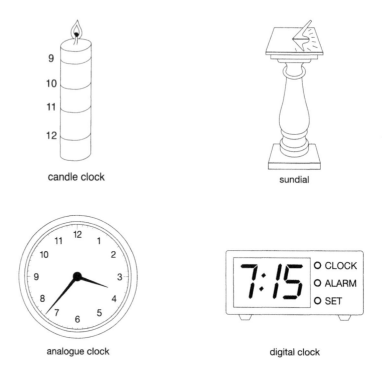

Figure 2.10 *Some of the devices used to measure time, now and in the past.*

Your reaction time, which is of the order of 0.1 second, introduces an error in measurements of time made on a stop watch or stop clock. Whenever possible, repeat timings and average the results.

Things to do

Write a paper entitled 'Time keeping through the ages'. In your paper pay attention to some or all of the following: day and night, phases of the Moon, the seasons, shadows, the pendulum, quartz crystals, the maser and atomic clocks.

2.4 Measuring temperatures

A number of devices are available for measuring temperature. Our discussion in this section, however, will be restricted to the liquid-in-glass thermometers of which Galileo's thermoscope (see page 10) was the forerunner.

The most common liquid-in-glass thermometer consists of a bulb filled with mercury. This bulb is attached to a long

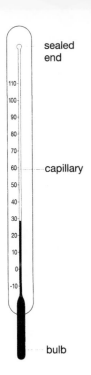

Figure 2.11 *A liquid-in-glass thermometer.*

Things to do

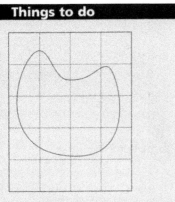

Estimate the area of the shape above, given that each small square has an area of 1 cm².

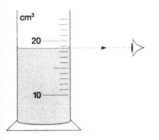

Figure 2.12 *The measuring cylinder is used for measuring or pouring out volumes of liquids.*

capillary of narrow cross-section. The thermometer is sealed at its upper end. Common laboratory mercury-in-glass thermometers measure temperatures in the range $-10\,^{\circ}\mathrm{C}$ to $110\,^{\circ}\mathrm{C}$.

We shall return to the subject of thermometers in Chapter 10.

2.5 Areas and volumes

Areas and volumes are examples of derived quantities. **Area** is the extent or measure of a surface. The SI unit of area is the square metre (m^2). The areas of regular surfaces can be calculated from formulae.

Table 2.1 Some useful formulae for finding areas	
The area of a ...	**is given by ...**
rectangle	length × width
square	$[\text{length}]^2$
circle	$\pi \times (\text{radius})^2$
cylinder	$2\pi \times \text{radius} \times \text{height}$
sphere	$4\pi \times \text{radius}^2$

The areas of irregular surfaces have to be estimated.

Volume is the capacity or the amount of space occupied in three dimensions. The SI unit of volume is the cubic metre (m^3).

The volumes of regular solids may be determined from formulae.

Table 2.2 Some useful formulae for finding volumes	
Solid	**Volume**
cuboid	length × width × height
cube	$[\text{length}]^3$
cylinder	$\pi \times (\text{radius})^2 \times \text{height}$
sphere	$\frac{4}{3} \times \pi \times (\text{radius})^3$

You can measure the volumes of liquids with common laboratory equipment such as the measuring cylinder, the burette and the pipette.

The **measuring cylinder** is used for measuring or pouring out volumes of liquids. Stand the measuring cylinder on a horizontal surface when taking readings. Your eye should be level with the bottom of the meniscus of the liquid surface (Figure 2.12).

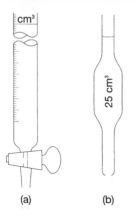

Figure 2.13 *(a) A burette; (b) a pipette.*

The **burette** (Figure 2.13(a)) can be used to deliver any volume of liquid up to its maximum stated capacity.

Pipettes of the type shown in Figure 2.13(b) deliver a predetermined volume of liquid. Always use a pipette filler to draw liquid into the pipette.

Measuring the volumes of irregular solids

You cannot measure the volume of an irregular solid directly.

Things to do

(i)   0 5 10

20 mm 30

(ii)   0

20

mm   15

(i) is the length of a small cylinder as measured with vernier callipers.

(ii) is the diameter of the cylinder as measured with a micrometer screw gauge.

Work out **a** the surface area and **b** the volume of the cylinder.

How to find the volume of a solid which is denser than water and which does not dissolve in water

1 Partly fill a measuring cylinder with water. Record the volume of the water.

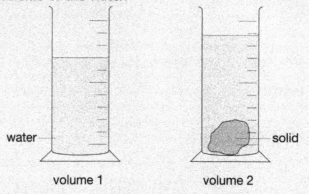

water ⟍   solid

volume 1   volume 2

2 Carefully drop the object into the cylinder so that it is completely immersed in water.
3 Read the new water level and record it.
4 The difference between the two readings is the volume of the irregular object.

Alternatively, you can use the method illustrated below if the object is too big to fit in a measuring cylinder.

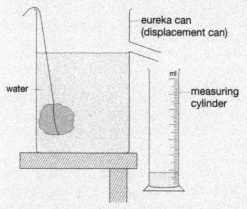

The volume of water displaced and collected in the measuring cylinder equals the volume of the irregularly shaped object.

Thinking it through

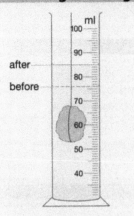

What is the volume of the stone?

How to find the volume of an irregularly shaped object that floats in water

You can do this using a eureka can and measuring cylinder as shown below. You will also need a heavy 'sinker' to which the floating object can be attached.

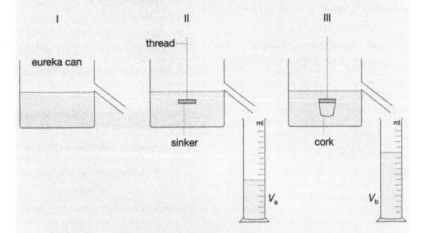

1 Fill the can with water and allow it to overflow. The water is now exactly at the spout.
2 Submerge the sinker alone and collect the overflow volume (V_a) in the measuring cylinder (II).
3 Refill the eureka can.
4 Attach the object to be measured to the sinker and lower both into the water (III). The volume of water displaced (V_b) equals the volume of the sinker plus the object.
5 Volume of object $= V_b - V_a$.

Table 2.3 The densities of some useful substances

Substance	Density (in $kg\,m^{-3}$)
helium	0.17
air	1.2
gasoline	800
water	1000
aluminium	2700
steel	7920
copper	8900
glass	2500
soft wood	500–600
gold	19 300

2.6 Density

Experiment shows that the greater the volume of a given material, the greater the mass:

$$mass(m) \propto volume(V) \qquad (1)$$

or

$$m = \rho \times V \qquad (2)$$

The constant of proportionality ρ is known as the **density** of the material. Density is defined as the mass per unit volume of a given substance. The units of density are $kg\,m^{-3}$ (SI) or $g\,cm^{-3}$.

Rearranging equation (2) gives:

$$\rho = \frac{m}{V}$$

Density is a useful property for comparing materials and for deciding which material is best for a given job. Polystyrene has a very low density. It is widely used as fillers for pillows, matresses and furniture, and in packaging. The very low density of helium coupled with its unreactive nature makes it ideal for filling balloons.

How to find the density of a solid

1 Find the mass of the solid using a suitable balance.
2 Find the volume of the solid using the appropriate formula, if the solid is regular, or the 'displacement method', if the solid is irregular.
3 Divide the mass by the volume.

How to find the density of a liquid

A. An approximate method

1 Find the mass, m_1, of an empty container such as a measuring cylinder.
2 Find the mass, m_2, of the container plus a known volume (100 cm³) of the liquid.
3 Subtract m_1 from m_2 to get the mass of the liquid.
4 Divide the mass of the liquid by the volume of the liquid.

B. Using a density bottle

The density bottle is a specially designed bottle (see Figure 2.14). It has a ground-glass stopper with a hole. This bottle holds a definite volume of liquid.

1 Find the mass (m_1) of the density bottle, including the stopper.
2 Fill the density bottle with the liquid, L, that you want to measure.
3 Insert the stopper; liquid will overflow through the hole so that there is no air left in the bottle.
4 Wipe away all traces of liquid from the outside of the bottle.
5 Find the mass (m_2) of the density bottle plus the liquid L.
6 Pour out liquid L.
7 Rinse the density bottle with water.
8 Fill the density bottle with water and insert the stopper.
9 Find the mass (m_3) of the density bottle plus water.

The mass of the water $= (m_3 - m_1)$. The density of water is $1\,\mathrm{g\,cm}^{-3}$. So:

$$\text{volume of water} = (m_3 - m_1)\ \mathrm{cm}^3$$

This is also equal to the volume of liquid L.

$$\text{Mass of liquid } L = (m_2 - m_1)\ \mathrm{g}$$

$$\text{Density of liquid } L = \frac{m_2 - m_1}{m_3 - m_1}$$

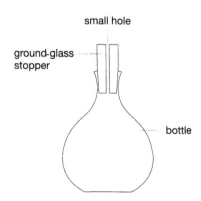

small hole

ground-glass stopper

bottle

Figure 2.14 *A density bottle.*

Thinking it through

Devise an experiment to show that carbon dioxide is more dense than air.

Things to do

Determine the relative densities of all the materials in Table 2.3.

Relative density

The relative density ρ_r measures how many times a given material is denser than water.

$$\text{Relative density } (\rho_r) = \frac{\text{density of material}}{\text{density of water}}$$

Relative density is a ratio. It is a dimensionless quantity, i.e. it has no units.

Relative density may also be defined as:

$$\frac{\text{mass of a given volume of material}}{\text{mass of the same volume of water}}$$

Problem

In an experiment to determine the density of a liquid using a density bottle, the following data were collected:

mass of the density bottle empty = 33.1 g

mass of bottle plus liquid L = 77.2 g

mass of bottle plus water   = 88.2 g

What is the density of liquid L in (i) $g\,cm^{-3}$ (ii) $kg\,m^{-3}$?

2.7 Using electrical measuring instruments

Electrical measuring instruments include the galvanometer, the wattmeter (this measures electrical power), the joulemeter (this measures electrical energy), the ohmmeter (this measures resistance), the ammeter and the voltmeter. Often, the ammeter, the voltmeter and the ohmmeter are combined in one instrument known as the multimeter.

The galvanometer is a device which detects a small current. The scale reading on the galvanometer is directly proportional to the current flowing through it.

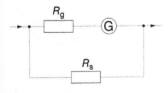

Figure 2.15 *You may think of the ammeter as a galvanometer with a small resistor (R_s) placed across it. $R_s > R_g$, where R_g is the resistance of the galvanometer.*

The ammeter

The ammeter measures current. The SI unit of current is the ampere (symbol: A). Ammeters may be of the analogue type or of the more accurate but more expensive digital type.

Note

The symbol for the ammeter is

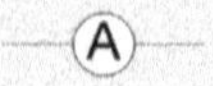

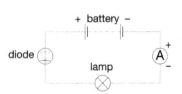

Figure 2.16 *An arrangement of a battery, a diode, a lamp and an ammeter.*

How to use an ammeter

- Ammeters are delicate instruments and should be handled with care. The positive terminal (+), often colour-coded red, must be connected directly or indirectly to the positive terminal of the cell or battery. The negative terminal (–), usually colour-coded black, is connected to the wire which goes back to the negative terminal of the cell or battery.
- Ideally, the effective resistance of the ammeter is so small that it has little effect on the current that it is measuring.

The voltmeter

The voltmeter measures the potential difference or voltage across components in electrical circuits.

Figure 2.17 *You can think of the voltmeter as a galvanometer to which is connected a series resistor of high value (R_m). This resistor is known as a* ***multiplier.***

How to use a voltmeter

- Like the ammeter, the voltmeter has a positive terminal (+) and a negative terminal (–) and must be connected correctly.
- When you want to measure the potential difference (or drop) across a circuit component, connect the voltmeter in parallel with it.
- Ideally, the voltmeter has such a large (effective) resistance that it has little effect on the potential difference or voltage that it is measuring.

Note

The symbol for the voltmeter is

In Figure 2.18, the voltmeter is connected to measure the potential difference across the diode. The positive terminal of the voltmeter is connected to the positive side of the battery. The negative terminal is connected to the negative side of the battery.

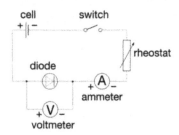

Figure 2.18 *The voltmeter is connected to measure the potential difference across the diode. The positive of the voltmeter is connected to the positive side of the cell or battery.*

Checklist

After studying Chapter 2 you should be able to:

- read a vernier
- use vernier callipers to make accurate measurements
- use a micrometer screw gauge to make accurate measurements
- recall the difference between mass and weight
- recall the relationship linking weight and mass
- recall that mass is independent of location
- recall that weight varies on different planets and also with latitude

- select instruments for measuring mass and weight
- define area, calculate and estimate areas, stating the appropriate units
- recall the standard formulae for finding areas
- define volume, stating its SI units
- use standard formulae to determine the volume of regular solids
- determine the volumes of irregular solids
- use standard laboratory glassware to determine volumes of liquids
- define density, stating its units
- solve problems involving density
- recall appropriate methods for determining density in the laboratory
- define and determine relative density
- recall the uses of basic electrical measuring instruments
- use basic electrical instruments appropriately

Questions

1 When a micrometer screw gauge is tightly closed, it gives the reading shown in Figure (a). When a metal sphere is measured, the reading shown in Figure (b) is obtained. Determine the diameter of the sphere. Calculate the area and the volume of the sphere.

(a)

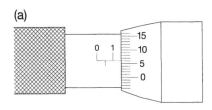

(b)

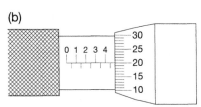

2 List the steps involved in using a micrometer screw gauge. Make sure you use the correct names of each part of the gauge.

3 When NASA measured the weight of Neil Armstrong at their Florida base, it was 600 N. When Neil and his co-astronauts landed on the Moon, his weight was measured again and found to be 100 N. Given that he has a mass of 61 kg, and that this mass remains constant, find the value of the acceleration due to gravity on the Moon (g_{moon}). Use $g_{earth} = 9.8\,\text{N}\,\text{kg}^{-1}$.

4 Godzilla entered the city of Manhattan, and wreaked havoc in the downtown district of South Street Sea Port. When he was finally captured, and the city cleaned, one area consisting of a footprint was left. From the diagram, estimate the size of this area.

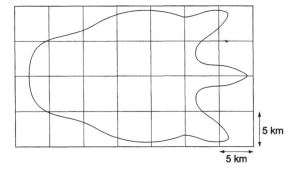

5 Would a circular pool of depth 5 m and diameter 2.5 m filled with water to a height of 4.5 m be able to hold four sea lions each of volume $0.5\,m^3$ without overflowing? Show the reasoning behind your answer.

6 An irregularly shaped block of material of density $272\,kg\,m^{-3}$ is placed in a cylinder containing liquid of density $150\,kg\,m^{-3}$. Will the material sink or float? Describe the methods which could be used to determine the volume of the block of material. The volume of the block of material was found to be $24\,cm^3$. Calculate the mass of the block.

7 Discuss the types of balance which could be used to measure an object's mass, listing the advantages and disadvantages of each.

8 When the height of a metal cylinder of mass 426 g was measured using vernier callipers, the reading shown in Figure (a) was obtained. Figure (b) shows the value of the diameter of the cylinder. Calculate the cylinder's **a** volume, **b** surface area and **c** density.

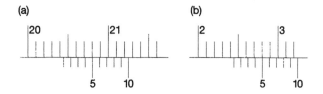

9 The two main types of thermometers used are mercury-in-glass and alcohol-in-glass thermometers. Which of the two would be more appropriate to measure the temperature of the following?
a boiling water (100°C)
b melting hexane ($-95\,°C$)
c melting octane ($-57\,°C$)
d boiling eisocane ($344\,°C$)

Note:	Mercury:	Boiling point	$357\,°C$
		Melting point	$-39\,°C$
	Alcohol:	Boiling point	$78\,°C$
		Melting point	$-115\,°C$

10 An empty bottle has a mass of 41 g. When this bottle is filled with water its mass is 79 g, and when it is filled with a certain liquid its mass is 52 g. What is the relative density of the liquid?

11 Leisel is in the middle of her end of year Physics lab exam, in which she is required to measure the potential difference across two resistors. Unfortunately she cannot remember which electrical measuring instrument should be used, or where it should be placed. Can you help Leisel?

12 An experimenter records the length of wire in an electrical circuit as $1.33 \pm 0.01\,m$ and the potential difference across that length of wire as $1.65 \pm 0.05\,V$.
a What is the smallest scale division on the instrument used to measure length?
b What is the smallest scale division on the voltmeter?
c Calculate the potential difference per metre of wire in the circuit.

13 Hollow iron pipes are made in lengths of 5 metres. The external diameter of the pipes is 30 mm, while the internal diameter is 20 mm. Each 5 metre length of iron pipe has a mass of 15.4 kg.
a Determine (i) the volume of iron in each length of pipe (give your answer in m^3), (ii) the density of iron in $kg\,m^{-3}$.
b If the iron corrodes at the average rate of 10% per year, calculate the mass of uncorroded iron after 3 years.
c Describe how you would determine the internal and external diameters of the pipes.
d You are provided with a burette and a millimetre scale. Describe how you would obtain, graphically, the internal diameter of a boiling tube. State all assumptions.

3 Experimenting

3.1 Galileo and the scientific revolution

During the first half of the 17th century, Galileo (1564–1642) pioneered a new approach to the study of natural phenomena.

Galileo advanced the ideas that:

- theories about natural phenomena should be subjected to carefully controlled experiments in which measurement plays a vital role. These experiments should be such that they could be reproduced easily by other scientists;
- all real (scientific) knowledge should be expressed in mathematical terms.

Galileo, as much as any one else, contributed to the ways of knowing and finding out what we now call the 'scientific method'. Galileo was the prime mover in the scientific revolution.

3.2 Planning and carrying out experiments

During your physics course you will carry out many experiments: some routine, others as part of the School Based Assessment (SBA) requirements if you are pursuing the CXC Physics course. It is important that you develop a systematic approach to practical work in physics.

Sometimes you will be given instructions for experiments. At other times you will have to design the investigation and plan the most sensible sequence of steps for carrying it out.

Among the things you will be expected to do are:

- clearly state the question or define the problem to be investigated;
- formulate the hypothesis or scientific theory to be tested.

These help you to clarify the purpose of the investigation and plan how to do it.

Consider the hypothesis: '*If the length of a simple pendulum is decreased, while the mass of the bob and the angle of swing are held constant, then the period (i.e. the time for one complete to and fro motion) will increase.*'

This hypothesis:

- describes, in broad outline, the experiment to be carried out;
- suggests the measurements to be made;
- clearly identifies:
  - the **manipulated variable**, which in this case is the length of the pendulum,
  - the variables to be controlled, i.e. the mass of the bob and the angle of swing,
  - the **responding variable**, which in this case is the period of swing.

The data from the experiment will show that a decrease in the length of a pendulum results in a corresponding *decrease* in the period. So the data do not support the original hypothesis. The hypothesis must be modified.

The hypothesis is a major tool in scientific research. It may lead to important discoveries even when not correct itself.

A controlled experiment is one in which only one factor (the manipulated variable) is allowed to change at a time. All other conditions (factors) are fixed. This enables the experimenter to determine the effect of the changing factor.

How to carry out an experiment

1 Read instructions carefully.
2 Be clear about all steps. Ask questions when in doubt. Do not follow instructions like a robot.
3 Use time sensibly. For example, while waiting for water to boil organize tables or work out appropriate scales for graphs and so on.
4 Ensure that your working space is not cluttered.
5 Do not be too quick to dismantle apparatus. You may have to cross-check readings.

How to write up an experiment

1 Make your presentation as attractive as possible.
2 Give the date of the experiment or investigation.
3 Give it a title or state the aim of the experiment clearly.
4 Number your experiments.
5 Make a table of contents.
6 Where necessary, draw clearly labelled diagrams of the experimental set-up.
7 Describe as a numbered list all the steps in your procedure.
8 Use the passive voice (e.g. 'the water was heated' instead of 'I heated the water') and the past tense, except when writing up 'plan and design' activities (see Appendix).

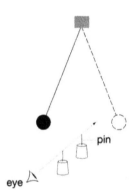

Figure 3.1 *An example of a controlled experiment. The simple pendulum may be used to find a value for g – the acceleration due to gravity.*

In Figure 3.1, what is the purpose of the pins? Explain the positioning of the eye in relation to the two pins.

Thinking it through

The investigation with the simple pendulum is an example of a controlled experiment. A student proposes to design and carry out a controlled experiment to investigate the relationship between the angle of swing and the period of a pendulum. Formulate a possible hypothesis which he could test.

Things to do

Investigate all the factors you can think of that may affect the period of swing of a simple pendulum.

3.3 Collecting and organizing data

The raw data collected during an experiment or investigation must be organized in a log, chart or a table. Prepare as much of the table or chart as possible before starting the investigation. Tables 3.1 and 3.2 are examples of acceptable formats.

Table 3.1 **Results from an investigation of the extension of a copper wire when different weights (loads) are attached to it**

Weight (N)	Extension (mm)
0.0	0.00
1.0	0.35
2.0	0.70
3.0	1.05
4.0	1.85
5.0	3.25

Table 3.2 **Results from an electrical method for determining the specific heat capacity of a liquid**

	Experiment 1	Experiment 2
Current through the coil	1.5 A	1.2 A
Voltage across the coil	6 V	5 V
Mass of liquid collected	210 g	325 g
Time	30 s	60 s
Inlet temperature	18 °C	10 °C
Outlet temperature	20 °C	20 °C

Note

A graph may show at what point a particular relationship ceases to be linear and starts to curve.

Graphs reveal systematic errors. Points which are way off the line or curve may be the results of systematic errors. These can be rechecked experimentally.

3.4 Graphs

Graphs:

- may be the best way of 'averaging' results;
- are a convenient way of presenting experimental results;
- show the connection between measured quantities, i.e. between the responding variable and the manipulated variable;
- often reveal 'information' that is not obvious from just looking at the raw data.

Note

If you do not have a flexicurve or a French curve, you may use your elbow as a pivot when drawing curves.

Note

A **line of best fit** is one that follows the trend of the data but does not necessarily pass through each point.

The line of best fit reduces random errors.

When the graph is a curve, it is often possible to choose the variables so that a straight-line graph is obtained.

For example, the relationship between the distance (x) travelled by a trolley and time (t) may be of the form:

$x = at^2$, where a is a constant.

A plot of x versus t will be a curve, but a plot of x versus t^2 will be a straight-line graph.

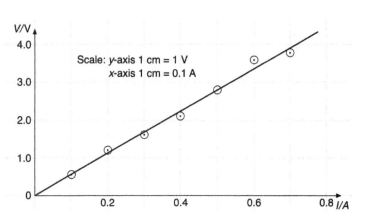

Figure 3.2 *A line of best fit on a graph of potential difference (V) against current (I).*

Extracting additional information from graphs

Wherever possible, we prefer to draw straight-line plots. Straight-line graphs make it easier to determine the links between the manipulated and the responding variables. Recall that a straight line may be represented by the mathematical relationship (equation):

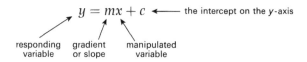

Note

The unit of the gradient equals

$$\frac{\text{vertical units}}{\text{horizontal units}}$$

A straight-line graph provides information in the form of:

- the gradient;
- the values of the intercepts on the x-axis and y-axis.

The gradient

The gradient or slope is defined as the change in the quantity plotted on the y-axis divided by the corresponding change in the quantity plotted on the x-axis. A straight-line graph has a gradient which may be positive or negative.

Example 1

In an experiment to investigate the relationship between the displacement of an object which is moving in a straight line and the time which has elapsed since motion started, the following results were obtained.

Table 3.3								
Displacement (m)	0	120	240	305	480	600	720	840
Time (s)	0	20	40	60	80	100	120	140

The data of Table 3.3 are plotted as a graph in Figure 3.3.

Clearly, one of the readings was incorrectly taken when obtaining the data of Table 3.3. Identify this reading.

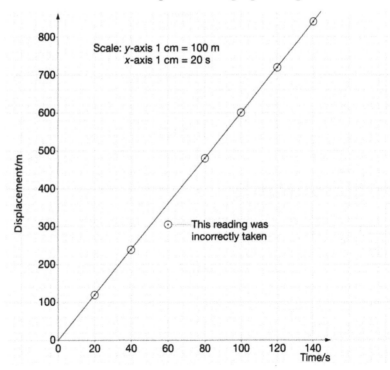

Figure 3.3 *The graph of displacement (y-axis) versus time (x-axis).*

Note

The gradient, s, of the graph of Figure 3.3 is given by:

$$s = \frac{840 - 0}{140 - 0} = \frac{840}{140}\ \text{m s}^{-1}$$

$$= 6.0\ \text{m s}^{-1}$$

Advice: Use as much of the graph as possible when determining the gradient (but make sure that the points you use can be read off the graph accurately).

Example 2

A car decelerating uniformly had velocities at the times shown in Table 3.4.

Table 3.4

v (m s⁻¹)	21.8	17.9	14.1	10.2	6.2	2.0
t (s)	2.0	4.0	6.0	8.0	10.0	12.0

Figure 3.4 gives the graph of the velocity (v) of this car against time (t). The graph of Figure 3.3 has a positive gradient. But the graph of Figure 3.4 has a negative gradient.

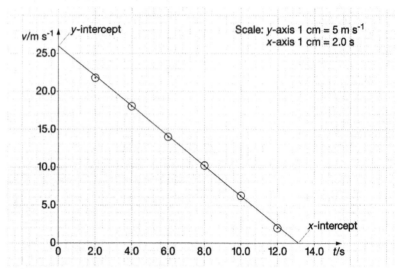

Figure 3.4 *A graph of velocity versus time.*

The intercepts of straight line graphs

What is the significance of the intercepts on the x-axis and y-axis in Figure 3.4?

A graph which does not pass through the origin has an intercept on the y-axis and an intercept on the x-axis.

Drawing curves

If the plotted points do not lie on a straight line, draw the smoothest possible curve through them.

Things to do

Convert the data in the graph of Figure 3.5 into a table.

Note

The gradient of a curve at any point is the tangent to the curve at that point. The gradient of the curve may be found by placing a plane mirror across the curve. Your teacher will show you how to do this.

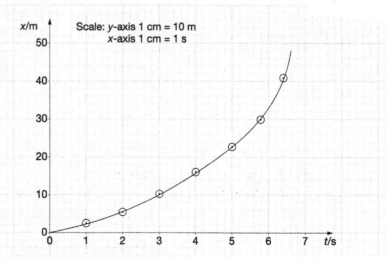

Figure 3.5 *Displacement of a trolley versus time.*

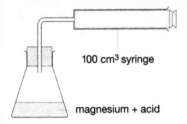

100 cm³ syringe

magnesium + acid

Figure 3.6 *Hydrogen gas is given off when magnesium reacts with the acid. The gas given off is collected in the syringe. Figure 3.7 is a plot of volume of gas versus time.*

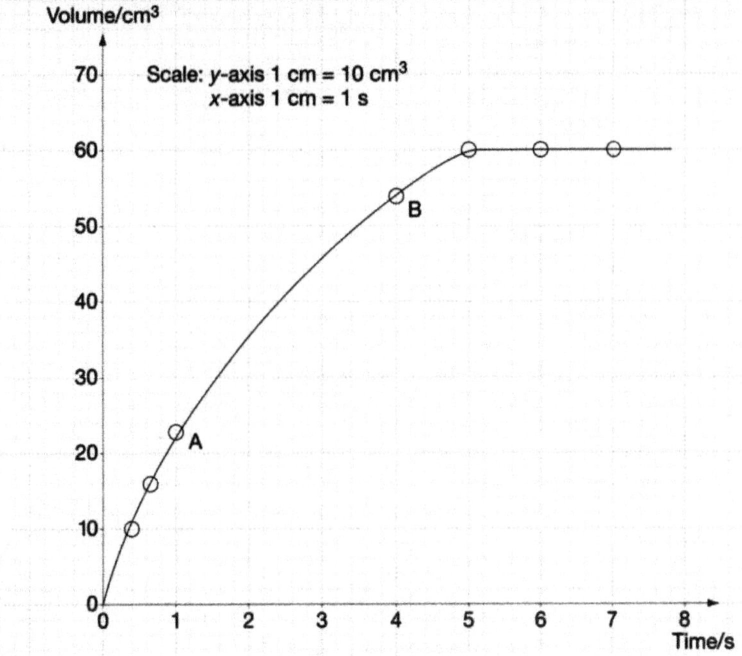

Volume/cm³

Scale: y-axis 1 cm = 10 cm³
x-axis 1 cm = 1 s

Figure 3.7 *A graph of volume versus time.*

For the graph of Figure 3.7 too few readings were taken. You cannot be sure of the 'curve' between points A and B. Take as many readings as possible in the parts of a graph that curve.

Making predictions from experimental data

Study Figure 3.4 again. Now answer the following questions.

1 Predict the velocity of the vehicle at the end of the seventh second.
2 At what time will the velocity be $15.0\,\mathrm{m\,s^{-1}}$?

Your answers should be (1) $12\,\mathrm{m\,s^{-1}}$ and (2) $5.6\,\mathrm{s}$.

These are examples of **interpolation** – the predicted values lie within the range of the experimental data, i.e. $(2.0–21.8)\,\mathrm{m\,s^{-1}}$ and $(2.0–12.0)\,\mathrm{s}$. The values of the intercepts on the y-axis and on the x-axis in Figure 3.4 are examples of **extrapolated values**. Extrapolated values lie outside the range of the experimental data.

Breaking the scale

A student determined the thicknesses of several sheets of duplicating paper. His results are shown in Table 3.5.

Table 3.5						
Thickness (t) (mm)	23.0	33.5	43.5	53.5	63.0	74.0
Number of sheets (n)	20	30	40	50	60	70

The student's graph is shown in Figure 3.8.

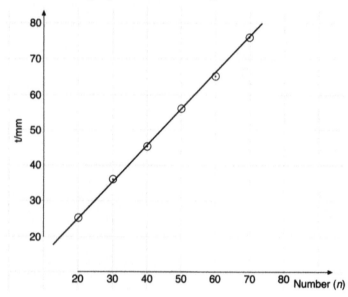

Figure 3.8

Advice

Read through the entire question before you decide to break scales.

Note

The word **discrepant** means different, unexpected ... inconsistent with accepted patterns or trends. Unexpected results sometimes play important roles in scientific discoveries.

'New knowledge very often has its origin in some quite unexpected observation or chance occurrence during an investigation. The importance of this factor in discovery should be fully appreciated and research writers ought deliberately to exploit it ...' Beveridge in *The Art of Scientific Investigation* (p. 40).

This student broke the scale on both axes, i.e. he did not start the scales at the origin. It is permissible to break the scale on both axes, provided certain conditions are satisfied:

- if you do not have to extrapolate from the graphical data, e.g. having to find the intercepts;
- if starting from the origin results in poor use of graph space.

3.5 Concluding: making generalizations based on experimental results

Important questions to be asked here include:

- Do you have confidence in the experimental data?
- Have you made several determinations of each measurement? Why is this question important?
- Have you checked *discrepant* readings (see Note)?
- What is the final experimental result?
- What are the possible sources of errors?
- What have you done to reduce the effects of these errors?
- Have you identified the errors over which you had no control?
- Have you identified relationships or trends which are obvious or which are indicated by the results?
- Do the results agree with expected theory? If not, what further experiments do you plan to carry out?
- Have you limited your conclusions/generalizations to the results of the experiment?

Checklist

After studying Chapter 3 you should be able to:

- approach experimental work in physics systematically
- carry out routine practical exercises
- define the terms 'hypothesis', 'variables' and 'controlled experiments'
- use hypotheses as the bases for planning and designing scientific investigations
- present experimental data in appropriate formats
- draw properly labelled graphs, straight lines and curves
- extract information from graphs (i.e. interpolate and extrapolate)
- make generalizations from experimental data

Questions

1 Using suitable examples, explain the meaning of the following terms:
a manipulated variable, **b** controlled experiment, **c** hypothesis, **d** interpolation, **e** extrapolation.

2 List the advantages of presenting scientific data graphically.

3 The electrical resistance of a wire varies with length as indicated in the table.

Resistance/Ω	1.9	2.5	2.8	3.1	3.6	4.1
Length/m	5.0	6.0	7.0	8.0	9.0	10.0

a Plot a graph of resistance (y-axis) versus length (x-axis), starting both scales from the origin.
b Is the graph linear or non-linear?
c Determine the gradient of the graph.
d What is (i) the resistance of 12 m of this wire, (ii) the length of wire of resistance 3.3 Ω?
e What is the value of the intercept on the x-axis? Explain the significance of this value.

5 The relationship connecting displacement (s), acceleration (a) and time (t) may be expressed as follows: $s = ut + \frac{1}{2}at^2$, where u is the initial velocity of the object. Rearrange this equation so that you will be able to plot a straight-line graph.

6 In an attempt to improve their end of year average, Jennifer, Lisette and Richard decided to conduct an extra credit assignment. This involved measuring the count rate of a radioactive source over a period of time. The following results were obtained.

Count rate/c.p.s.	160	110	80	60	40	30	20
Time/min	0	2	4	6	8	10	12
Count rate/c.p.s.	15	10	8	5	4	25	
Time/min	14	16	18	20	22	24	

a Plot a graph of count rate against time.
b From this graph find the gradient at time (i) 3 min (ii) 21 min.
c What do the gradient values tell you about the count rate with respect to time?

7 You are given the following hypothesis: 'Ice melts faster in a solution of sugar than in a solution of salt'. Plan and design an experiment to test this hypothesis.

Multiple-choice questions

1 Which of the following is the correct symbol for a base SI unit?

A W   B J   C kg   D V

2 Radio 107 FM broadcasts on a frequency of 107 MHz. This is the same as

A 1.07×10^8 Hz   B 1.07×10^6 Hz
C 1.07×10^4 Hz   D 1.07×10^2 Hz

3 How many significant figures are there in the measurement 0.0075 V?

A 1   B 2   C 3   D 4

4 When determining the diameter of a wire using a micrometer, a student obtained the following readings: Zero reading = 0.03 mm Reading with wire in position = 0.37 mm The diameter of the wire, in mm, is:

A 0.40   B 0.385 C 0.355 D 0.34

5 1 kilowatt hour = 3.6 MJ.
1 kilowatt hour converted to joules is:

A 3.6×10^{-6} J   B 3.6×10^{-3} J
C 3.6×10^3 J   D 3.6×10^6 J

6 Which of the following is not a fundamental quantity?

A length   B current
C temperature   D energy

Questions 7–9

Use the relationship density = mass ÷ volume to answer these questions. A copper object has a volume of 0.47 cm^3 and a mass of 4.18 g.

7 The mass of the object in kg is:

A 4.18×10^{-6} B 4.18×10^{-3}
C 4.18×10^3 D 4.18×10^6

8 The volume of the object in m^3 is:

A 4.7×10^5 B 4.7×10^3
C 4.7×10^{-7} D 4.7×10^{-6}

9 The density of copper, in kg m^{-3}, is:

A 8.93×10^3 B 8.89×10^3
C 8.9×10^3 D 9.0×10^3

10 A mirror behind the pointer of a voltmeter serves to

A reflect light onto the pointer
B make the scale more easily visible
C eliminate zero error
D reduce parallax error.

11 Momentum is defined as the product of the mass and velocity of an object. The unit of momentum is therefore:

A m^{-1} s kg   B kg m s
C kg^{-1} m s   D kg m s^{-1}

12 A room measures 15.3 m × 10.6 m × 10.1 m. What is the volume of the room?

A 1638 m^3 B 1630 m^3
C 1.64×10^3 m^3 D 1.6×10^3 m^3

13 Which of the following is most suitable for measuring the depth of a test tube?

A micrometer   B vernier callipers
C metre rule   D 15 cm rule

14 7.6 mA expressed in microamperes (μA) is:

A 7.6×10^{-2} B 7.6×10^{-3}
C 7.5×10^3 D 7.6×10^9

Questions 15–17

A measuring cylinder contains 75 cm^3 of water. When 100 steel ball bearings are added to the water the volume increases to 140 cm^3. The 100 ball bearings have a combined mass of 501 g.

15 What is the average volume of the ball bearings in cm^3?

A 0.65   B 0.75 C 105   D 210

16 What is the average diameter of the ball bearings?

A 1.0747 cm   B 1.074 cm
C 1.07 cm   D 1.1 cm

17 The density of the steel used to make the balls, in g cm^{-3}, is:

A 8.0   B 7.7   C 7.692 D 7.6

Questions 18 and 19

An object weighs 225 N on another planet. The mass of the object on Earth is 75 g.

18 What is the value of the acceleration due to gravity on the other planet in N kg^{-1}?
A 0.33   B 3.0   C 10   D 300

19 What does this object weigh on Earth?
A 75 N   B 225 N
C 300 N   D 750 N

20 Which of the following is an example of a dimensionless quantity?
A relative density   B linear momentum
C work   D energy

21 The diagram below shows the face of an ammeter.

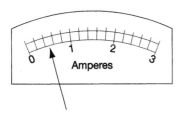

What is the reading of the ammeter?
A 0.20 A   B 0.3 A
C 0.40 A   D 0.5 A

22 The diagram below shows a magnified section of a millimetre scale.

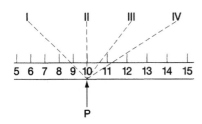

What is the correct reading of the arrow point (P) which is placed beside the scale?
A I   B II   C III   D IV

Questions 23–25

The following diagram shows the scale of a voltmeter and the position of the pointer when no current is flowing.

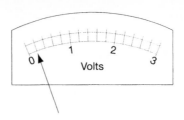

23 The smallest division on the scale corresponds to:
A 0.1 V   B 0.2 V   C 0.5 V   D 1.0 V

24 This voltmeter has a zero error of:
A +0.05 V   B +0.2 V
C −0.05 V   D −0.1 V

25 What would the meter read when placed across a device which has a potential difference of 1.4 V across it?
A 1.6 V   B 1.5 V
C 1.35 V   D 1.2 V

Questions 26 and 27

I The unstretched length of a spring is 7.0 cm.
II The spring extends by the same amount for every newton load added to it.
III The length of the spring when it is loaded with 25 N is 12.0 cm.

26 By what length is the spring stretched for each newton load added to it?
A 5 cm   B 2 cm   C 0.5 cm   D 0.2 cm

27 What is the length of the spring when a 15 N load is attached?
A 3 cm   B 4 cm   C 10 cm   D 15 cm

Questions 28–30

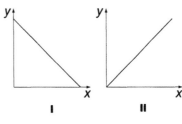

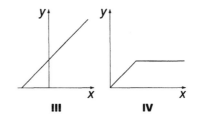

Which of the graphs I–IV:

28 has a positive gradient, but does not pass through the origin?

   A I   B II   C III   D IV

29 has a negative gradient?

   A I   B II   C III   D IV

30 has a positive gradient, but a negative intercept on the x-axis?

Questions 31–35

A bottle weighs 35 g when empty, 95 g when filled with water and 119 g when filled with an organic liquid.

31 What is the capacity (volume) of the bottle?

   A 60 cm^3 B 84 cm^3

   C 130 cm^3 D 154 cm^3

32 The relative density of the organic liquid is:

   A 0.37   B 0.71 C 1.4   D 3.4

33 What would be the mass of the bottle if it were filled with another organic liquid of relative density 0.80?

   A 131 g   B 120 g

   C 110 g   D 83 g

34 A student wishes to measure accurately 25.0 cm^3 of a liquid. Which of the following should he choose for the task?

   A 50 cm^3 measuring cylinder

   B 25 cm^3 conical flask

   C 25 cm^3 pipette

   D 50 cm^3 burette

35 What is a good estimate of the area enclosed on this map?

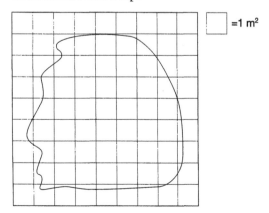

$\square$ =1 m^2

A 35 m^2 **B** 45 m^2

C 49 m^2 **D** 53 m^2

Questions 36 and 37 concern the graph which shows the length of a spiral spring under different loads.

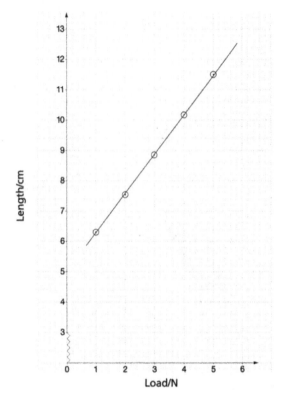

36 What is the gradient of this graph?

   A 1.5 cm N^{-1} B 1.3 cm N^{-1}

   C 0.78 cm N^{-1} D 0.63 cm N^{-1}

37 The unstretched length of the spring is:

   A 5.0 cm   B   5.6 cm

   C 7.6 cm   D   8.0 cm

Structured and free-response questions

1 The winner of the 100 metre race was timed in 9.86 seconds.

   a How many significant figures are there in the (i) length, (ii) time?

   b The timing of this race was to the nearest (i) tenth of a second, (ii) hundredth of a second, (iii) thousandth of a second. Which of (i) to (iii) is true?

c Given that speed = distance ÷ time, what is the average speed of the winner of this race?

2 Write the following numbers in standard form. **a** 0.043 **b** 326 **c** 30 000.

3 Write each of the following numbers correct to three significant figures.
a 7.256 **b** 4278.9 **c** 57.823.

4 List five fundamental quantities.

5 Which of these are derived quantities? velocity, length, time, power, potential energy

6 Which of these are sub-multiples? mega, nano, kilo, milli, micro

7 **a** State one advantage of a linear scale over a non-linear scale.
b State two advantages of a digital scale over an analogue scale.

8 For the instrument shown, what is the **a** precision **b** range **c** reading?

9 Name three sources of error.

10 Suggest two ways by which experimenters may reduce errors.

11 **a** What does SI mean?
b List two advantages of SI over older systems of measurements.
c What is meant by the phrase 'the accuracy of a reading'?

12 What is meant by the following?
a cm³ **b** 1 μg **c** 1 ks **d** 1 mA **e** 1 nm

13 A block of aluminium measures 12 cm × 12 cm × 12 cm. What is its volume in cubic metres?

14 **a** What is meant by the phrase 'a line of best fit'?
b List three bits of information that may be obtained from a straight line graph.

15 The table shows distances of a bus from a station at different times.

Distance (s)/m	13	22	37	66	85	118
Time (t)/s	1	2	3	4	5	6
Time²(t²)/s²						36

a Copy and complete the table by filling in values of time². The distance s and the time t are related by the equation:

$$s = bt^2 + D, \text{ where } b \text{ and } D \text{ are constants}$$

b Plot a graph of s against t^2.
c Use the graph to (i) identify the reading which was incorrectly recorded, (ii) find a value for b [state the unit of b], (iii) find the distance of the bus from the station at time = 0.

16 You are provided with a 25 cm³ measuring cylinder and a 1 dm³ bottle. Describe how you would calibrate the bottle to measure volumes of liquids to the nearest 20 cm³.

17 Data on the energy requirements of different people are given below.

	15-year-olds		Adults	
normally active	Boys	12 000 kJ	Male	12 200 kJ
	Girls	9500 kJ	Female	10 500 kJ
very active	Adult male	15 200 kJ		
	Adult female	12 600 kJ		

a Present the above data in another format.
b Make as many deductions as possible from the data.

18 The pressure exerted by a fixed mass of gas at constant temperature was varied and the corresponding volume determined. The results are shown in the table.

Pressure (P)/kPa	50	75	125	150	175	200
Volume (V)/m³	0.10	0.066	0.04	0.033	0.0285	0.025
$\frac{1}{V}$/m⁻³						

a Copy and complete the table by filling in values of $1/V$.
b Plot a graph of pressure (P) [y-axis] against volume (V) [x-axis].
c Use the graph to determine the value of the volume (V) when the pressure is $100\,kPa$. What is the corresponding value of $1/V$?
d Comment on the value of (pressure × volume) for this set of readings.
e Is this an example of a controlled experiment? Discuss, identifying the manipulated, the responding and the controlled variables.
f Without drawing a graph, state the relationship between pressure and $1/V$.
g Will the graph of pressure $1/V$ pass through the origin? Explain.

19 Sodium hydrogencarbonate is sold in waterproof containers which measure $5.0\,cm \times 10.0\,cm \times 13.0\,cm$. Each full container has a mass of $454\,g$.
a What is the mass of each container in kilograms?
b What is the volume of each container in cm^3 and in m^3?
c Obtain an estimate for the density of sodium hydrogencarbonate, stating all assumptions made.

20 The vertical height and the corresponding horizontal distances of a projectile are given in the table below.
a Plot a graph of vertical height (y-axis) against horizontal distance (x-axis), starting both axes from zero [let the range of the x-axis extend to 90 cm].
b Use the graph to answer the following: (i) What is the maximum height gained by the projectile? (ii) What is the range (the maximum horizontal distance) of the projectile? (iii) Should the graph pass through the origin?
c Are all readings of vertical height given

to the same number of significant figures? Explain.
d In what region of the graph should the greatest number of readings be taken? Explain.

21 I The only micrometer screw gauge available to you has a zero error of $+0.25\,mm$.
II You have a box of copper rods with diameters in the range $3.25\,mm$ to $8.65\,mm$.
III You need a copper rod of uniform diameter of $5.60\,mm$ or nearest.
a Describe how you would use the micrometer to determine (i) the diameter of a suitable rod, (ii) whether the rod has a uniform cross-sectional area.
b You do not have access to a balance, but you have a metre rule. Describe how you would determine the length of copper rod of diameter $5.60\,mm$ which has a mass of $0.14\,kg$.

22 Since the world is round, one could get to the east by sailing west. Using this statement as an example, discuss the main features of a hypothesis.

23 The following table shows the variation of the velocity of a space shuttle with time.

Velocity /$m\,s^{-1}$	500	690	820	970
Time/s	1	2	3	4
Velocity /$m\,s^{-1}$	1140	1290	1450	1600
Time/s	5	6	7	8

a Using a scale of $2\,cm = 1\,s$ on the x-axis, and $2\,cm = 100\,m\,s^{-1}$ on the y-axis, plot a graph of the data given.
b Find the gradient of the graph.
c What is the significance of the gradient?
d Were any of the points accurate? If so, which?

Horizontal distance/cm	10.0	20.0	30.0	35.0	40.0	45.0	50.0	60.0	65.0	70.0
Vertical height/cm		6.8	13.0	16.2	16.8	17.0	16.5	14.7	9.6	7.0	5.0

4 Forces, moments, stability and machines

4.1 Forces

A **force** is a push or pull: any action which changes or tends to change the state of rest or state of motion of an object. Forces may be used to compress (squeeze), twist, bend, stretch, move, rotate . . . objects.

A force applied to an object may change its:

- shape;
- size;
- motion.

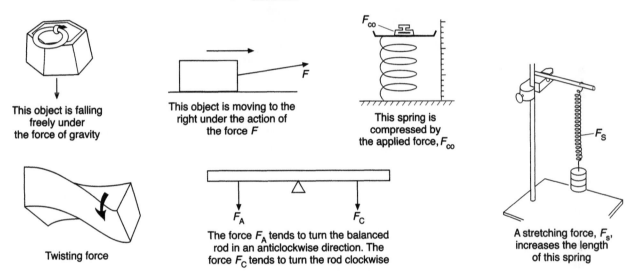

Figure 4.1 *Examples of force at work.*

This object is falling freely under the force of gravity

This object is moving to the right under the action of the force F

This spring is compressed by the applied force, F_{co}

Twisting force

The force F_A tends to turn the balanced rod in an anticlockwise direction. The force F_C tends to turn the rod clockwise

A stretching force, F_s, increases the length of this spring

Contact and non-contact forces

Forces may be broadly classed as **contact** and **non-contact** (see Figure 4.2). With contact forces the objects interacting are physically in contact (they are touching). Non-contact forces act at a distance.

- The regions within which non-contact forces act are called **fields** (force fields).
- The closer the objects or fields are, the greater the force which exists between them.

Force diagrams

Force diagrams, or free-body diagrams, show all the forces acting on a body in a given situation. Forces are represented by arrows.

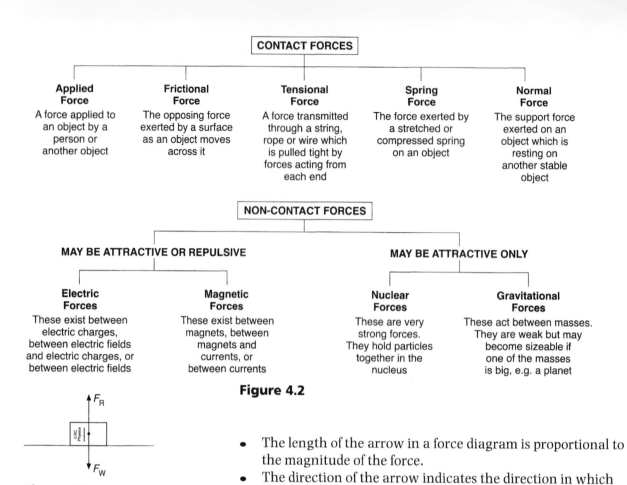

Figure 4.2

Figure 4.3

Figure 4.4

Figure 4.5

- The length of the arrow in a force diagram is proportional to the magnitude of the force.
- The direction of the arrow indicates the direction in which the force is acting.

Examples

1 A book resting on a table. The forces are the weight F_W of the book, which acts vertically down, and the reaction F_R of the table on the book, which acts vertically upwards (Figure 4.3)

$$F_W = F_R$$

2 A force is applied to a book which is resting on a table, causing it to move to the right (Figure 4.4).

4 An acrobat is suspended motionless from a bar which hangs from the ceiling by two ropes (Figure 4.5).

F_W = weight of the acrobat, which acts vertically down.

F_{T1} and F_{T2} are the tensions in the rope and they act vertically up. The values of F_{T1} and F_{T2} depend on the position of the acrobat on the bar and on the position of the ropes on the bar.

$$F_W = F_{T1} + F_{T2}$$

frictional force

Figure 4.6 *Frictional force between your shoe and the ground acts backwards. This pushes you forwards.*

air resistance

weight

Figure 4.7 *Friction prevents this parachute from falling freely. Because of air resistance the parachute eventually reaches a constant (terminal) velocity.*

4.2 Frictional forces

A ball which is rolling along the ground eventually comes to a stop. So too does a pendulum which has been set in motion or a bicycle which is free-wheeling along the street . . . Clearly, a force is opposing motion in each of these situations. This force is the force of friction.

Friction is not an 'active' force. Friction arises as a reaction to 'applied forces'.

The force of friction always acts in a direction opposite to the direction of motion.

Frictional forces act so as to slow down motion between moving objects and prevent motion between stationary objects.

Friction can be a nuisance

Friction is responsible for the wear and tear of the moving parts of machinery. It also produces heat. This wear and heating reduce the efficiency and shorten the working life of machines. Heat is produced whenever work is done against friction (test this statement by rubbing your hands together).

Friction can be useful

Although friction may be a nuisance, there are situations in which friction is highly desirable. Some examples:

- without friction car wheels would simply skid;
- without friction we would need suction pads on our feet to help us walk;
- friction allows a nail to be hammered into a block of wood. Explain why.

Thinking it through

The frictional force *F* on a ball moving through a column of air is given by the relationship:

$$F = \rho A v^2$$

where ρ = density of the air, A = cross-sectional area of the ball, v = velocity of the ball.

Predict:
1. the relative change in *F* if both the radius and the velocity of the ball are doubled;
2. the qualitative change in *F*, at constant *A* and *v*, if the air becomes heated up.

Figure 4.8 *An electron microscope photograph of a razor blade, a 'smooth' surface.*

Figure 4.9 *A streamlined shape.*

Causes of friction

Many solid surfaces appear to be smooth to the unaided eye but, if you look at them under a microscope, they are in fact jagged and bumpy. Two surfaces which are in contact are, in reality, in close contact at only a few points. The conventional view is that friction increases with the roughness of the surfaces in contact but this is not necessarily so. Recently, it has been shown that the extent of friction depends mostly on the molecular (cohesive) forces in the region of real contact.

In fluids (liquids and gases) friction arises as follows. An object which is moving through a fluid pushes some of the fluid molecules aside. Since a force is required to achieve this, the moving object slows down.

What factors determine the extent of friction?

Frictional forces depend on the type of surfaces involved and, in particular, on the cohesive forces between points of contact. Frictional forces increase if:

- the area of the surfaces in contact increases;
- the weight of the moving object increases (with greater weight there is an increase in the number of points of contact);
- the velocity of the moving object increases.

Ways of reducing friction

Since friction is sometimes a nuisance, we need to find ways of reducing it. Some suggested ways are the use of

- smoother surfaces or fluids of lower viscosities (see Note);
- grease and other lubricants;
- rollers, wheels and ball bearings;
- objects, e.g. cars and planes, with streamlined shapes.

Things to do

1 Describe and account for the width and tread pattern in a racing car tyre when compared to that of an ordinary car.
2 Discuss how the shapes of different animals' bodies are designed to reduce friction.
3 Identify parts of a car where friction is undesirable and should be reduced to a minimum.
4 List as many situations as possible where friction is desirable.
5 A steel ball is travelling downwards through liquid at constant velocity. Given that the oil exerts a force of 4 N on the ball, what is the weight of the ball? Hence, find the mass of the ball.

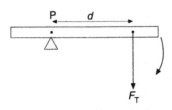

Figure 4.10 *The force F_T exerts a turning effect on the rod.*

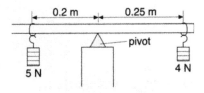

Figure 4.11 *A system in which there is no tendency to rotate in either direction.*

4.3 Moments

In Figure 4.10, the force F_T exerts a turning effect on the rod about the point P. F_T is at a perpendicular distance d from the turning point P.

It can be shown that the effect of a turning force increases if:

- the size of the force increases, and
- the perpendicular distance from the point of application of the force to the turning point (pivot or fulcrum) increases.

The **moment of a force** is measured by multiplying the force by the *perpendicular* distance of the line of action of the force from the fulcrum or pivot. The unit of moment of a force is the **newton metre (N m)**.

$$\text{moment} = F \times d$$

A moment which tends to rotate an object or system in a clockwise direction is called a clockwise moment.

A moment which tends to rotate an object or system in an anticlockwise direction is known as an anticlockwise moment.

The system of Figure 4.11 shows no tendency to rotate in either a clockwise or anticlockwise direction. Why?

The moment of the 5 N weight (anticlockwise)
$$= (5 \times 0.20)\,\text{N m} = 1\,\text{N m}$$

The moment of the 4 N weight (clockwise)
$$= (4 \times 0.25)\,\text{N m} = 1\,\text{N m}$$

For this system, then, clockwise moment = anticlockwise moment. This system remains balanced (or in **equilibrium**) unless disturbed.

For systems in equilibrium the sum of the clockwise moments about a point or axis equals the sum of the anticlockwise moments. This is commonly known as the **principle of moments**.

Using the principle of moments to solve problems

A uniform rod with the fulcrum at its centre of gravity

The weight of the rod acts through its centre of gravity (see section 4.4), which in this case is at the pivot or fulcrum. The moment due to the weight of the rod is therefore zero.

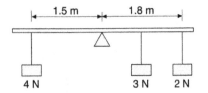

Example

A uniform rod 4 metres long is balanced at its centre. A 4 N weight is attached 1.5 m to the left of the fulcrum. A 2 N weight is attached 1.8 m to the right of the fulcrum. Where must a 3 N weight be placed for the system to remain balanced?

Solution

The 4 N weight 1.5 m to the left of the fulcrum gives an anticlockwise moment of value 6.0 N m. (1)

The 2 N weight, 1.8 m to the right of the fulcrum, gives a clockwise moment of value 3.6 N m. (2)

Since (1) > (2) the 3 N weight should be placed to the right of the fulcrum for balance.

Let the distance of the 3 N weight from the fulcrum be d metres.

From the principle of moments:

$$(4 \times 1.5)\,\mathrm{N\,m} = (2 \times 1.8 + 3 \times d)\,\mathrm{N\,m}$$
$$6.0\,\mathrm{N\,m} = (3.6 + 3d)\,\mathrm{N\,m}$$
$$3d = 2.4$$
$$d = 0.8$$

The 3 N weight should be placed 0.8 m to the right of the fulcrum.

A uniform rod which is not balanced at its centre of gravity

In this situation, the moment due to the weight of the rod is not zero.

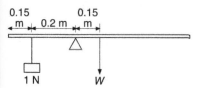

1 N   W

Examples

1 A metre rule is balanced at the 35 cm mark by a 1 N weight placed at the 15 cm mark. What is the weight of the ruler?

Solution

Note that the weight of the ruler is taken as acting through its centre of gravity.

From the principle of moments:

$$(1 \times 0.2)\,\mathrm{N\,m} = (W \times 0.15)$$

$$W = 1.33\,\mathrm{N}$$

Note

When a number of parallel forces act on a body so that it is in equilibrium:

- the principle of moments applies;
- the sum of the forces in one direction equals the sum of the forces in the opposite direction;
- moments can be taken about any point.

2 As shown in the force diagram below a uniform rod of weight 10 N and length 5 m is supported by ropes from points P and Q. The rope at P is attached 1.0 m from end M and the rope at Q is attached 0.25 m from end O. Find the tensions T_P and T_Q.

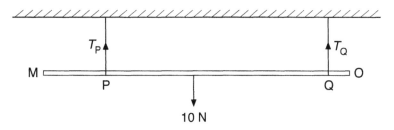

Solution

The weight of the rod acts through its centre of gravity, i.e. the 10 N weight acts at a distance 1.5 m from P and distance 2.25 m from Q.

$$\text{the distance } PQ = (1.5 + 2.25)\,\mathrm{m} = 3.75\,\mathrm{m}$$

$$T_P + T_Q = 10\,\mathrm{N} \text{ (because the system is in equilibrium)}$$

Taking anticlockwise moments about P:

$$\text{moment due to } T_P = 0$$
$$\text{moment due to } T_Q = (T_Q \times 3.75)\,\mathrm{N\,m}$$
$$\text{clockwise moment} = \text{moment due to the weight of the rod}$$
$$= (10 \times 1.5)\,\mathrm{N\,m}$$

From the principle of moments:

$$3.75 \times T_Q = 10 \times 1.5$$

$$T_Q = 4\,\mathrm{N}$$

$$T_P = 10 - 4 = 6\,\mathrm{N}$$

Problems

1 Verify that $T_P = 6\,N$ in Example 2 on page 45 by taking moments about Q.
2 Use the values of the forces and distances shown below to calculate the distance x.

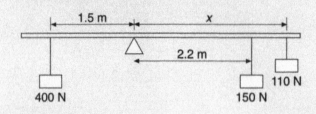

Moments and levers

A lever is any device which turns about a fulcrum or pivot.

Have you noticed that a long-handled spanner turns a stubborn nut more easily than a short-handled spanner? The longer spanner exerts a bigger force (**effort**) on the nut. Devices such as spanners, bottle openers, nutcrackers are examples of levers.

In a lever an effort is applied to overcome a resistive force called a **load**. Levers are **force multipliers**.

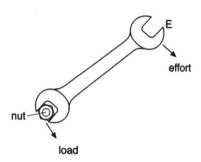

Figure 4.12 *The spanner. The effort is applied at E. The nut is the load. The fulcrum is at the centre of the nut.*

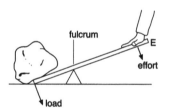

Figure 4.13 *The crowbar. If the distance from the fulcrum to the load is 0.15 m and the distance from the fulcrum to the effort is 2.4 m, a force applied at E is multiplied 16 times.*

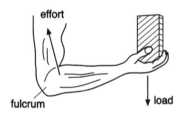

Figure 4.14 *The forearm. The fulcrum is the elbow joint. The biceps muscles provide the effort.*

Things to do

1 Draw diagrams of a bottle opener, a nutcracker and a pair of scissors. In each case label the fulcrum and the load.

2 You are provided with a long plank. Describe how you would use this to move a heavy rock. Carefully explain the underlying physics.

- Given two known masses, describe how you would determine the mass of a non-uniform rod.
- Describe how you would use a beam balance to determine the mass of an object.

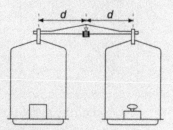

- Why is the beam balance above suspended from its centre?
- The box in the diagram below contains a rock. Explain why the box does not topple.

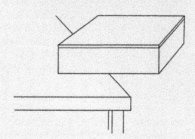

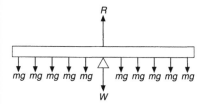

Figure 4.15 *A heavy balanced rod resting on a pivot.*

4.4 Centre of gravity

In Figure 4.15 a heavy balanced rod is resting on a pivot. Each particle of the rod exerts a downward force as shown. The sum of these downward forces is balanced by an upward force, or reaction, R at the pivot. We can represent the downward forces by a single force W acting through the centre of the rod.

$$W = \text{sum of the weights (forces) of the particles}$$
$$= \text{weight of the rod}$$

The centre of the rod P is the **centre of gravity** or **centre of mass** of the rod.

The centre of gravity or centre of mass of a body is the point through which the total weight of the body is considered to act.

Note

The centre of gravity of a regularly shaped object is its geometrical centre.

A **lamina** is any thin flat object.

How to locate the centre of gravity of regularly shaped flat objects

1

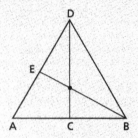

The centre of gravity of a square lamina is the point of intersection of the diagonals.

2

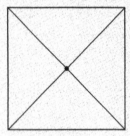

The centre of gravity of a triangular lamina is the point of intersection of the lines which join the vertices of the triangle to the midpoints of the opposite sides. C and E are the midpoints of AB and AD respectively. B and D are vertices.

How to locate the centre of gravity of irregularly shaped objects

You can find the centre of gravity of an irregularly shaped lamina or flat board by:

- trial and error. For example, by adjusting the lamina on a pencil point or other support until it balances;
- using a plumbline as described below.

1 Obtain a suitable lamina (this could be any shape cut from a piece of bristol board).
2 Pierce three widely spaced holes (A, B and C, say) near the edges of the lamina.
3 Support the lamina from hole A, using a pin held in a clamp as shown in Figure 4.16.
4 Hang a plumbline from A and draw the vertical, i.e. the line next to the string.
5 Repeat the procedure, using holes B and C in turn.
6 The point where the drawn verticals meet is the centre of gravity. As a test, place the experimentally determined centre of gravity of the lamina on the point of a pencil and see if it balances.

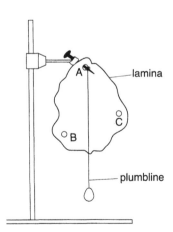

Figure 4.16

Centre of gravity and stability

Consider the three solids A, B and C in Figure 4.17.

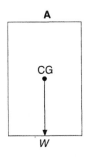
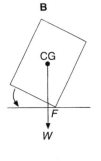
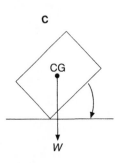

Here the weight of block A acts vertically down. There is no turning effect on the block, which remains resting on its base.

Block B is slightly tilted. Point F acts as a pivot. The weight, W, of the block falls inside the base. The anticlockwise turning effect results in the block falling back on its base.

Block C is tilted further to the right than block B. W acts outside the base. Block C topples under the influence of the clockwise turning effect.

Figure 4.17 *Centre of gravity and stability.*

Block A is in neutral equilibrium, block B is in stable equilibrium and block C is in unstable equilibrium.

A body which stays in its new position when slightly displaced is in **neutral equilibrium**.

A body is in **unstable equilibrium** if, when it is given a slight displacement, its centre of gravity falls.

A body is in **stable equilibrium** if, when it is given a slight displacement, its centre of gravity rises.

The stability of an object is determined by:

- the position of its centre of gravity. **The lower the centre of gravity, the more stable the object**.
- the surface area of its base. **The wider the surface area of the base, the more stable the object**.

continued

Figure 4.18 *A tightrope walker improves his stability by using a long pole.*

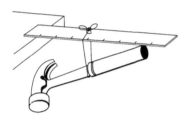

Figure 4.19 *This system does not topple. Why? The centre of gravity is normally right in the middle of the ruler. With the arrangement as shown, the centre of gravity is near the head of the hammer. The ruler balances from its tip.*

- Describe how you would use an adjustable inclined plane to show that stability depends on position of centre of gravity and on area of base.
- How would you improve the stability of a tall, narrow, hollow plastic toy?
- Look at the tightrope walker in Figure 4.18. Explain how the long pole improves his stability.

Some delicate balancing acts

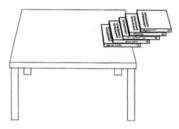

Figure 4.20 *These books are balanced over the edge of the table, with the top book way over the edge. Why is this system stable? The portion of the weight directly over the table is more than that of the part hanging over the edge. The centre of gravity of the system is not outside the table.*

Figure 4.21 *Explain as fully as you can why this system does not topple.*

Thinking it through

Discuss the physics of the following situations:

- the decision to place the heaviest items in the top drawer of a filing cabinet;
- placing a door handle on the same side of the door as the hinges.

4.5 Machines

The non-scientist thinks of a machine as a device which has moving parts and which makes work easier. Strictly speaking, however, a machine is any device which enables a force to be applied at one point to overcome another, usually larger, force at another point.

The applied force is called the effort, E.

The force moved or overcome is known as the load, L.

Since early times people have used machines to help them do work (Figure 4.22).

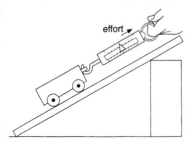

Figure 4.23 *An inclined plane. A hill and a ramp are examples of inclined planes. Give one other example.*

Figure 4.24 *A pencil sharpener. Identify the load and the effort in this machine.*

Figure 4.22 *The ancient Egyptians used rollers to move very large loads.*

Figures 4.23 and 4.24 show some common simple machines.

Machines serve many purposes.

- Machines multiply forces:

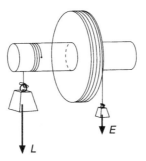

Figure 4.26 *A wheel and axle. This is a continuous lever. A small effort applied to the wheel gives a large force on the axle.*

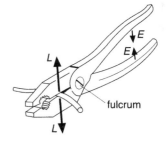

Figure 4.27 *Pliers. Explain how this machine is able to multiply the applied force (effort).*

Figure 4.25 *A block and tackle. A relatively small effort is needed to lift the engine out of the car.*

- Machines transfer energy:

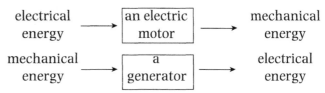

You can think of a screw as an inclined plane wrapped round a bar. What is meant by the pitch of a screw? Consider two screws: one of small pitch and the other of large pitch. Which of the two screws would need less effort for a given load?

- Machines transfer energy from one place to another:

| energy from the burning of gasoline in a motor car's engine | → | energy transferred via rods, shafts and axles . . . | → | to the wheels of the car |

- Machines increase speed. Observe that the rear wheel of a bicycle moves faster than the pedal which drives the sprocket.
- Machines change the direction of a force. Think of two common examples of machines that change the direction of a force.
- Machines can multiply distance (Figure 4.28).

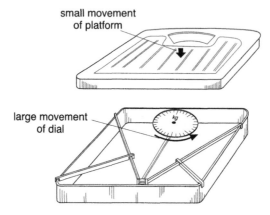

small movement of platform

large movement of dial

Figure 4.28 *Some machines, such as these bathroom scales, multiply distance.*

4.6 The efficiency of machines

Most machines multiply force, but they do not multiply the work done or the energy transferred. That is, the principle of conservation of energy is not violated (see page 96).

Friction affects the performance of all machines. Also, the moving parts of machines are not weightless. Work has to be done to move them. This means that no machine is 100% efficient and less energy or work is got out of machines than is put in.

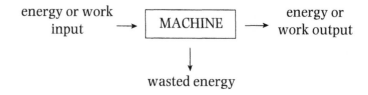

energy or work input → MACHINE → energy or work output

wasted energy

Problems

1 A machine is used to cut sheets of metal. Each sheet is cut in 0.025 s. Given that the efficiency of the machine is 90% and that the input power to the machine is 2.5 kW, calculate **a** the energy supplied by the machine and **b** the work done by the machine in cutting five sheets of metal.

2 List the factors which affect the efficiency of a machine.

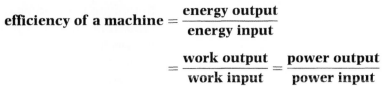

$$\textbf{efficiency of a machine} = \frac{\textbf{energy output}}{\textbf{energy input}}$$

$$= \frac{\textbf{work output}}{\textbf{work input}} = \frac{\textbf{power output}}{\textbf{power input}}$$

Efficiency may also be expressed as a percentage.

Efficiency can also be given in terms of mechanical advantage and velocity ratio.

Mechanical advantage is the ratio of the load moved (overcome) to the effort applied.

$$\text{M.A.} = \frac{\text{load}}{\text{effort}}$$

For example, if an effort of 12 N is needed to overcome a load of 60 N, then the mechanical advantage is 5.

Mechanical advantage is dimensionless (i.e. it has no unit). The mechanical advantage in any situation is affected by the amount of friction involved (why?).

Velocity ratio is the ratio of the distance moved by the effort to the distance moved by the load.

$$\text{V.R.} = \frac{\text{displacement of the effort per second}}{\text{displacement of the load per second}}$$

$$= \frac{\text{displacement of effort}}{\text{displacement of load}}$$

Velocity ratio, like mechanical advantage, is dimensionless. However, velocity ratio is not affected by friction.

It can be shown that

$$\% \textbf{ efficiency} = \frac{\textbf{mechanical advantage}}{\textbf{velocity ratio}} \times \textbf{100}$$

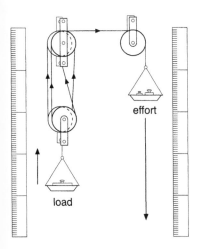

Figure 4.29

How to determine the efficiency of a pulley system

1 Set up the pulley system as shown in Figure 4.29, but with no weights on the 'effort' pan.
2 Record the initial position of the load and the effort, using the ruler.
3 Place increasing masses (weights) on the effort scale pan until the load just rises steadily, i.e. without any acceleration.
4 Record the value of the effort (weight) E which just gets the load rising steadily.
5 Record d_L and d_E – the distances moved by the load and effort respectively in the same time.

6 Determine the mechanical advantage and the velocity ratio from the results you obtain.
7 Repeat for different loads and obtain the corresponding values of E, d_L and d_E.
8 For each set of values calculate the mechanical advantage and velocity ratio.
9 Plot a graph of mechanical advantage versus velocity ratio. Comment on the shape of the graph and find the efficiency (from the gradient).

Checklist

After studying Chapter 4 you should be able to:

- define a force
- discuss, with examples, contact and non-contact forces
- draw and use force diagrams
- discuss, with examples, friction and frictional forces
- define the moment of a force and recall its unit
- define clockwise and anticlockwise moments
- recall and use the principle of moments
- define a lever
- discuss applications of levers
- define the centre of gravity
- locate the centre of gravity of regularly shaped flat objects
- locate the centre of gravity of irregularly shaped flat objects
- discuss equilibrium
- define stability
- recall the factors that determine stability
- discuss practical applications of stability
- define machines
- recall examples of simple machines
- recall the many purposes machines serve
- define and determine efficiency
- define and determine mechanical advantage and velocity ratio

Questions

1 a Explain what is meant by each of the following terms: moment, principle of moments, stability.

b A uniform rod of length 5 m and weight 20 N is supported at the 1 metre mark and at the 4 metre mark. A painter of weight 750 N stands at the 3 metre mark. Calculate the reaction at each support.

2 a Describe how you would determine the centre of gravity of an irregularly shaped object.

b Distinguish between neutral and stable equilibrium.

c What is the connection between the position of the centre of gravity and the stability of an object? Illustrate your answer by reference to a double decker bus.

d Mark the approximate position of the centre of gravity and describe the state of equilibrium for each of the following systems: (i) a basketball resting on a table, (ii) an empty milk bottle standing on its neck.

3 a What is meant by each of the following terms: lever, machine, mechanical advantage, velocity ratio, efficiency?

b A crowbar 2 m long is used to lift a rock weighing 800 N. If the fulcrum is 0.25 m from where the bar touches the rock, how much effort must be applied to move the rock? Draw a suitable force diagram.

c Why is the efficiency of machines less than 100%?

4

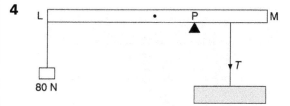

80 N

A uniform beam LM has a weight of 60 N and is 5 m long. It rests on a pivot at P. PM = 1.6 m. An 80 N weight hangs from L. A vertical rope attached midway between P and M prevents the beam from rotating. Calculate the tension, T, in the rope.

5 Consider the diagram below which shows a nutcracker.

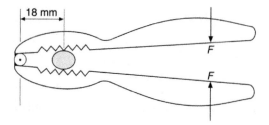

18 mm

a Why is the jaw of the nutcracker serrated?

b Discuss the advantage of making the handle of the nutcracker longer.

c On a copy of the diagram, label the fulcrum, load and effort.

d A force of 100 N is applied to the handle and a force of 650 N is needed to crack the nut. (i) What is the mechanical advantage of the nutcracker? (ii) What is the distance between the fulcrum and the point of application of the 100 N force? Take moments about the fulcrum.

6 a Describe how you would determine the mass of a metre rule, using the principle of moments, given the following: a knife edge, a metre rule and a 100 g mass.

b A seesaw is 3 metres long. It is balanced when child A sits 0.8 m from the fulcrum and child B sits 0.6 m from the fulcrum. Child A has a weight of 300 N. What is the weight of child B?

7 a What factors limit the efficiency of a machine?

b A man exerts an effort of 250 N through a distance of 4 m to lift a load of 1000 N through a height of 0.75 m. Calculate (i) the mechanical advantage, (ii) the velocity ratio, (iii) the efficiency (%).

5 Kinematics – the motion of objects

Note

Kinematics is concerned with the motion of objects and *not* with the forces that cause the motion.

Kinematics is concerned with describing the motion of objects using words, diagrams, numbers, graphs and equations. As you study this chapter and the others which follow, keep firmly in mind that physics is a mathematically based science.

The terms *scalar, vector, distance, displacement, speed, velocity* and *acceleration* are widely used in physics, especially to describe the motion of objects. It is in your best interest to acquire as clear an understanding of these terms as possible.

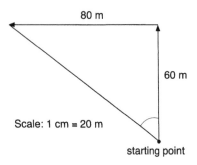

80 m
60 m
Scale: 1 cm ≡ 20 m
starting point

Figure 5.1 *From the scale, you are 100 m from your starting point in the direction shown (53.1° west of north).*

5.1 Scalars and vectors

Some physical quantities have size (magnitude) only. These are called **scalar** quantities. Examples are mass, time, speed, distance, volume and energy. Scalar quantities are added by ordinary arithmetic. For example, $5.2\,\text{kg} + 3.8\,\text{kg} = 9.0\,\text{kg}$.

By contrast, if you walk 60 m due north and then 80 m due west you are not 140 m from your starting point: you cannot simply add the quantities arithmetically in this situation. 60 m due north is a displacement. Quantities such as displacement, velocity, acceleration, force and momentum have both size (magnitude) and direction. They are **vector** quantities.

Note

In vector diagrams:

- give a scale;
- draw an arrow of precise length, consistent with the scale;
- indicate clearly the direction of the vector.

25 N
a force of 25 N due east
30 N
30°
Scale:
1 cm ≡ 10 N
a force of 30 N at 30° to the horizontal

Figure 5.2

Vectors are represented on diagrams by straight lines. The length of the line represents the magnitude of the vector. An arrow on the line indicates the direction or line of action. In Figure 5.3, the vectors **a** and **b** are *parallel* vectors. They act in the same direction on the object. The combined effect or **resultant** of these two vectors is **a** + **b**.

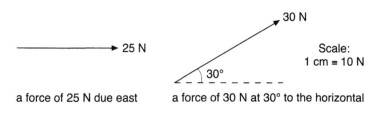

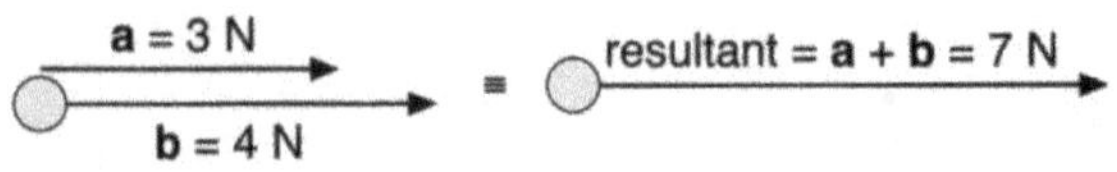

a = 3 N
b = 4 N
≡
resultant = a + b = 7 N

Figure 5.3 *Parallel vectors.*

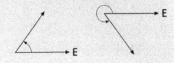

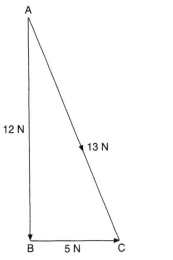

Figure 5.5 Scale drawing also gives a value of 13 N for the resultant. The angle between the 12 N force and the resultant is 22.6°.

The vectors **c** and **d** in Figure 5.4 are *anti-parallel*. Their combined effect = **d** − **c**. Why?

$$c = 3\,N \qquad d = 6\,N \qquad \equiv \qquad d - c = 3\,N$$

Figure 5.4 *Anti-parallel vectors.*

Consider the vector AB of magnitude (size) 12 N and the vector BC of magnitude 5 N; the two vectors being at right angles to each other (Figure 5.5). The resultant, AC, can be obtained from the relationship:

$$AC^2 = AB^2 + BC^2$$
$$= 144 + 25$$
$$= 169\,N^2$$
$$AC = 13\,N$$

Adding vectors using the parallelogram law

The resultant of any two vectors, irrespective of the angle between them is found by constructing a parallelogram of vectors. To do this, construct the parallelogram as you would in mathematics. The diagonal of the parallelogram gives the magnitude and direction of the resultant.

Example

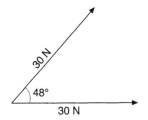

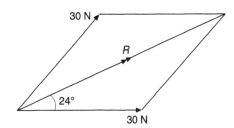

Resultant = 52 N. The resultant makes an angle of 24° with the 30 N horizontal force.

Solution

a Draw the vectors to scale, ensuring that the angle between them is accurately measured.

b Complete the parallelogram, and draw the diagonal.

c Measure the length of the diagonal and the angle between the diagonal and the horizontal vector.

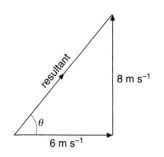
Figure 5.6 *Vector addition.*

Resolving vectors

A vector which is directed at an angle may be resolved into two mutually perpendicular directions (directions at right angles to each other) – e.g. one vertical and the other horizontal.

Relative velocity

Objects may move within a medium or substance which is itself moving with respect to an observer. Consider a plane moving eastwards at 200 km h^{-1} which experiences a tail wind of 25 km h^{-1}. The velocity of the plane relative to an observer on the ground is $(200 + 25)$ km h^{-1}. If instead the plane runs into a head wind of 40 km h^{-1}, the velocity of the plane relative to a ground level observer is $(200 - 40)$ km h^{-1}.

Suppose a boat moves northwards across a river at 8 m s^{-1} while the water moves to the east at 6 m s^{-1}. What is the velocity of the boat relative to an observer on the opposite bank?

To determine the resultant and its direction, use the method of vector addition (Figure 5.6).

From Pythagoras' theorem:

$$R^2 = 6^2 + 8^2$$
$$R = 10 \text{ m s}^{-1}$$

The direction of the resultant is angle θ, the anticlockwise angle between the due east direction and the resultant.

From trigonometry:

$$\theta = \tan^{-1}\frac{8}{6} = \tan^{-1} 1.33 = 53°$$

5.2 Describing linear motion with words

Distance and displacement

Distance and displacement may seem to be the same 'thing' but they have very different meanings. Whereas distance is a scalar quantity, i.e. it has only magnitude, displacement is a vector quantity. It is described by both its magnitude and its direction.

Essentially, distance is concerned with how much ground an object has covered. Displacement, on the other hand, is concerned with the change in the position of the object. Displacement takes the direction of the movement into account.

Figure 5.7 shows someone starting at point A, moving to point B, then to C, then to D and back to A.

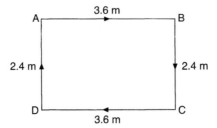

Figure 5.7

The person has walked a distance of

$$(3.6 + 2.4 + 3.6 + 2.4)\,m = 12.0\,m$$

However, there is no displacement – displacement is 0, as the person has not changed position relative to the starting point A.

Speed and velocity

Speed and velocity have different meanings. Speed is a scalar quantity. Velocity is a vector quantity. Speed is 'indifferent' to direction. Velocity is direction 'dependent'.

The speed of an object is an indication of how fast it is moving. If the speed of an object varies during its motion, you need to compute its **average speed**.

$$\text{average speed} = \frac{\text{total distance travelled}}{\text{total time taken}}$$

The **instantaneous speed** refers to the speed at any instant.

The average speed is the average of all the instantaneous speeds.

Consider two cars, each moving due west at $5\,m\,s^{-1}$. These objects have the same speed and the same velocity since they are moving in the same direction. If one car is moving due east at $10\,m\,s^{-1}$ and the other moving due west at $10\,m\,s^{-1}$, both cars have the same speed but their velocities are different because they are heading in different directions.

The vector quantity velocity is the 'rate at which an object changes its position', i.e. velocity is rate of change of displacement.

Instantaneous velocity is the velocity at any given instant. **Average velocity** is defined as:

$$\frac{\text{change in total displacement}}{\text{total time taken for the change}}$$

Examples

1 A motor cyclist travelled a distance of 20 km in 30 minutes, followed by 10 km in 18 minutes, followed by 15 km in 12 minutes. What is the average speed of the motor cyclist in **a** $km\,h^{-1}$ **b** $m\,s^{-1}$?

Solution

a Total distance travelled $= (20 + 10 + 15)$ km

$$\text{Time taken} = (30 + 18 + 12)\ \text{min}$$
$$= 1\ h$$

$$\text{Average speed} = \frac{\text{total distance}}{\text{total time}}$$
$$= 45\ km\,h^{-1}$$

b To convert $km\,h^{-1}$ to $m\,s^{-1}$ multiply by 5 and divide by 18:

$$1\ km\,h^{-1} = 1000\ m/(60\ s \times 60\ s) = \frac{5}{18}\ m\,s^{-1}$$

$$\text{Answer} = 12.5\ m\,s^{-1}$$

2 A motor car travelled due east at $60\ km\,h^{-1}$ for $\frac{1}{2}$ h and then due north at $120\ km\,h^{-1}$ for $\frac{1}{4}$ h.

  a What is the displacement of the car relative to its starting point?

  b For how long was the car travelling?

  c What is the average velocity of the car?

Solution

The car travelled 30 km due east and then 30 km due north.

a Displacement $= R$ km $= 42$ km

b The total time of travel $= \frac{3}{4}$ h

c Average velocity $= 56\ km\,h^{-1}$

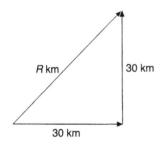

R km   30 km

30 km

Acceleration

An object is accelerating only if its velocity is changing (increasing or decreasing). If the velocity of an object remains constant then it has zero acceleration.

Table 5.1 shows the velocity of an object over 6 seconds.

Table 5.1

Velocity/$m\,s^{-1}$	0	8	16	24	32	40	48
Time/s	0	1	2	3	4	5	6

Figure 5.8 *The parachutist falls eventually with terminal velocity (constant velocity), i.e. he is not accelerating.*

You can see that the velocity of this object is changing by a constant amount each second. The object is moving with *constant* acceleration.

A body is said to have constant or uniform acceleration if it experiences equal changes in velocity in equal successive time intervals.

For the moving object in Table 5.2 the velocity is changing every second but the change is not constant or regular. This object is not moving with constant acceleration. Examples of objects which move with varying acceleration include a swinging pendulum or a weight attached to a spring which is pulled down and released.

Table 5.2						
Velocity/m s^{-1}	0	14.14	18.48	19.3	19.6	19.74
Time/s	0	1	2	3	4	5

Acceleration is defined as the 'rate at which an object changes its velocity'.

$$\text{acceleration } (a) = \frac{\text{change of velocity } (\Delta v)}{\text{time } (t)}$$

Acceleration is a vector quantity. Its sign depends on whether the object is speeding up or slowing down. If it is positive, the object is speeding up and the acceleration is in the same direction as the velocity. If it is negative, the object is slowing down or decelerating and the acceleration is in the opposite direction to the velocity.

Thinking it through

1 Look at the data in the tables below. For each set of data determine, explaining how you arrived at your answer, if

- the object is moving with constant or varying acceleration;
- the object is accelerating or decelerating.

a	Velocity/m s^{-1}		−8	−6	−4	−2	0
	Time/s	0	0.5	1.0	1.5	2.0	

b	Displacement/m	0	7.5	10.6	13.0	14.5	15
	Time/s	0	1.0	1.5	2.0	2.5	3.0

c	Displacement/m	0	3.5	14	31.5	56	87.5
	Time/s	0	1	2	3	4	5

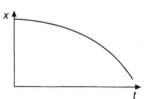
5.3 Describing linear motion with graphs

Displacement–time graphs

A graph of displacement (x) versus time (t) is shown in Figure 5.9.

The object whose motion is described by Figure 5.9

- starts from rest;
- is not accelerating: there are equal changes of position in equal intervals of time – the object has constant velocity.

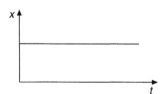

Figure 5.9 *Displacement versus time.*

Figure 5.10 *This object is moving slowly, with constant velocity.*

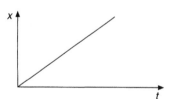

Figure 5.11 *This object is moving quickly, with constant velocity.*

Figure 5.12 *This object is moving back towards its starting point, slowly, with constant velocity.*

Figure 5.13 *Here, the displacement remains constant over time. This object is stationary. Velocity equals zero since slope is zero.*

Figure 5.14 *This graph is a curve. The slope is increasing. Therefore, velocity is increasing with time. This object is accelerating.*

Figure 5.15 *This graph is also a curve. Velocity is decreasing with time. This object is decelerating.*

Velocity–time graphs

- The *slope* of the velocity–time graph for an object indicates its acceleration ($a =$ [change in v]$/t$).
- The *area* under a velocity–time graph for an object gives the total distance travelled by the object ($s = v \times t$).

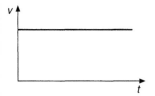

Figure 5.16 *This object is moving with zero acceleration (constant velocity). The slope of this graph is zero.*

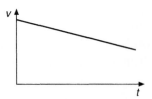

Figure 5.17 *This object is decelerating. The slope is negative.*

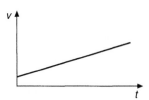

Figure 5.18 *This object is accelerating. The slope is positive.*

The velocity–time graph of Figure 5.19 describes the motion of an object over 7 seconds.

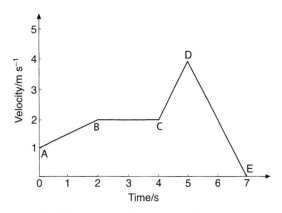

Figure 5.19 *Velocity versus time.*

From A to B, the object moves with constant acceleration of $0.5\,\mathrm{m\,s^{-2}}$ ($\Delta v = 1\,\mathrm{m\,s^{-1}}$ and $\Delta t = 2\,\mathrm{s}$).

From B to C, the object moves with constant velocity ($2\,\mathrm{m\,s^{-1}}$); its acceleration is 0.

From C to D, the object is again accelerating.
Acceleration $= 2\,\mathrm{m\,s^{-2}}$.

From D to E, the object is uniformly decelerated.
Acceleration $= -2\,\mathrm{m\,s^{-2}}$.

Copy the diagram on graph paper. Divide the figure into triangles and squares (or trapezia) and show that the total area under the graph is $14\,\mathrm{m^2}$.

Note

The unit of velocity ($\mathrm{m\,s^{-1}}$) multiplied by the unit of time (s) gives the unit of distance. The area under a velocity–time graph is the total distance covered.

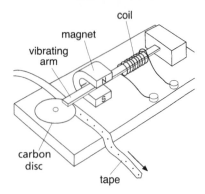

Figure 5.20 *A ticker timer.*

5.4 Describing linear motion with diagrams

Ticker tape diagrams and **vector diagrams** are commonly used to describe the motion of objects. In this section the focus will be on ticker tape diagrams.

A ticker device is shown in Figure 5.20. A long tape is attached to a moving object and threaded through the device which imprints on it a trail of dots at regular intervals of, say, every 0.02 seconds. The details of the motion of the object can be deduced from the pattern of dots or 'ticks' on the tape.

The spacings between dots represent changes in the position of the object as a function of time. Widely spaced dots indicate that the object is moving quickly. If the spacings between dots are small, then the object is moving slowly. For which of Figures 5.21(a) and (b) is the object moving faster?

When dots are evenly spaced as in Figures 5.21(a) and (b) the object is moving with constant velocity: acceleration is zero.

Uneven spacing of the dots (Figures 5.21(c) and (d)) indicates that velocity is not constant over equal intervals of time, i.e. the object is accelerating or decelerating.

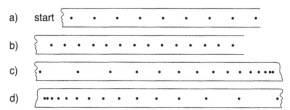

Figure 5.21 *The object whose motion is described by tape (c) is decelerating. The object whose motion is described by tape (d) is accelerating.*

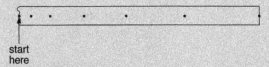

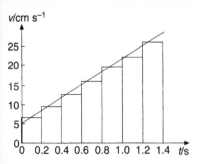

Figure 5.22 *A ticker tape chart. Use the tape chart to determine the acceleration of the object. Explain the steps in your calculation.*

Ticker tapes can be used directly to make charts of velocity versus time. Cut the tape into 10-dot sections. (Note that if the object is accelerating the sections get longer.) Use the sections to make a tape chart like the one in Figure 5.22.

Thinking it through

Carefully analyse the tape charts below. Describe in detail the motion depicted in each. Which tape corresponds to:

a uniform velocity; **b** uniform acceleration; **c** uniform deceleration; **d** variable acceleration; **e** zero acceleration?

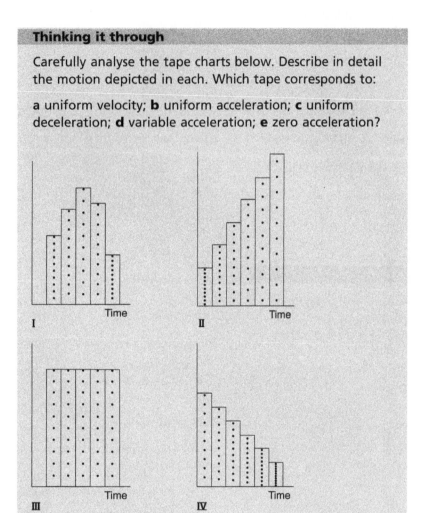

5.5 Describing linear motion with equations

Let the initial velocity of an object be u m s^{-1}
Let the final velocity of an object be v m s^{-1}
Let the acceleration (assumed constant) be a m s^{-2}
Let the interval for the motion be t s
Let the displacement be s m

By definition, acceleration equals the rate of change of velocity.

Example

An object accelerates uniformly at $5\,\mathrm{m\,s^{-2}}$ for 4 s. What velocity change does it undergo? If the object starts from rest what is its final velocity?

Solution

$$v - u = a \times t$$
$$= (5 \times 4)\ \mathrm{m\,s^{-1}}$$

Since object starts from rest,

$$\text{initial velocity} = 0\ \mathrm{m\,s^{-1}}$$
$$\text{final velocity} = 20\ \mathrm{m\,s^{-1}}$$

Problem

An object accelerating uniformly for 3 seconds reached a final velocity of $30\,\mathrm{m\,s^{-1}}$. If its initial velocity was $12\,\mathrm{m\,s^{-1}}$, what is:

a its average velocity;
b its acceleration;
c the distance travelled in the 3 seconds?

Note

Equations of linear motion

$$s = \left(\frac{u+v}{2}\right)t$$

$$v = u \pm at$$

$$s = ut \pm \tfrac{1}{2}at^2$$

$$v^2 = u^2 + 2as$$

These are used only when acceleration is uniform.

Therefore,

$$\text{acceleration } (a) = \frac{\text{change of velocity } (v - u)}{\text{time } (t)}$$

or

$$v = u + at$$

More generally,

$$v = u \pm at \tag{1}$$

where the $-$ sign is used to indicate deceleration,

Distance covered (displacement) = average velocity × time

$$s = \frac{u+v}{2} \times t$$

$$s = \left[\frac{u + (u + at)}{2}\right] \times t$$

$$= ut + \tfrac{1}{2}at^2$$

More generally,

$$s = ut \pm \tfrac{1}{2}at^2 \tag{2}$$

By combining equations (1) and (2) it can be shown that:

$$v^2 = u^2 \pm 2as \tag{3}$$

Examples

1 An object accelerates uniformly from $10\,\mathrm{m\,s^{-1}}$ to $30\,\mathrm{m\,s^{-1}}$ in 5 s. What is:

    a the acceleration of the object;
    b the distance covered in the 5 s?

Solution

$u = 10\,\mathrm{m\,s^{-1}}, v = 30\,\mathrm{m\,s^{-1}}, t = 5\,\mathrm{s}, a = ?, s = ?$

a Using the equation $v = u + at$, we get

$$30 = 10 + a \times 5$$
$$5a = 20$$
$$a = 4\ \mathrm{m\,s^{-2}}$$

b Using $v^2 = u^2 + 2as$, we get

$$30^2 - 10^2 = 2 \times 4 \times s$$
$$s = 100\ \mathrm{m}$$

2 A car is travelling at $60\,\mathrm{m\,s^{-1}}$ when the traffic light turns red. The car decelerates at $6\,\mathrm{m\,s^{-2}}$.

    a How long does the car take to stop?

    b How far from the lights must the driver start braking in order not to overshoot the lights?

Solution

$u = 60\,\mathrm{m\,s^{-1}}, v = 0\,\mathrm{m\,s^{-1}}, a = -6\,\mathrm{m\,s^{-2}}, t = ?, s = ?$

Using $v = u + at$, we get $t = 10\,\mathrm{s}$.
Using $s = ut + \frac{1}{2}at^2$, we get $s = 300\,\mathrm{m}$.

Problem

1 A bus leaves one stop, accelerates at $2\,\mathrm{m\,s^{-2}}$ for 3 seconds, then travels at constant velocity for 4 minutes, then decelerates at $5\,\mathrm{m\,s^{-2}}$ to the next stop. What is the distance between the two stops?

2 How far does an object move in the third second if it starts from rest with a uniform acceleration of $8\,\mathrm{m\,s^{-2}}$? (*Hint:* find the distance travelled after $4\,\mathrm{s}$ and the distance travelled after $3\,\mathrm{s}$.)

5.6 Motion in a vertical plane – free-fall and the acceleration due to gravity

Note

A moving object which is acted upon only by the force of gravity is said to be in **free-fall**.

So far in this chapter we have been studying motion in a horizontal plane. The focus will now be on motion in a vertical plane and on free-falling objects in particular.

Objects in free-fall:

- experience no air resistance or other frictional forces;
- accelerate downwards at the rate of about $10\,\mathrm{m\,s^{-2}}$. The numerical value for this acceleration is commonly called **acceleration due to gravity** (g);
- can be described by the following equations of motion (see page 66:

$$v = u + gt$$
$$h = ut + \tfrac{1}{2}gt^2$$
$$v^2 = u^2 + 2gh$$

where h is the height through which the object falls.

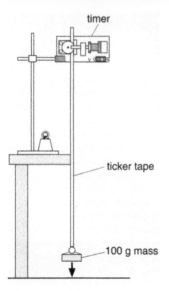

Figure 5.23

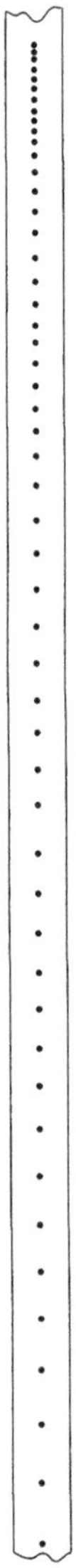

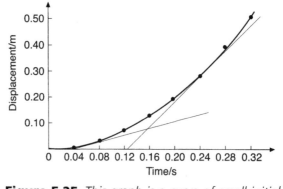

Figure 5.24

How to obtain and analyse data on objects in free-fall

1. Clamp a ticker timer about 2 metres above the desk or ground as shown in Figure 5.23.
2. Thread 2 metres of tape into the timer; attach a 100 g mass to the leading end of the tape.
3. Switch on the timer and allow the mass to fall freely downwards.
4. Examine the pattern of dots on the tape.
5. Describe the pattern fully. The pattern of dots should be similar to that of Figure 5.24.
6. Construct a suitable tape chart. (You may find it convenient to use two-dot strips.) This tape chart is essentially a graph of velocity versus time. (Explain why this is so; if not sure, consult your teacher. Keep in mind that the spacings between ticks (dots) represent a specific time interval. This time interval depends on the frequency of the timer.)
7. Use your tape chart to obtain a value for the acceleration due to gravity. Explain the basis of your calculation.

A student performed this experiment and carefully analysed data from a tape chart. She presented the data as shown in Table 5.3 and plotted graphs of displacement versus time (Figure 5.25) and velocity versus time (Figure 5.26).

Table 5.3

Displacement/m	Time/s	Displacement/m	Time/s
0.008	0.04	0.200	0.20
0.032	0.08	0.280	0.24
0.072	0.12	0.392	0.28
0.128	0.16	0.512	0.32

Figure 5.25 *This graph is a curve of small initial slope and large final slope. Velocity is small initially but increases as the mass falls. The mass is accelerating downwards.*

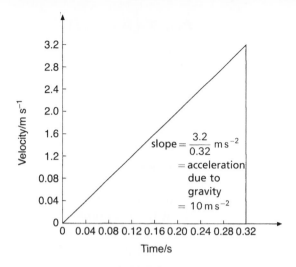

Figure 5.26 *This graph is a straight line. The object is falling with constant acceleration. It starts with zero velocity and finishes with large velocity. Calculation shows that the slope has a value of $10\,m\,s^{-2}$. The object, therefore, fell freely.*

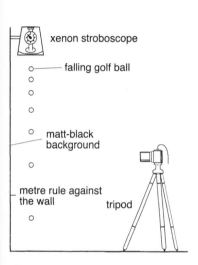

Figure 5.27

Note

A stroboscope is a light that flashes at regular intervals.

How to use multi-flash photography to investigate free-fall

1 Hold a 2 metre black rule (white scale) against a black background.
2 Set up a stroboscope as shown in Figure 5.27. Adjust the frequency to obtain a convenient number of flashes per second.
3 Switch on the strobe. Simultaneously release the golf ball and open the shutter of the camera.
4 Close the shutter when the ball strikes the table.
5 Repeat to obtain several photographs of the falling ball.

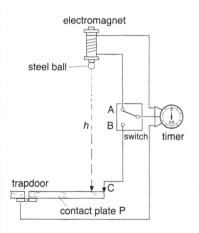

Figure 5.28

How to find a value for the acceleration due to gravity

In this experiment a value for g is obtained by timing the direct fall of a steel ball.

1 Set up the apparatus of Figure 5.28. When the current is switched on, the electromagnet holds the steel ball.
2 Open the two-way switch. This releases the steel ball (why?) and automatically starts the timer.
3 The steel ball falls through a height h metres and strikes the contact plate P. This breaks the circuit and stops the timer. Note the time of fall.
4 Vary h and find the corresponding times of fall.
5 The equation $h = ut + \frac{1}{2}gt^2$ can be used to find a value for g. $2h = gt^2$ since the object starts from rest and $u = 0$. A graph of $2h$ (y-axis) against t^2 gives a straight line, the slope of which is g.

Thinking it through

Two fourth-formers attempted to find a value for *g* as follows:

- One student climbed to the top of the science lab and released a steel ball.
- His co-worker started a stop watch at the moment she got a signal that the ball was being released. She stopped the watch the instant the ball hit the ground.
- They repeated the above procedure three times and took the average time for the ball to reach the ground as the true time of fall.
- Explain how they could **a** measure the height through which the ball fell and **b** determine a value for *g* from their readings.
- Discuss the weaknesses in their experiment. What precautions should they have taken?

Things to do

In an experiment carried out by a group of fifth-formers the data in Table 5.4 were obtained.

Table 5.4

Height, h/m	0.50	0.75	1.00	1.25	1.50	1.75	2.00
$2h$/m							
Time, t/s	0.32	0.39	0.45	0.50	0.55	0.59	0.64
t^2/s^2							

1 Copy and complete the table by filling in values of $2h$ and t^2.
2 Plot a graph of $2h$ versus t^2.
3 Find a value for *g*.

5.7 Motion in two dimensions – projectiles

So far we have been discussing motion in one dimension only. Many objects, however, move in two dimensions. Projectiles are common examples of objects which move in two dimensions. Many sports involve throwing balls, i.e. they involve the physics of projectiles.

A projectile is any object which, given an initial velocity, continues in motion by its own inertia and is influenced only by the downward force of gravity. A spacecraft in orbit (Figure 5.29) is a type of projectile.

Figure 5.29 *The motion of an orbiting space station is influenced only by the Earth's gravity.*

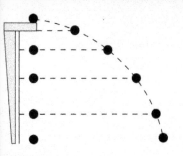

Figure 5.30 *Analysing the motion of a projectile.*

We can analyse the motion of projectiles in terms of their horizontal and their vertical motions.

- The vertical motion of an object projected horizontally is identical to that of an object dropped from rest. In Figure 5.30 the object dropped from rest reaches the ground at the same time as the object projected horizontally at the same time.
- Projectiles travel horizontally at constant speed. The horizontal speed of a projectile is the same at the end as at the beginning of its range.

Let us pretend that gravity can be turned on or off at the flick of a switch.

Hypothetical case 1

Gravity switch OFF. An object is projected horizontally to the right as in Figure 5.31(a) at $30\,\mathrm{m\,s^{-1}}$.

In the absence of gravity, the object will continue to move with the same speed and in the same direction. The motion here is consistent with Newton's first law of motion (see Chapter 6).

(a)

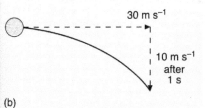

(b)

Figure 5.31

Hypothetical case 2

Gravity switch ON. An object is again projected to the right at $30\,\mathrm{m\,s^{-1}}$. The object will continue to move to the right at $30\,\mathrm{m\,s^{-1}}$, while falling simultaneously under gravity (see Figure 5.31(b)). Since no force acts in the horizontal direction there can be no acceleration in this direction. (According to Newton's law, a force is needed for acceleration but not for motion at constant velocity. See page 76.) With the gravity switch ON, there is simultaneous motion in two directions and the trajectory of the object is as shown in Figure 5.31(b).

Table 5.5 summarizes the motion of projectiles.

Note

The vertical motion of a projectile is influenced only by the force of gravity.

When thinking about projectiles or when solving problems on projectiles it is useful to divide the motion into its vertical and horizontal components.

The path of a projectile through the air is a curved one called a **parabola**.

Table 5.5	The motion of projectiles	
	Horizontal	**Vertical**
Velocity	constant	changing
Acceleration	zero	$10\,\mathrm{m\,s^{-2}}$
Forces	none	gravity

This table tells us there is no acceleration in the horizontal direction (acceleration $= 0$). It follows that velocity remains constant in this direction.

There is a constant acceleration of 10 m s^{-2} in the vertical direction. Velocity, therefore, increases uniformly with time in this direction in accordance with the equation for free fall: $v = gt$.

To summarize: A projectile is an object on which the only force acting is gravity. Gravity determines the vertical acceleration of projectiles. Since there are no horizontal forces, projectiles move with constant horizontal velocity.

5.8 Going round the bend – circular motion

No net force acts on an object which is moving in a straight line at constant speed. Consider an object which is moving with uniform circular motion (that is, in a circle with constant speed). It moves with constant speed but not with constant velocity. Why? Because its *direction* is constantly changing.

In Figure 5.32, at position 1 the velocity vector is directed northwards. At position 2, the velocity is directed eastwards. Since the velocity is changing, the object is accelerating. The acceleration is directed towards the centre of the circle and is described as 'centripetal'.

If an object in circular motion is accelerated towards the centre of the circle, there must be a centripetal force causing this acceleration. This centripetal force changes the direction of the object without changing its speed.

The centripetal force, F_c, which keeps an object moving in a circle of radius r is given by the equation:

$$F_c = \frac{mv^2}{r}$$

where v is the linear speed of the object.

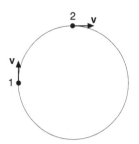

Figure 5.32 *Velocity vectors.*

Note

'Centripetal' means 'seeking the centre'.

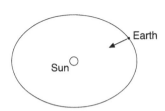

Figure 5.33 *The Earth orbits the Sun. The gravitational force of attraction between Earth and Sun provides the centripetal force required for the circular motion. Gravity keeps the Moon in near-circular motion around the Earth.*

Figure 5.34 *This car is going round the bend. The frictional force acting on the tyres of the car provides the required centripetal force.*

Figure 5.35 *The rubber bung is whirled in a circle. Identify the centripetal force.*

Note

Weightlessness is a condition which is experienced when all contact forces are removed.

Roller coaster riders who are temporarily lifted out of their seats experience weightlessness.

Astronauts who are orbiting the Earth experience the sensation of weightlessness.

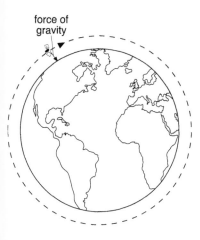

Figure 5.36 *Centripetal force.*

Things to do

Design experiments to:

a show that a force must be exerted on an object to keep it moving in a circular path;

b show that the acceleration of an object moving in a circular path depends on the square of the speed of the object and inversely on the square of the radius of the circular path.

It follows that F_c must increase if:

- the mass of the object is increased;
- the speed of the object is increased;
- the radius of the circle is decreased.

You can determine the speed of an object in circular motion from the relationship:

$$v = \frac{2\pi r}{T} = \frac{\text{distance travelled round circle}}{\text{time}}$$

where T is the time for one complete orbit of the object. T is called the **period** of the orbit. If the distance of an object from the centre of the Earth is known then its period can be calculated. Objects, natural or artificial, which orbit the Earth are known as **satellites**.

If a satellite is put into orbit with too low a speed it falls back to Earth. If the speed is too great it flies off into space. However, with the correct speed it orbits the Earth and is held in motion by a centripetal force – the force of gravity which is directed towards the centre of the Earth (see Figure 5.36).

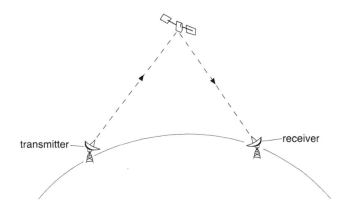

Figure 5.37 *A geostationary communications satellite link.*

If a satellite is placed in orbit directly above the equator, so that its radius of orbit is 42 000 km, then that satellite takes 24 hours to circle the Earth. Such a satellite is called a geostationary satellite. Geostationary satellites play important roles in telecommunication especially in telephony and in television transmissions. Six suitably positioned geostationary satellites could provide adequate coverage for the entire globe.

In Figure 5.37, a microwave beam 'laden with information' is directed towards a geostationary satellite (the up-link). The satellite receives the beam and redirects it back to an Earth station at a different frequency (the down-link).

Figure 5.38 *The predominant physics principles underlying rides at amusement parks are those of projectile motion and circular motion.*

Checklist

After studying Chapter 5 you should be able to:

- define scalar and vector quantities, giving examples
- determine the resultant of any two vectors
- resolve any vector directed at an angle into two mutually perpendicular directions
- recall and use the concept of relative velocity
- understand the differences between displacement and distance, and between speed and velocity
- recall and use the equation: average speed $=$ distance/time
- recall and use the equation: average velocity $=$ displacement/time
- define acceleration
- recall that $s \propto t$ for objects moving with zero acceleration
- recall that acceleration $= \Delta v / \Delta t$
- describe linear motion graphically
- recall that velocity is the gradient of a displacement–time graph
- recall that speed is the gradient of a distance–time graph
- recall that acceleration is the gradient of a velocity–time graph
- recall that the total distance covered is the area under a velocity–time or speed–time graph
- use ticker tape printouts to describe the motion of objects
- recall and use the equations of linear motion
- discuss motion in a vertical plane
- discuss projectile motion
- discuss circular motion
- define centripetal force
- recall practical applications of circular motion

Questions

1 Describe how you would determine experimentally the average speed of a sprinter in a 100 metre race. Explain how his speed at the 5th second may be determined.

2 A ball strikes a wall at $20 \, m \, s^{-1}$ and rebounds at $10 \, m \, s^{-1}$. What is the velocity change? (*Hint:* velocity is a vector quantity.)

3 A car is travelling east at $45 \, m \, s^{-1}$ when the street lights turn red. If the car decelerates at $5 \, m \, s^{-2}$, how long does it take to stop? How far away from the street light must the car start braking in order to stop in time?

4 A motor cyclist travels due north. His velocity was recorded at intervals up to 60 seconds as shown in the table.

Velocity/$m \, s^{-1}$	0	12	24	36	48	48	48
Time/s	0	10	20	30	40	50	60

  a Plot a graph of velocity versus time.
  b Use the graph to determine:
    (i) the initial acceleration;
    (ii) at what time the motor cyclist first reaches maximum velocity;
    (iii) the total distance covered by the motor cyclist in the 60 seconds.

5 A ball strikes a wall at $12 \, m \, s^{-1}$ and rebounds with velocity $8 \, m \, s^{-1}$ after an interval of $0.1 \, s$. What acceleration does the ball experience?

6 What is the acceleration of an object which, starting from rest, reaches a velocity of $25 \, m \, s^{-1}$ in $5 \, s$?

7 The tape chart shown was obtained from a moving trolley. The timer was connected to a $50 \, Hz$ mains supply. (This means that dots were made every 0.02 seconds.) The tape was cut into 5-dot strips.

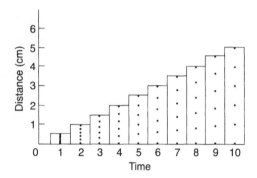

Calculate:
a the average speed during the 5th second;
b the average speed during the 10th second;
c the acceleration of the trolley.

8 A car starts from rest, and moves initially with an acceleration of $5 \, m \, s^{-2}$ for $5 \, s$. It then travels at constant velocity for $50 \, s$, finally coming to rest with uniform retardation in $10 \, s$. What is the retardation? Draw a velocity–time graph and determine the total distance covered.

9 An Olympic gold medallist accelerates from rest to $10 \, m \, s^{-1}$ in $1.25 \, s$. What is his acceleration and the distance travelled in this time?

10 A car, starting from rest, travels $100 \, m$ in $8 \, s$. Assuming constant acceleration, find the average speed for these $8 \, s$. What is the acceleration and the distance covered in the first 5 seconds?

6 Exploring the link between forces and motion

Note

The branch of mechanics which deals with the motion of objects under the action of forces is called **dynamics**.

In Chapter 5 we studied motion, but paid no attention to the forces which cause motion. In this chapter we focus on the links between forces and motion.

The Greek philosopher Aristotle (384–322 BC) believed that inanimate objects did not move by themselves but were always set in motion by something else from outside – what we would now call a force. If you push a book sideways on a table it moves. If you stop pushing, it stops. The harder you push the faster it moves. Observations like these led Aristotle to conclude that the velocity of an object is proportional to the force that makes it move. Aristotle thought that $v \propto F$. But this idea does not explain why, for example, when you throw or roll a ball it goes on moving after you stop pushing it.

Note

Newton based his laws of motion on the observations of Galileo (1564–1642) and others.

Almost 2000 years later, Newton, the 17th century mathematician and scientist, described how objects moved using three laws – commonly referred to as **Newton's laws of motion**.

No net force

$$\sum F = 0$$

acceleration of object $= 0$

| Object at rest stays at rest velocity $= 0$ | Moving objects continue to move with constant velocity |

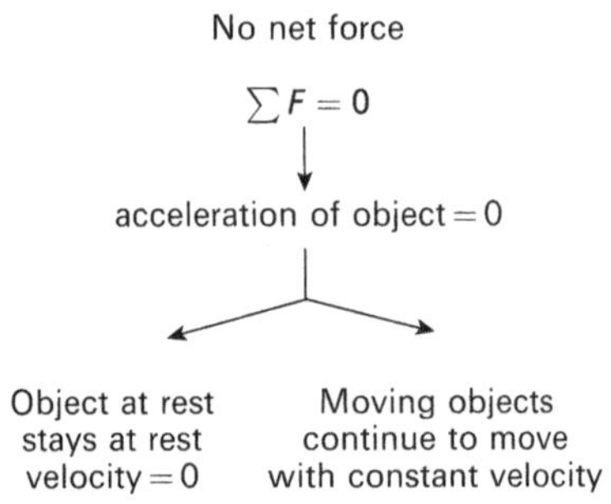

Figure 6.1 *Newton's first law of motion.*

6.1 Newton's first law of motion

An object at rest tends to stay at rest and an object in motion tends to stay in motion with the same velocity, unless the object is acted upon by an unbalanced force.

Stated another way, when *no net (resultant) force* acts on an object, if it was initially at rest it would remain at rest, and if it was moving with constant velocity it would continue to move at that constant velocity. Newton's first law of motion may be summarized as in Figure 6.1.

Essentially, the first law tells us that for an object in motion no additional force is required to keep it moving. But does this not seem at variance with our everyday experience? Before Galileo's and Newton's time it was believed that all moving objects eventually come to rest.

Note

$\sum F$ means the sum of the forces, here added as vectors, i.e. the net or resultant force.

To account for the behaviour of objects at rest and of moving objects, when no net force is acting, Galileo introduced the concept of 'inertia'. He also reasoned that moving objects (in the absence of a net or unbalanced force) come to rest eventually because of friction – that force which opposes motion.

Figure 6.2

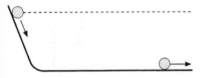

Figure 6.3 *The ball continues for ever, if friction is removed.*

Note

When an object is standing still it takes a force to get it moving. Once an object is moving it takes a force to stop it.

All objects resist changes in their state of motion.

All objects have inertia.

Mass is a quantitative measure of inertia.

The greater the mass of an object the greater its inertia.

Figure 6.4 *The bottle and coin trick.*

From his experiments with inclined planes, Galileo deduced that if there was no friction at all, then a ball released at the left side in Figure 6.2 would reach the same height at the right.

He further argued that if friction could be completely eliminated, a ball released as in Figure 6.3 would continue in motion for ever. Newton used Galileo's ideas as a platform for formulating the first law of motion.

Inertia

Inertia is the resistance an object has to change in its state of motion. Imagine two objects of identical dimensions, one made of clay and the other of lead. The objects are painted black so that they cannot be distinguished by looking, and are lying on a table. What single test could you use to tell which object is made of clay and which of lead?

One possible solution is to apply the same force (push) to each object. The object which moves more readily is made of clay; the other is made of lead. The clay object has *less inertia* than the object made of lead. The greater the inertia of an object the more it resists changes in its state of motion.

You must be wondering: is there a connection between the inertia of an object and its mass? Indeed, there is. The inertia of an object is a quantity which is solely dependent upon its mass.

Some applications of Newton's first law

The bottle and coin trick

Place a small coin on a stiff card over the opening of a bottle (Figure 6.4). When the card is flicked horizontally, the coin drops into the bottle.

Explanation: The coin remained in its original position – at rest. But the card no longer holds it up, so it falls into the bottle.

Getting ketchup out of a bottle

A fourth-former suggests the following way of dislodging ketchup from the bottom of a bottle. Turn the bottle upside down. Thrust it downwards vigorously, then halt it abruptly. The ketchup comes out.

Explanation: Initially, the bottle is thrust downward and so is the ketchup. When the bottle is 'halted' abruptly, the ketchup continues to move downward relative to the bottle.

Thinking it through

What is the link between Newton's first law of motion and the following?

- the use of car head rests to reduce whiplash injuries during collisions
- the sensation experienced by a person in a descending elevator which stops suddenly

Watch it!

Consider what happens to you if the maxi-taxi or car in which you are travelling stops suddenly. On braking, the wheels of the vehicle lock. The force of the road on the wheels provides a net force ($\sum F \neq 0$) which causes the vehicle to slow down and stop. If you are not wearing a seat belt you experience no net force – for you $\sum F = 0$. You continue forward, sliding along the seat . . . moving towards the front of the vehicle.

Install air bags – save a life or two!

Inertia plays a key role in road safety. Using seat belts and front and side air bags may reduce the extent of injuries during crashes. Within 0.04 second after the collision the air bag is fully inflated with nitrogen gas. (Why do you think nitrogen is used and not another gas, such as oxygen or carbon dioxide?)

Why should the seat in a car be pushed way back? Why should passengers sit with their backs right against the car seat? Why should passengers be buckled with minimal belt slack?

Thinking it through

Use the concept of inertia to explain each of the following.

- When each of A or B is displaced from its equilibrium position, system A vibrates rapidly while system B vibrates slowly.

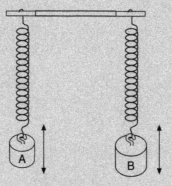

- The head of a hammer can be tightened onto its wooden handle by hitting the base of the hammer on a hard surface.
- If someone jumps off a fast moving vehicle they may be seriously injured.

- If the tablecloth is pulled firmly the jug stays in position.

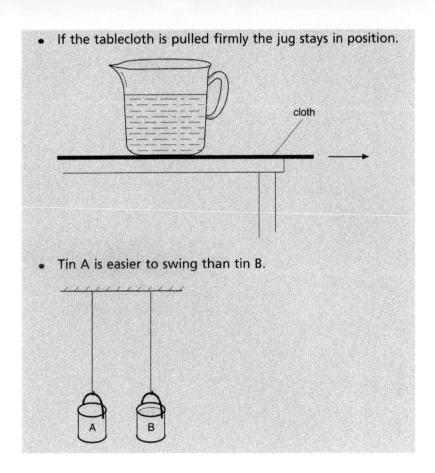

cloth

- Tin A is easier to swing than tin B.

A   B

6.2 A first look at Newton's second law

Newton's first law is concerned with the effect of balanced forces on the motion of objects. By contrast, Newton's second law predicts the behaviour of objects for which all existing forces are *not* balanced.

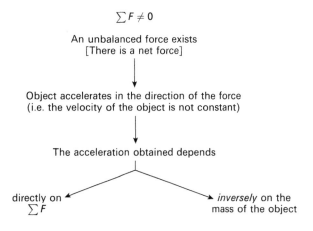

$$\sum F \neq 0$$

An unbalanced force exists
[There is a net force]

Object accelerates in the direction of the force
(i.e. the velocity of the object is not constant)

The acceleration obtained depends

directly on
$\sum F$

inversely on the
mass of the object

Figure 6.5 *A statement of Newton's second law.*

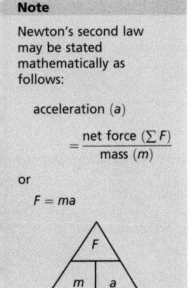

Note the following:

- Newton's second law does not imply that motion always results when a force is applied: e.g. the shape may change.
- Acceleration may involve a change of direction or a change of velocity.
- Equal forces which are not acting along the same line cause motion.
- When equal but opposite forces act along the same line, they cancel each other's effect and no motion occurs.

How to use ticker timers and trolleys to explore the relationship between unbalanced force and mass

1 Set up the apparatus as shown using a friction-compensated runway.

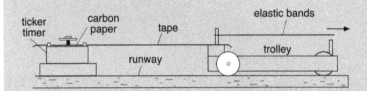

2 Use identical trolleys (assume that each trolley has unit mass). Trolleys may be placed on top of each other to obtain different masses.
3 Apply forces by stretching identical elastic bands, each stretched by the same length. The tension in one elastic band may be taken as a unit of force.
4 Attach tape from the timer to the trailing end of the trolley.
5 Make tape charts, as described in Chapter 5, for the following situations:
   a number of trolleys fixed but number of elastic bands varied;
   b number of trolleys varied, but number of elastic bands fixed.

Example

A constant force changes the speed of a 100 kg sprinter from $4.0\,\mathrm{m\,s^{-1}}$ to $6.0\,\mathrm{m\,s^{-1}}$ in $1.0\,\mathrm{s}$. Calculate

a the acceleration of the sprinter during this time

b the magnitude of the force.

Solution

a $v = u + at$

   $6 = 4 + a$

   $a = 2\,\mathrm{m\,s^{-2}}$

b $F = m \times a$

   $= 2 \times 100$

   $= 200\,\mathrm{N}$

When you carefully analyse the tape charts you should see that:

$a \propto \sum F$ when the mass is constant

$a \propto \dfrac{1}{m}$ when the applied force $\sum F$ is constant

By combining the above results, we get

$\sum F \propto ma$

or

$$\sum F = kma$$

If you are using SI units, $k = 1$, so $F = ma$.

Problem

A toy vehicle of mass 0.75 kg rests on a friction-compensated horizontal surface. At $t = 1$ s a force of 3 N is applied to the vehicle. This force is maintained for 3 s, then removed.

a What is the acceleration of the vehicle between $t = 1$ s and $t = 3$ s?

b What is the maximum velocity reached by the vehicle?

c Sketch the velocity–time graph for the first 6 s.

d Calculate the distance covered by the vehicle in this time.

6.3 Momentum and impulse

The extent of damage done to two colliding cars depends on

- the masses of the cars;
- the velocities of the cars.

A 10 tonne truck travelling at $10\,\mathrm{m\,s^{-1}}$ may cause greater damage than the same truck travelling at $5\,\mathrm{m\,s^{-1}}$. A car travelling at $20\,\mathrm{m\,s^{-1}}$ may cause greater damage than the truck travelling at $1\,\mathrm{m\,s^{-1}}$. **Momentum** is that physical quantity which takes into account both the mass of an object and its velocity. **Linear momentum** (p) of an object is the product of the mass (m) of the object and its velocity (v). Momentum is measured in $\mathrm{kg\,m\,s^{-1}}$.

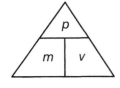

$$\boldsymbol{p} = \boldsymbol{mv}$$

Momentum is a vector quantity. It is fully described by both direction and magnitude.

Example

An object of mass 4 kg is travelling due east at $5\,\mathrm{m\,s^{-1}}$. Another object of mass 2 kg is travelling due west at $3\,\mathrm{m\,s^{-1}}$. What is the momentum of each object?

Solution

Let us assign the direction due east as positive. Then the direction due west is negative.

$$\text{Momentum of the } 4\,\text{kg car} = 4\ (\text{kg}) \times 5\ (\text{m}\,\text{s}^{-1})$$
$$= +20\ \text{kg}\,\text{m}\,\text{s}^{-1}$$
$$\text{Momentum of the } 2\,\text{kg car} = -2\ (\text{kg}) \times 3\ (\text{m}\,\text{s}^{-1})$$
$$= -6\ \text{kg}\,\text{m}\,\text{s}^{-1}$$

Remember that an object with momentum can be stopped if a force is applied against it for some time. Recall further that an unbalanced force always changes the velocity of an object. If velocity is changed, then so too is momentum. So there is a connection between force and change of momentum.

Looking at the equations:

$$\sum F = m \times a = m \times \frac{\Delta v}{t}$$
$$\Rightarrow \quad \sum F \times t = m \times \Delta v$$

The product $\sum F \times t$ is known as **impulse** (symbol I).

impulse = change in momentum

We will call this the **impulse–momentum change** equation.

In collisions, objects experience an impulse. Impulse is equal to the change of momentum.

From the impulse–momentum change equation we can see that:

- the longer the time over which a collision occurs, the smaller the force which acts on the object involved;
- to increase the effect of the force on an object involved in a collision, the time must be decreased.

We shall discuss the practical significance of the impulse–momentum change equation later.

> **Note**
>
> We can view impulse I as the average force applied to an object in order to change its momentum.
>
> Impulse has the unit newton second (N s).

Impact forces (Newton's second law revisited)

Impact forces occur when one object strikes another; for example, when a footballer kicks a football or when a gymnast falls on a trampoline.

Consider an object of mass m moving initially with a velocity v_1.

Let a retarding force act on the object such that, after t seconds, its velocity reduces to v_2.

Then the initial momentum of the object $= m \times v_1$ and the momentum of the object after t seconds $= m \times v_2$

> **Note**
>
> Forces which result from changes in momentum are known as impact forces.
>
> Impact forces are greatest when the duration of the impact is short.

The change in momentum during this time $= m \times v_2 - m \times v_1$

$$= m(v_2 - v_1)$$

The rate of change of momentum $= \dfrac{\text{change in momentum}}{\text{time}}$

$$= \dfrac{m(v_2 - v_1)}{t}$$

$$= \text{mass} \times \text{rate of change of velocity}$$

$$= \text{mass} \times \text{acceleration}$$

$$= \text{force}$$

So **force is equal to the rate of change of momentum**. This is an alternative statement of Newton's second law.

Example

An object of mass 10 kg, initially travelling at $20 \, \text{m s}^{-1}$, is acted upon by a force which reduces its velocity to $5 \, \text{m s}^{-1}$ in 5 seconds.

Determine:

a   the deceleration of the object
b   the average force acting on it
c   the initial momentum of the object
d   the momentum of the object after 5 seconds
e   the rate of change of momentum of the object.

Solution

a $a = \dfrac{v_2 - v_1}{t} = \dfrac{5 - 20}{5} = -3 \, \text{m s}^{-2}$

b $F = ma = 10 \times (-3) = -30 \, \text{N}$

c $p_1 = mv_1 = 10 \times 20 = 200 \, \text{kg m s}^{-1}$

d $p_2 = mv_2 = 10 \times 5 = 50 \, \text{kg m s}^{-1}$

e   rate of change of momentum

$$= \dfrac{50 - 200}{5} \left(\dfrac{\text{kg m s}^{-1}}{\text{s}} \right) = -30 \, \text{kg m s}^{-2}$$

$$= -30 \, \text{N}$$

Note that the answer for **b** is the same as the answer for **e**.

The conservation of linear momentum

Provided that the vector sum of the external forces acting on a system is zero (i.e. $\sum F = 0$), the total linear momentum of that system remains constant during collisions. This is a statement of the **principle of conservation of linear momentum**.

Problem

An object of mass 5 kg travelling from left to right at $6 \, \text{m s}^{-1}$ collides with an object of mass 4 kg which is travelling at $2.5 \, \text{m s}^{-1}$ from right to left. The 4 kg object bounces back at $2 \, \text{m s}^{-1}$. Determine

a   the momentum of the 5 kg object before collision
b   the momentum of the 4 kg object before collision
c   the total momentum of the system before collision
d   the momentum of the 4 kg object after collision
e   the momentum of the 5 kg object after collision
f   the velocity and direction of the 5 kg object after collision.

Note

The law of conservation of linear momentum holds for collisions involving any number of objects or particles.

In a system of interacting particles, momentum is redistributed but the total momentum is conserved.

Figure 6.6 *Head-on collision between two objects.*

Momentum is conserved for all interactions in which $\sum F = 0$.

Consider the head-on collision between two objects X and Y (Figure 6.6).

Objects moving to the right are assigned positive velocities (velocity is a vector quantity). Objects moving to the left are assigned negative velocities.

$$\text{Momenta before collision } \sum p_b = m_X v_X - m_Y v_Y$$
$$\text{Momenta after collision } \sum p_a = -m_X v'_X + m_Y v'_Y$$

Since momentum is conserved, $\sum p_b = \sum p_a$ and

$$m_X v_X - m_Y v_Y = -m_X v'_X + m_Y v'_Y$$

This is a mathematical statement of the principle of conservation of momentum as it applies to the system of Figure 6.6.

You can use the experimental set-up of Figure 6.7 to show that momentum is conserved during collisions. The trolleys (1) have the same mass, (2) are on a friction-compensated surface. From the pattern of dots in the tape of Figure 6.8, you can deduce that the velocity is halved when the trolleys collide and stick together. The velocity of trolley A before collision $= 80 \, \text{cm s}^{-1}$ and the velocity of the combined trolleys $= 40 \, \text{cm s}^{-1}$. (Explain how these values are arrived at.)

$$\text{momentum before collision} = \text{momentum after collision}$$
$$m \times v = (2m) \times \tfrac{1}{2}v$$
$$= m \times v$$

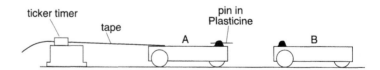

Figure 6.7 *Before collision: trolley A is on the move and trolley B is stationary. After collision: both trolleys stick together and move on with a common velocity.*

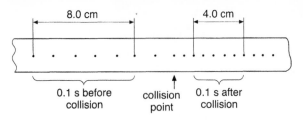

Figure 6.8 *The tape from a collision experiment.*

Momentum and collisions

Collisions play a central role in many areas of physics. Collisions may be head-on as in Figure 6.6 or glancing as in Figure 6.9.

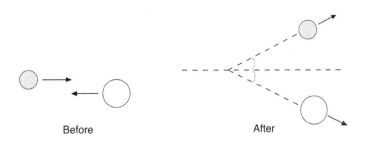

Before   After

Figure 6.9 *An example of a glancing collision.*

Collisions may be elastic or inelastic.

- *Elastic:* momentum conserved, kinetic energy conserved and total energy conserved.

In elastic collisions, the relative velocity before equals the relative velocity after collision. Particles approach at a given speed and, after collision, they recede with the velocity vector having the same magnitude but opposite sign.

$$(v_1 - v_2) = -(v_1 - v_2)$$

The collision of two billiard balls may be taken as elastic. During this collision there is no deformation of the particles (the balls) and there is no loss of kinetic energy.

- *Inelastic:* momentum conserved, kinetic energy not conserved, total energy conserved.

Example

A bullet is fired into a stationary block of wood mounted on wheels. The bullet becomes embedded in the block. Block plus bullet move off in the forward direction with velocity v. (This is an example of an inelastic collision.) Derive an expression for the velocity of the bullet, v_b.

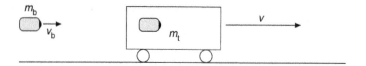

Solution

$$\text{Momentum before collision} = m_b \times v_b + 0$$

$$\text{Momentum after collision} = (m_b + m_t) \times v$$

$$m_b \times v_b = (m_b + m_t) \times v \quad \text{(conservation of momentum)}$$

$$\therefore \quad v_b = \frac{v(m_b + m_t)}{m_b}$$

Practical situations involving impulse and momentum

As you work through the examples below remember that the force which an object experiences during a collision depends:

- inversely on the collision time, i.e. $F \propto \dfrac{1}{t}$;
- directly on the velocity change during collision, i.e. $F \propto \Delta v$.

The operating equation is

$$\text{force} \times \text{time} = \text{momentum change}$$

$$= \text{mass} \times \text{velocity change}$$

1 Riding the punch

A boxer throws a jab at the head of his opponent, who relaxes his neck and allows his head to move backwards on impact.

This action minimizes the effect (force) of the jab by extending the impact time. The bigger the impact time, the smaller the effect of the jab.

2 Another take on air bags

The use of air bags may increase collision time by a factor of about 100. This means that the impact force is reduced to one-hundredth of its value as compared to a collision in a car without air bags.

Figure 6.10 *The perfect follow-through.*

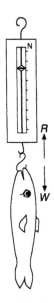

Figure 6.11 *The force exerted by the balance on the fish must equal the force exerted by the fish on the balance – otherwise the fish would fall on the ground!*

3 The follow-through in sports such as cricket and lawn tennis

How hard the ball is struck depends on the force imparted by the batsman or player but a good follow-through plays an important role. It ensures that the ball leaves the bat or racket with greater velocity because the contact time is increased.

$$F \times t = m \times \Delta v$$

$$\Delta v = \frac{Ft}{m}$$

Since F and m are constant, an increase in the value of t increases Δv and hence v.

Thinking it through

Use the momentum–impulse principle to explain the following:

- A raw egg smashes when thrown against a wall. Another raw egg, thrown against a bed sheet held as shown does not break.
- Motor vehicles have crumple zones.
- The dashboards in motor vehicles are padded.
- The best way of catching a cricket ball in the outfield is to use outstretched arms and carry the ball for a metre or so before 'stopping its momentum'.

6.4 Newton's third law of motion

For every action there is an equal and opposite reaction.

This is a statement of Newton's most famous law. However, this law is often misunderstood. Make sure you understand what it means.

A fish hanging on a spring balance exerts a downward force, equal to its weight W, on the balance. The balance pulls the fish upwards with a force R, equal but opposite to W. W is an action force and R is a reaction force.

There is a pair of forces acting on any two interacting objects. The magnitude of the force on the first object *equals* the magnitude of the force on the second. The direction of the force on the first object is *opposite* to the direction of the force(s) on the second object.

Newton's third law could be restated as follows:

Forces exist in pairs – 'equal and opposite action–reaction pairs'.

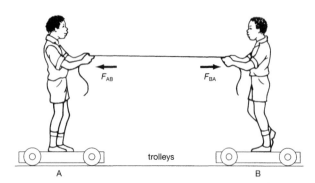

Figure 6.12 *Newton's third law.*

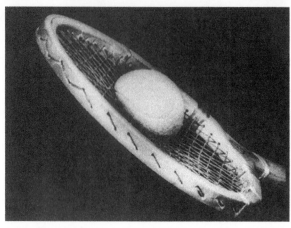

Figure 6.13 *The force of the tennis ball on the racquet deforms the racquet strings. The force of the racquet on the ball drives the ball back in the opposite direction.*

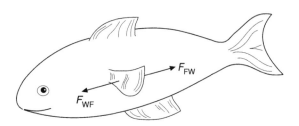

Figure 6.14 *The fish uses its fins to push water backwards. The water, in turn, reacts by pushing the fish forwards.* $F_{FW} = -F_{WF}$ *(F = fish; W = water).*

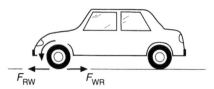

Figure 6.15 *How does this car move? The wheels of the car spin backwards. The road is pushed backwards. The road reacts by pushing the car forwards.* $F_{WR} = -F_{RW}$ *(W = wheels; R = road).*

Consider Figure 6.12. If boy A pulls boy B, both boys move towards each other. If boy B pulls boy A, both boys move towards each other. A pair of action–reaction forces exists between boy A and boy B. These forces act in opposite directions and are equal in magnitude. $F_{AB} = -F_{BA}$.

A clearer way of stating Newton's third law is: if body A exerts a force on body B, then body B exerts a force of the same size on body A, but in the opposite direction.

Things to do

Collect information by observing or experimenting in each of the following situations:

- kicking a football;
- participating in a tug of war;
- rowing a boat;
- playing a straight drive in cricket;
- making a backhand stroke in lawn tennis.

Draw or sketch each situation, clearly indicating the action and reaction forces involved. Use arrows to indicate the direction of the forces involved.

Note

The third law holds between two objects whether they are at rest or in motion.

(a)

(b)

Blast off – how do rockets work?

If you tape a balloon to a straw which can slide along a string (Figure 6.16), blow it up and then let go, the balloon moves along the string. When you let go, air escapes from the balloon through the narrow neck. The backward moving air stream produces an action force. The reaction force drives the balloon forwards.

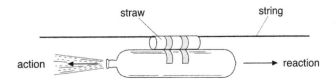

Figure 6.16 *A balloon 'rocket'.*

In a conventional rocket, cold liquid oxygen and cold liquid hydrogen are fed into a combustion chamber where they are ignited. The heat produced causes the gases to expand rapidly. The hot gases are forced out of the exhaust nozzle at very high speeds, creating an action force. The corresponding reaction force lifts the rocket off the launch pad.

The mass of gases which exit each second from the exhaust nozzle may be small but the exhaust gases come out at very high speeds. The rate of change of momentum (Newton's second law) in this situation is quite significant and so is the resultant force. The rocket is propelled at high speeds.

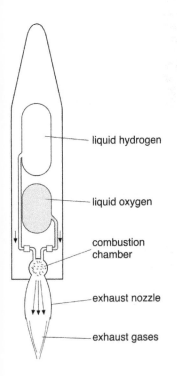

liquid hydrogen

liquid oxygen

combustion chamber

exhaust nozzle

exhaust gases

Figure 6.17 *Many physical principles from mechanics, heat, and electricity and magnetism are involved in rocket launches.*

Checklist

After studying Chapter 6 you should be able to:

- recall Newton's laws of motion
- recall that friction causes moving objects to come to a stop

- define and discuss inertia
- recall some practical applications of inertia
- recall some practical applications of Newton's first law of motion
- recall Newton's second law of motion
- recall that Newton's second law can be written as $F = ma$
- define momentum
- recall and use the linear momentum equation $p = mv$
- recall and use the impulse–momentum change equation $F \times t = m \times v$
- recall that force = rate of change of linear momentum
- recall and use the equation $F = \dfrac{(m \times \Delta v)}{t}$
- recall and use the principle of conservation of linear momentum
- define an elastic collision
- define an inelastic collision
- discuss practical situations involving impulse and momentum
- recall Newton's third law of motion
- discuss practical applications of the third law

Questions

1 If the greatest decelerating force which may be applied to a vehicle is numerically equal to its weight, find **a** the least distance and **b** the least time, in which the vehicle travelling on a horizontal road can be brought to rest from an initial velocity of $25\,\mathrm{m\,s^{-1}}$.

2 An unbalanced force of 25 N acts on a mass of 10 kg. What is the acceleration of the object?

3 A car of mass 900 kg travelling at $15\,\mathrm{m\,s^{-1}}$ crashes into a wall and comes to rest in 0.3 s. Calculate **a** the average deceleration of the car and **b** the average force exerted on the wall by the car.

4 An object of mass 20 kg moving at $10\,\mathrm{m\,s^{-1}}$ due west collides with and sticks to an object of mass 10 kg moving due east at $5\,\mathrm{m\,s^{-1}}$. What is the common velocity of the objects after collision?

5 A 3 kg trolley is travelling at $3\,\mathrm{m\,s^{-1}}$, when an object of mass 1.5 kg is dropped onto it. What is the new velocity v_n of the trolley?

6 A rifle is 2000 times as heavy as the bullet it fires. A bullet leaves the rifle at a velocity of $100\,\mathrm{m\,s^{-1}}$. What is the recoil velocity of the rifle?

7 A cart (X), initially at rest and having mass 0.75 kg, is pulled with a 1.5 N force for 1 second. Another cart (Y), also initially at rest and having mass 0.75 kg, is pulled with a 3.0 N force for 0.5 s.
   a Which cart has the greater acceleration?
   b Which cart experiences the greater impulse?
   c Which cart has the greater momentum change?

8 What is the effect on the momentum of an object if its velocity is quadrupled? What is the momentum change?

7 Work, energy, power – a trio of scalar quantities

Work and energy are important topics in physics; so too is power. Work in physics has quite a different meaning to work in everyday usage. In physics, work is done only when an object is moved in the direction of an applied force. Energy is the ability to do work.

7.1 Work

An object which has the ability to do work is said to have energy. Work is done: things happen whenever energy transfers take place.

Energy and work are measured in **joules (J)**, named after James Prescott Joule (1818–1889).

The joule is defined as the energy transformed or the work done when a force of 1 newton displaces an object a distance of 1 metre in the direction of the force.

Mathematically, work (W) is expressed by the equation:

$$W = F \times d \cos \theta$$

where F is the force applied, d is the displacement and θ is the angle between the force and the displacement vectors. When force and displacement are in the same direction,

$$\theta = 0, \quad \cos \theta = 1 \quad \text{and} \quad W = F \times d$$

For the rest of this section we shall consider only situations in which force and displacement are in the same direction.

Examples of situations in which work is done include:

- a shopper pushing a cart down the aisle of the supermarket;
- a field event athlete throwing a javelin;
- a fourth-former carrying a bag of books;
- two men pushing a stalled car.

By contrast, no work is done when a boy pushes on a wall. In this situation there is no displacement.

A waiter holding a tray of dishes (Figure 7.1) while moving horizontally across the room does no work on the tray. Here, the force vector and the displacement vectors are at right angles, i.e. $\theta = 90°$ and $\cos \theta = 0$.

Figure 7.1 *A vertical force does no work on a horizontally displaced object.*

Examples

1 A 30 N force is applied horizontally, causing a block to be displaced horizontally 5 m along a frictionless floor. How much work is done?

Solution

$$W = F \times d \cos \theta = F \times d, \quad \text{since } \theta = 0$$
$$= (30 \times 5) \, \text{N m} = 150 \, \text{J}$$

2 An object of weight 40 N is pulled up at constant speed a vertical distance of 10 m. How much work is done?

Solution

Force to move the object $= 40 \, \text{N}$

$$\therefore \quad \text{Work done } (W) = 40 \times 10 = 400 \, \text{J}$$

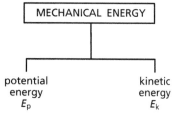

Figure 7.2

7.2 Mechanical energy

The energy acquired by objects on which work is done is known as **mechanical energy**. Mechanical energy is energy possessed by an object due to:

- its stored energy (potential energy);
- its motion (kinetic energy).

Potential energy

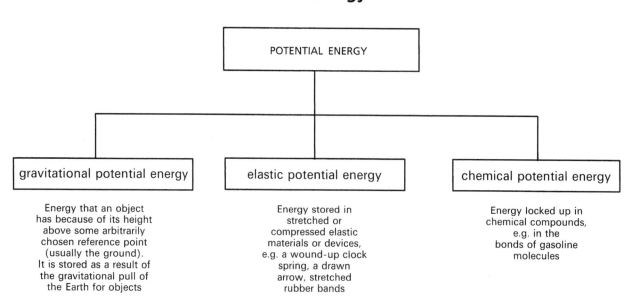

Figure 7.3 *Types of potential energy.*

For the remainder of this section the focus will be on gravitational energy and elastic potential energy.

Gravitational potential energy

Gravitational potential energy, E_p, of an object may be expressed mathematically as follows:

$$E_p = \text{mass (kg)} \times \text{acceleration due to gravity (N kg}^{-1})$$
$$\times \text{ height (m)}$$
$$= m \times g \times h$$

Gravitational potential energy, then, is dependent on two variables:

- mass;
- height above reference position.

We assume that the acceleration due to gravity (g) is constant. Is this assumption valid?

Example

What is the gain in gravitational energy of an object of mass 50 kg which is lifted a vertical height of 40 m?

Solution

$$E_p = m \times g \times h = 50 \times 10 \times 40 \ (\text{kg N kg}^{-1} \text{m}) = 20\,000\,\text{J}$$

Elastic potential energy

The amount of potential energy stored in materials is related to the extent of deformation of these materials. The more the stretch or the greater the compression, the greater the elastic potential energy stored.

We can explain the deformation of materials in terms of the forces (usually large) between atoms and molecules. When materials are deformed the atoms or molecules are pushed (or pulled) out of their equilibrium or rest position and energy is stored. The stored energy is released when the atoms or molecules regain their rest positions.

Stretching and compressing

If a force is applied to an object that cannot move, the object stretches (extends) or compresses (Figure 7.4). The extent of deformation depends on the force applied.

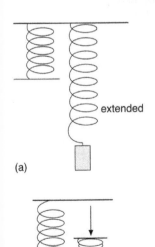

(a) extended

(b) compressed

Figure 7.4 *Stretching and compressing.*

Note

A spring which is neither stretched nor compressed has no stored elastic potential energy. The atoms in such a spring are said to be in equilibrium.

Note

For certain materials the applied force is directly proportional to the extension or compression.

Deformation is described as *plastic* if the atoms or molecules of the materials do not return to their equilibrium positions when the deforming forces are removed. A piece of modelling clay retains its new shape and size when stretched or compressed.

Deformation is *elastic* if the atoms or molecules return to their original positions when the deforming forces are removed.

How to investigate the stretchiness of a spring

1 Arrange a spiral spring as shown.

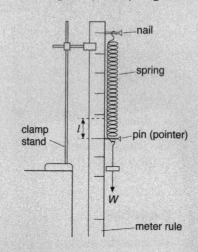

2 Attach a small weight W_0 to the lower end of the spring and record the original length of the spring. W_0 serves to separate the coils of the spring.
3 Add increasing weight W_1, W_2 to the spring, in addition to W_0, and record the corresponding extensions (increases in length) l_1, l_2, etc.
4 Plot a graph of extension (y-axis) against load (x-axis).

Figure 7.5 is a graph of extension against load for a given spring. Note that, for the loads used, extension is directly proportional to load. Under the conditions of the experiment the spiral spring is obeying **Hooke's law**. Hooke's law states that **extension is directly proportional to load or applied force, provided the elastic limit is not exceeded**.

The behaviour of a wide range of materials under applied forces may be investigated using apparatus similar to that above.

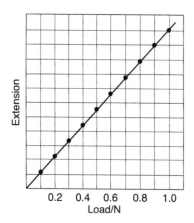

Figure 7.5 *Graph of extension against load for a spiral spring.*

Example

A rubber band is 14.5 cm long. When a mass of 120 g is attached to one end, it stretches to 17.5 cm. If the band obeys Hooke's law, $F = kx$ where x is the extension, what is the constant k?

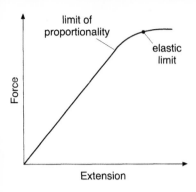

Figure 7.6 *Beyond the limit of proportionality, extension is no longer proportional to the applied force. Beyond the elastic limit, the material will not return to its original length when the force is removed.*

Solution

$$F = \text{weight of mass} = mg = 0.12 \text{ (kg)} \times 10 \text{ (m s}^{-1})$$

$$= 1.2 \text{ N}$$

$$x = 17.5 - 14.5 = 3 \text{ cm} = 0.03 \text{ m}$$

$$\therefore \quad 1.2 = k \times 0.03$$

$$\therefore \quad k = \frac{1.2}{0.03} = 40 \text{ N m}^{-1}$$

Some practical aspects of elastic potential energy

For all materials, there is a point, called the **limit of proportionality**, beyond which Hooke's law is not obeyed. If a material is stretched further than this point, it reaches its **elastic limit** (Figure 7.6). Beyond the elastic limit the material forms a 'neck'. Further stretching may cause the material to break.

Making materials stronger

Concrete, a brittle material, is made tougher by reinforcing it with steel. Energy passes through the reinforcing steel rods so that cracks do not develop readily in the concrete.

Plaster can be made tougher by reinforcing it with paper. The resulting material is called 'plasterboard'. Concrete and plasterboard are just two examples of a rapidly increasing range of man-made materials called **composite materials**. Steel, an alloy of iron, is not a composite material since its constituents are combined in such a way that they are indistinguishable. Concrete, however, is a composite material because the individual constituents, cement and gravel, are easily distinguishable.

Things to do

- Obtain about 25 g of Plasticine.
- Roll the Plasticine into a sphere.
- Determine the diameter of the sphere (explain how you did this).
- Stretch the Plasticine, a little at a time, measuring the 'length' after each stretch.
- By what factor can the Plasticine be stretched before it breaks? Explain how you arrived at your answer.

Cardboard and metal sheets can be made stronger by corrugating them. How does corrugating them make them stronger?

Things to do

1 Materials which are capable of large plastic deformation can easily be shaped. These materials are described as **ductile**. Identify five ductile 'materials' that are commonly used in the home.

2 Wood, which is usually strong in one direction only, may be made stronger by plying. Carefully examine a sheet of plywood and determine how its greater strength is achieved.

Kinetic energy

Kinetic energy is the energy that an object has because it is
moving.

$$\text{kinetic energy (in joules)} = \tfrac{1}{2} \times \text{mass} \times \text{speed}^2$$

A doubling of speed leads to a quadrupling of kinetic energy; a
doubling of mass leads to a doubling of kinetic energy.

Examples

1 An object has mass 60 kg and moves at $4\,\mathrm{m\,s^{-1}}$.

$$\text{kinetic energy} = \tfrac{1}{2} \times 60\ (\mathrm{kg}) \times [4\ \mathrm{m\,s^{-1}}]^2$$

$$= 480\ \mathrm{kg}\ (\mathrm{m\,s^{-1}})^2 = 480\,\mathrm{J}$$

2 A mini mook of mass 800 kg which is travelling at $36\,\mathrm{km\,h^{-1}}$
has

$$\text{kinetic energy} = \tfrac{1}{2} \times 800\ (\mathrm{kg}) \times [10\,\mathrm{m\,s^{-1}}]^2$$

$$= 40\,000\,\mathrm{J}$$

$$(36\,\mathrm{km\,h^{-1}} = 10\,\mathrm{m\,s^{-1}})$$

The interchange between gravitational potential and kinetic energy

Consider an object held 80 metres above ground. This object
has $E_p = m \times g \times h = m \times 10 \times 80$ (where m is its mass in kg).
Let this object be released so that it falls freely. As the object
falls its gravitational potential energy decreases as its kinetic
energy, i.e. its speed, increases. At the instant the object strikes
the ground all its initial potential energy has been changed to
kinetic energy.

Gravitational potential energy = kinetic energy the instant
at the point of release   the object strikes the ground

$$m \times g \times h = \tfrac{1}{2}m \times v^2$$

$$800m = \tfrac{1}{2}m \times v^2$$

$$40\,\mathrm{m\,s^{-1}} = v$$

The kinetic energy with which an object strikes the ground is
dissipated as heat, sound, etc. This is an example of the
principle of conservation of energy which may be stated as
follows:

**Energy is neither created nor destroyed, only converted
to another form, i.e. the total energy present at every
stage in a chain of energy conversions is the same.**

Figure 7.7 *Water power.*

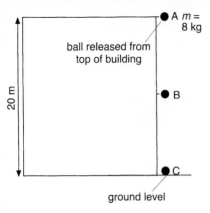

Figure 7.8 *What are the potential energy and the kinetic energy of the ball at points A, B and C?*

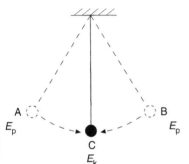

Figure 7.9 *A simple pendulum.*

Although energy is never lost as such, it eventually ends up in a 'diluted' form. Heat (or sound) is the final form in most sequences of energy transfers (energy chains). For this reason, heat is often described as a degraded form of energy.

Energy transformations

Energy is never 'lost', but it changes form. The process by which energy changes form is known as an **energy conversion** or as **transferring energy**.

When the bob of the simple pendulum in Figure 7.9 is displaced from the equilibrium position C to position B, it has gravitational potential energy relative to position C. If the bob is released its potential energy is transformed into kinetic energy which reaches a maximum at C.

As the bob overshoots C, its kinetic energy decreases while its potential energy again reaches a maximum at position A. This process is repeated cyclically and, in the absence of friction, the total mechanical energy of the system remains constant.

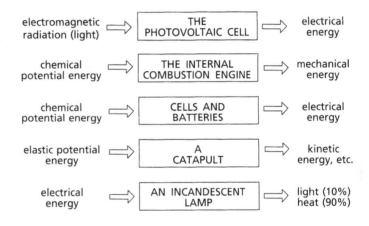

Figure 7.10 *Some energy transfers.*

7.3 Power

The power of persons or machines measures *how quickly* they convert energy from one form to another or how quickly they do work.

Power is the work done or energy converted per unit time.

$$\text{power} = \frac{\text{force (N)} \times \text{displacement (m)}}{\text{time taken (s)}} = \frac{\text{energy transformed (J)}}{\text{time taken (s)}}$$

Power is measured in **watts** (W), named after James Watt (1735–1819).

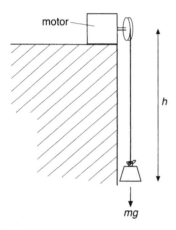

Figure 7.11 *Measuring the power output of an electric motor.*

1 watt of power is dissipated if 1 joule of work is done or 1 joule of energy is transformed in 1 second. Large amounts of power are measured in kilowatts (kW) or megawatts (MW).

How to determine the power output and the efficiency of an electric motor

1 Allow a small motor to raise a mass (m) through a given height (h) at a constant speed. (Why should the speed be constant?)
2 Record the time taken for this to happen as t seconds.
3 The work done by the motor $= m \times g \times h$, where g is the acceleration due to gravity. (N.B. the work done is equal to the increase in gravitational potential energy of the mass.)
4 The power output of the motor $=$ work done $\div$ time taken.
5 If the power rating of the motor is known, its efficiency can be calculated:

$$\text{efficiency} = \text{power output} \div \text{power rating}$$

The efficiency of the motor is less than 100% because of 'losses' such as the kinetic energy of the moving parts, heat produced (as a result of friction in the windings and bearings) and sound produced.

Example

A force of 500 N acting for 20 seconds displaces an object a distance of 25 metres. Calculate: **a** the work done and **b** the power output.

Solution

$$\text{Work done} = \text{force (N)} \times \text{displacement (m)}$$
$$= (500 \times 25)\,N\,m = 12.5\,kJ$$
$$\text{Power output} = \text{work done} \div \text{time taken (s)}$$
$$= (12\,500 \div 20)\,J\,s^{-1} = 625\,W$$

Thinking it through

Identify all the energy transfers which take place in each of the following situations:

• A vehicle on a full tank of fuel climbs up a rough mountainous road, until its tank is empty.
• The spring of a toy car is wound up and the car allowed to move over a rough floor.

- A battery is used to start a motor car engine.
- Energy from the Sun is used to run tape recorders on a spaceship.
- A bicycle generator is used to light a headlamp.
- Large propellers use wind energy to run a generator which, in turn, charges storage batteries.

Checklist

After studying Chapter 7 you should be able to:

- define work
- recall and define the unit of work (joule)
- recall and use the equation $W = F \times d \cos \theta$
- discuss mechanical energy
- discuss potential energy and its various types
- define gravitational potential energy (E_p)
- recall and use the equation $E_p = mgh$
- define and discuss elastic potential energy
- define and discuss elastic and plastic deformation
- recall Hooke's law
- define the limit of proportionality and the elastic limit
- define, with examples, composite materials
- discuss ways in which materials are made stronger
- define kinetic energy (E_k)
- recall and use the equation $E_k = \frac{1}{2}mv^2$
- recall the energy interchange $mgh = \frac{1}{2}mv^2$
- recall and use the principle of conservation of energy
- discuss energy transfers
- define power
- recall and use the equation $P = F \times d/t = E/t$
- recall and define the unit of power (the watt)

Questions

1 a Define kinetic energy and gravitational potential energy.

b An object falls from a cliff to the ground 320 metres below. Calculate the velocity with which it hits the ground.

c Does this velocity depend on the mass of the object? Explain.

2 Define the watt. How is power related to energy transfer?

3 List four energy transfers which take place at an oil-burning power station.

4 A man of mass 80 kg climbs up 15 steps each 25 cm high in 7.5 seconds. Calculate the power that the man develops.

5 A windmill is used to pump water from a large container. Each second, 250 kg of air moving at 25 m s^{-1} strikes the vanes of the windmill.

a Calculate the rate at which power arrives at the windmill.

b Calculate the mass of water pumped each second through a height of 2.5 m, assuming that the windmill has an efficiency of 80%.

6 The diagram shows a model of an electric power station. The lamp lights normally if the power input to it is 0.5 W.

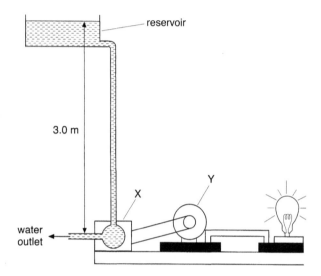

a Define power.

b List the energy conversions which take place in the system shown.

c If 0.1 kg of water flows through the turbine each second in order for the lamp to light, what is the overall efficiency of the system?

d Give the names of the devices X and Y.

e Identify, where, in this system, energy is lost. What happens to this 'lost energy'?

f Describe the function of both X and Y.

7 a Trace the energy transfers which take place when a bullet, fired from a gun, passes through the air, penetrates a target and eventually comes to rest.

b Name ONE device by which each of the energy transformations A–I below can take place.

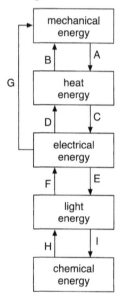

8 Extraction of energy from alternative sources – a Caribbean perspective

8.1 The classification of energy sources

One classification of energy sources is into conventional (fossil fuels) and alternatives (the others). Another, equally valid, classification is into *renewables* and *non-renewables*. The term renewable refers to those energy resources which are replaced as fast as they are used up. They can be used over and over again, without being depleted. The non-renewable resources are not replaced, once extracted and used.

We depend heavily on fossil fuels

We use energy all the time. Most of the energy we use today comes from oil, natural gas and coal – the fossil fuels. Indeed, we seem hooked on fossil fuels. The pie chart of Figure 8.1 highlights the present global dependence on fossil fuels. A significant proportion of the energy used in the Caribbean region is derived from oil and natural gas. This means that the countries of the region, Trinidad and Tobago excepted, have high energy import bills. These countries should strive to make greater use of the energy alternatives, especially 'solar energy'. Globally, too, we need to find alternatives to fossil fuels because:

- the fossil fuels are starting to run out (experts believe that oil and natural gas will run out in 50–60 years and coal in about 200–300 years);
- pollutants are produced when fossil fuels are burnt (these pollutants harm us and the environment);
- natural gas and some fractions of oil are too precious to burn (apart from their use as energy sources, natural gas and certain fractions of oil are important feedstocks (starting materials) for the production of a wide range of petrochemicals: drugs, fertilisers, detergents, synthetic rubbers, industrial alcohol, methanol, plastics, . . . to name a few).

Table 8.1 lists some of the primary pollutants formed when fossil fuels are burnt. These pollutants have a bad effect on the quality of our land, air and water. Climate may also be affected. For example, greenhouse warming which results from the presence of carbon dioxide and other gases may lead to a temperature rise of about 1 °C. This could result in the melting of the polar ice caps and the flooding and submergence of coastal towns and villages.

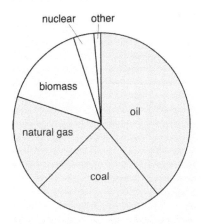

Figure 8.1 *Global energy use.*

Figure 8.2 *Motor car emissions cause pollution over Mexico City.*

Table 8.1

Table 8.1 | **Primary pollutants and their effects**

Pollutant	Sulphur dioxide	Oxides of nitrogen	Carbon monoxide	Carbon dioxide
Effects	Dissolves in moisture to form sulphuric acid – a major constituent of acid rain. Acid rain damages vegetation and marble and concrete in buildings.	Dissolve in moisture to form acids. Cause respiratory illnesses. Encourage excessive growth of algae. Cause the formation of smog.	Poisonous. Combines with a substance in the blood, preventing oxygen uptake.	A build-up of this may lead to a 'greenhouse effect'.

Note

The emissions of carbon dioxide and the oxides of sulphur and nitrogen disrupt the carbon cycle and contribute to acid rain. Some 25 billion tonnes of carbon dioxide are released into the air annually through the burning of fossil fuels.

What are the alternatives to fossil fuels?

Figure 8.3 shows the major alternatives to fossil fuels.

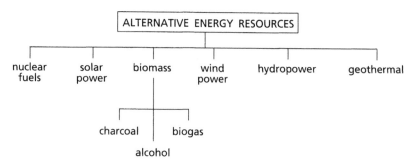

Figure 8.3 *Alternative energy resources.*

8.2 Solar energy

The flow chart of Figure 8.4 lists the major applications of solar energy in the Caribbean.

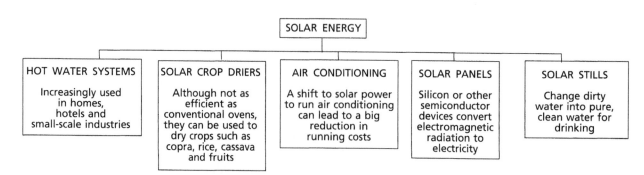

Figure 8.4 *Uses of solar energy in the Caribbean.*

Note

The solar still and the solar drier are examples of 'flat plate' collectors. Flat plate collectors do not focus the Sun's rays, are usually stationary but may be moved to follow the season.

The solar oven and the solar cooker are examples of focusing or concentrating collectors. These are moved during day to follow the Sun.

The receiving surfaces of all collectors should be blackened so as to absorb about 95% of incident radiation.

The initial cost of solar water heaters is high, but running costs are low.

The solar still

A solar still is a shallow rectangular 'box' which is insulated (why?). Suggest what materials might be suitable for use as insulation. The box is blackened on the inside. The top plate, made of glass or other transparent materials, slopes southwards in the northern hemisphere; the angle of slope is equal to the latitude. Cold water enters and is heated by radiant energy from the Sun. The water evaporates. The pure vapour condenses on the underside of the glass. The pure water is collected in tanks and stored for later use or is fed to taps for immediate domestic use.

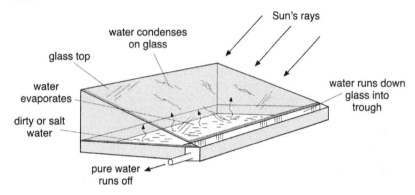

Figure 8.5 *A solar still.*

The solar water heater

Water is heated as it is pumped through blackened copper pipes in the absorbing panel. The heat is transferred to the hot water tank by a coil.

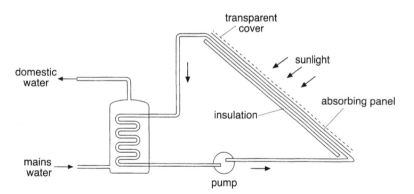

Figure 8.6 *A solar water-heating system.*

The solar drier

Most crops grown in warm, sunny regions can be dried using solar driers. Cold air enters the drier from the back and is heated by solar radiation. The warm air moves over the crop, removing moisture and drying it.

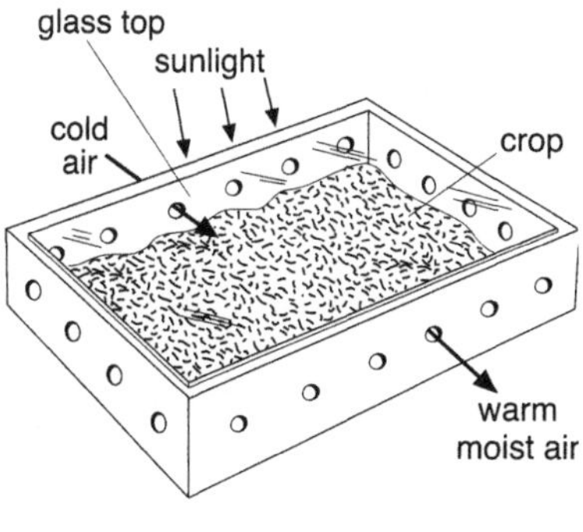

Figure 8.7 *A solar drier.*

This method of drying leads to some loss of food colour, but there is little loss in food quality.

Solar electricity

The solar electricity industry is still in its infancy compared to the solar heating industry. In the solar electricity industry, solar panels convert incident solar radiation to electricity. The solar panel or module consists of individual solar cells wired together. Each cell produces about 0.5 V at 2–4 amperes. Combinations of modules are called arrays.

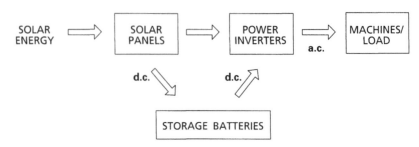

Figure 8.8 *Electricity from solar power.*

Solar power is used to

- power transmitters and other equipment on satellites;
- recharge batteries;
- power pumps for crop irrigation and watering stock;
- power pumps and blowers for solar heaters.

Solar energy, converted to electricity, may be stored in batteries and used in off-peak periods and at night.

For a.c. equipment, the direct current from the modules or arrays is passed through an inverter. Inverters convert d.c. to a.c.

Figure 8.9 *Solar panels on a roof.*

8.3 Biomass

Biomass may be viewed as 'indirect solar energy'. It is energy which can be obtained from plant or animal matter, and which came initially from the Sun.

Biogas

Biogas is produced by bacteria breaking down plant and animal wastes. This is most conveniently done in a biodigester.

Methane is the major constituent of biogas. Biogas may be used for cooking, heating and small-scale electrification. The sludge

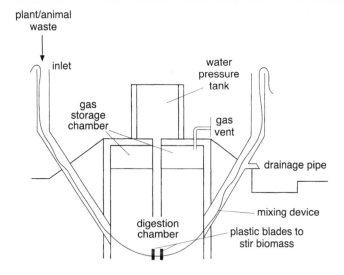

Figure 8.10 *A biodigester. The gas produced in this device could provide energy for homes or small farms. In India the gas is used to provide electricity for small remote villages.*

which remains from the decomposition may be used as fertilizers and in aquaculture.

Alcohol

Alcohol is produced when carbohydrates such as sugar undergo fermentation. Gasohol – a mixture of gasoline and alcohol – is used in some places as a fuel.

Wood and charcoal

Wood is burnt directly for cooking and domestic heating. Charcoal is made by heating wood in a kiln in a limited supply of air. Charcoal is cleaner burning than wood.

8.4 Wind power

Wind results from the uneven heating of the Earth's surface and atmosphere. In this sense, wind is an indirect form of solar energy. The wind has been harnessed by people for over 2000 years. Wind power has been used for centuries for such tasks as pumping water and grinding grain. Some countries in the Caribbean used wind power in the sugar industry and for other tasks.

Today, wind is used to drive windmills which, in turn, run generators to produce electricity. Since winds are seasonal and vary in speed, energy obtained from wind-driven machines may be stored in batteries for later use.

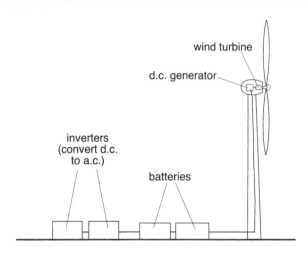

Figure 8.11 *Harnessing the wind.*

It is believed that some countries could generate up to 30% of their electricity from wind power. Recent designs of turbines have turned the wind into a reliable, clean source of electricity.

Is there a downside to harnessing the wind? Yes. There are negative impacts on land use. Noise pollution has an effect on humans and wildlife. Radio transmissions may be disrupted. Some experts believe that wind farms may have serious effects on bird populations. But this should be contrasted with the upbeat response of *Cable Network News* (1996) which reported on the building of wind farms in Northern California as follows: 'Computers track the wind patterns, automatically switching off and on at exact intervals, capturing the subtlest shifts in air currents. These new windmills have a design life that will carry them into the next century . . . Long after the last drop of oil comes from the ground, these windmills will be up . . . generating power.'

8.5 Water power

Water power was used in the distant past to turn mechanical water wheels and to mill grain. Today water power, or hydropower, is widely harnessed for electricity. Of all renewable energy sources, hydropower is the most efficient, economical and reliable. About 25% of the world's electricity is obtained from this source.

In a hydroelectric power plant, water falls on a turbine which turns the shaft of a generator, and electricity is produced. In some locations a dam is built on a river that has a big drop. Water is stored behind the dam and then allowed to fall, by gravity, onto the blades of the turbine.

Note

Water evaporates from the oceans, forms into clouds, falls out as rain and snow, gathers into streams and rivers. This constant movement provides opportunities for extracting energy.

Note

When rivers are dammed certain environmental problems are created. These include:

- blocked migration of fish
- destruction of habitats
- lower water quality
- flooding of areas behind dams.

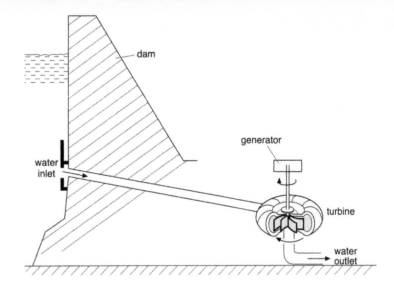

Figure 8.12 *The potential energy of the water in the reservoir is transferred to kinetic energy in the turbine and drives the generator.*

High waterfalls can be harnessed to produce electricity. The Niagara falls provide electric power for Canada and the north-eastern United States.

Guyana has enormous hydropower potential. St. Vincent and Dominica make significant use of this resource.

Hydropower:

- is clean;
- has relatively low running costs;
- is renewable.

Experts believe that, in the future, the trend will be to build small-scale hydropower plants that serve small or isolated communities.

8.6 Geothermal energy

Geothermal energy is heat which is produced by natural processes beneath the Earth's surface. It is the most stable energy resource.

This energy occurs at great depths in some places but may be found near the surface in other places. Geothermal resources are commonly found in earthquake and volcanic areas. About eighty countries have large reservoirs of underground heat. This heat is used to:

- generate electricity (many countries, including El Salvador, Mexico, Italy and New Zealand have large-scale geothermal electric plants);

Note

The areas with the greatest geothermal potential are in regions with active or geologically young volcanoes.

Geothermal energy provides one-third of El Salvador's electricity. Iceland gets 45% of all its energy from geothermal sources.

Hot spring water is used directly to heat greenhouses, to dry fish, de-ice roads, in oil recovery and to heat fish farms and spas. Geothermal heat pumps may 'use' the steady temperatures just underground to heat and cool buildings cleanly and inexpensively.

- heat buildings (many of the buildings in the city of Reykjavik, Iceland, are heated by underground water. Year-round crop growing is possible in this very cold land);
- provide energy for industrial processes (France, Hungary and New Zealand use geothermal water for industrial processes, e.g. in the extraction of valuable minerals);
- desalinate water, i.e. remove salts from sea water;
- warm greenhouses and soil.

St. Lucia is the only Caribbean country which has assessed its geothermal potential.

Much of the technology for the efficient harnessing of geothermal energy now exists. Geothermal power is likely to play an increasingly important role in our 'energy future'.

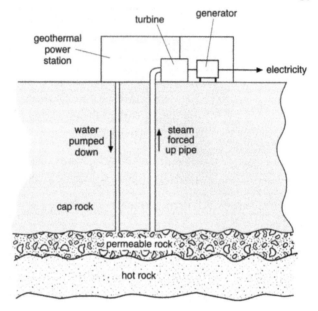

Figure 8.13 *A geothermal steam field.*

Figure 8.13 is a simplified diagram of a geothermal steam field. The basic requirements of such a field are:

- a good source of underground natural heat;
- an adequate water supply;
- an 'aquifer' or permeable reservoir rock;
- a cap rock – impermeable rock that water cannot pass through.

8.7 Conserving energy

Here are some ways in which we waste energy:

- Homes and business places are overlighted.
- Security lights are left on longer than is necessary.
- Inefficient incandescent lighting fixtures adorn our homes and business places (fluorescent lighting is at least three

Things to do

Using this list of ways in which we waste energy, and other sources of information, design a questionnaire to determine the extent to which energy is wasted:

1 at your home
2 in your school
3 in your neighbourhood.

Give out the questionnaire and, based on the results, make an action plan to conserve energy in your home and in your school.

times as efficient and more suitable for use in schools and commercial buildings).

- The refrigerator is overworked in a number of ways:
  - door seals do not fit snugly;
  - the door is open more often than necessary;
  - thermostat settings are often too high;
  - condenser coils are often covered with dirt, dust and even articles of underclothing;
  - too often, the refrigerator is sited too near stoves, etc., or air cannot circulate freely over the coils.

Additionally, we:

- cook on high heat settings throughout (this is unnecessary: why?);
- use open flames instead of the oven;
- do not make best use of oven heat by cooking as many dishes as possible at the same time;
- keep hot water temperatures unnecessarily high;
- run washing machines and, especially, the spin drier with less than maximum loads;
- run cars with poorly tuned engines, too low tyre pressure and with lots of junk in the trunk.

Checklist

After studying Chapter 8 you should be able to:

- define renewable and non-renewable energy sources
- discuss fossil fuels: their advantages and disadvantages
- discuss the primary pollutants of fossil fuels and their effects
- recall alternatives to fossil fuels
- discuss some practical applications of solar energy
- discuss biomass
- discuss wind power
- discuss water power
- discuss geothermal power
- discuss methods of conserving energy

Question

1 a What is meant by the term 'a non-renewable energy resource'?
   b List four non-renewable energy resources.
   c Which two non-renewable energy resources might most profitably be used in your country?
   d Identify four ways in which energy is wasted in the home.
   e On the basis of your answer to **d**, suggest ways by which energy might be saved in the home.
   f 'Pollution knows no frontiers.' Discuss this statement, making especial reference to the burning of fossil fuels.

9 Hydrostatics

Note

Pressure is produced when objects exert forces on each other.

Note

$1\,Pa = 1\,N\,m^{-2}$

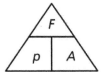

Figure 9.1 *This snow dweller has large paws (why?).*

Note

Pressure is a scalar quantity.

9.1 Pressure

An object exerts a pressure on any surface with which it is in contact. Pressure is related to force as follows:

$$\text{pressure } (p) = \frac{\text{force } (F)}{\text{area } (A)}$$

Pressure is the force acting perpendicular to unit surface area. The SI unit of pressure is $N\,m^{-2}$, also called the **pascal** (Pa).

Consider a box of dimensions $6\,m \times 4\,m \times 2\,m$ and mass $48\,kg$. The pressure which the box exerts on the ground depends on which face of the box is in contact with the ground: the faces of the box have areas of $24\,m^2$, $12\,m^2$ and $8\,m^2$.

The box weighs $[48\,kg \times 10\,N\,kg^{-1}] = 480\,N$. The box, therefore, will exert pressures of $20\,Pa$, $40\,Pa$ and $60\,Pa$, depending on which face is in contact with the ground.

A woman with high-heeled shoes exerts a greater pressure on the ground than one who is wearing flat shoes. An Inuit who is wearing an ordinary pair of shoes sinks into snow more easily than his friend who is wearing snow shoes.

If you hold two walnuts in your hand and press on them along a certain direction, the walnuts split open. However, if you squeeze one walnut with your hand you are not likely to open it, no matter how hard you press.

From the above examples, you can see that pressure is concerned not with force itself but with the distribution of force. Although pressure can produce a force, pressure is not a vector quantity because it has no directional characteristics.

Things to do

Design experiments:

1 to test the validity of the statement 'One pascal is the pressure equivalent of a dollar bill resting on a flat table';
2 to determine the pressure you exert on the floor of the classroom;
3 to determine whether a 10-tonne truck exerts a greater or lesser pressure on the highway than a car of mass 2000 kg.

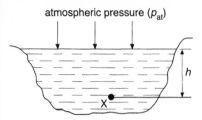

atmospheric pressure (p_{at})

h

X

Figure 9.2 *The pressure at any depth in a pond can be calculated.*

> **Note**
>
> The pressure exerted by an incompressible fluid is directly proportional to the depth in the fluid (h), its density (ρ) and the acceleration due to gravity (g). This pressure is independent of the total volume or the shape of the container.

9.2 Pressure in fluids

Fluids exert a pressure on objects with which they are in contact. A deep-sea diver experiences a greater pressure at the sea floor than at the surface because of the weight of water directly above him.

This pressure in a fluid depends on:

- depth below the surface;
- the density of the fluid.

For an object at a depth h below the surface of a pond (Figure 9.2), the pressure experienced equals atmospheric pressure + the pressure due to the column of water above the object. This latter pressure is called **excess pressure**.

The excess pressure p_{ex} on an object at point X below the surface is given by:

$$p_{ex} = \text{depth } (h) \times \text{acceleration due to gravity } (g) \times \text{density } (\rho)$$

So the total pressure at point X $= p_{at} + hg\rho$, where p_{at} is atmospheric pressure.

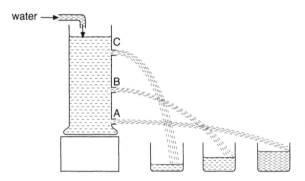

water

C

B

A

Figure 9.3 *The pressure in a fluid increases with depth. The water from the hole A travels further horizontally than that from hole B or hole C. The liquid jets from the holes describe parabolas – another example of projectile motion.*

Example

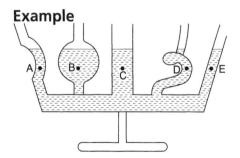

A B C D E

Figure 9.4 *The width or shape of the sections of this apparatus has no effect on the pressure at a given depth. The pressure at A, B, C, D and E is the same because they are at the same horizontal level.*

Figure 9.5 *Pressure at the same depth acts equally in all directions.*

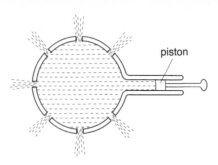

piston

Figure 9.6 *The pressure exerted on part of a liquid which is confined is transmitted equally to all parts of the confining vessel: the water is forced out of all the holes at equal speed.*

A diver working on an oil rig descends to a depth of 820 m. What pressure does he experience? (Density of sea water $= 1.03 \, \text{g cm}^{-3}$, atmospheric pressure $= 1.01 \, \text{kPa}$, $g = 10 \, \text{N kg}^{-1}$.)

Solution

$$\text{Density of water} = 1.03 \, \text{g cm}^{-3} = 1.03 \times 10^3 \, \text{kg m}^{-3}$$

$$\text{At bottom, } p = \rho g h + p_{\text{at}}$$

$$= [1.03 \times 10^3 \, (\text{kg m}^{-3}) \times 10(\text{N kg}^{-1}) \times 820 \, (\text{m})]$$

$$+ 1.01 \times 10^3 \, (\text{N m}^{-2})$$

$$= 8\,447\,010 \, \text{N m}^{-2}$$

$$= 8.45 \, \text{kN m}^{-2}$$

$$= 8.45 \, \text{kPa}$$

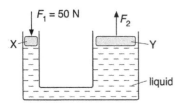

Figure 9.7 *The hydraulic principle illustrated.*

Draw a diagram of a hydraulic jack and explain how it is used to raise cars, etc., in garages.

9.3 Using liquid pressure to magnify forces

In Figure 9.7 a force of 50 N is applied to point X. X has a surface area A_1 of $0.01 \, \text{m}^2$.

$$\text{The pressure acting down at X} = \frac{F_1}{A_1}$$

$$= \frac{50 \, \text{N}}{0.01 \, \text{m}^2} = 5000 \, \text{Pa}$$

The pressure at X is passed on throughout the fluid and acts upwards at point Y. Since X and Y are at the same horizontal level, the pressure at Y is also equal to 5000 Pa. If the piston at Y has an area of $0.5 \, \text{m}^2$,

$$\text{the force acting up at Y} = \text{pressure} \times \text{area}$$
$$= 5000\,\text{N}\,\text{m}^{-2} \times 0.5\,\text{m}^2$$
$$= 2500\,\text{N}$$

You can see that a small force applied at X results in a larger force at Y. This is the principle on which the hydraulic jack and other hydraulic machines work.

Figure 9.8 *This mechanical digger and loader works on the hydraulic principle. Remember that liquids are incompressible and they 'pass on' any pressure which is applied to them.*

The braking system of a car (Figure 9.10) is an example of a hydraulic machine. When a force is exerted on the brake pedal, that force is transmitted throughout the brake fluid. The brake shoe (pad) pushes against the brake drum which, in turn, exerts a force which stops the wheel from turning. (What happens to the kinetic energy with which the car was travelling when it was brought to a stop by the brake?)

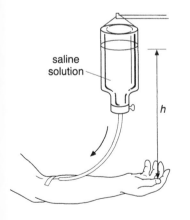

Figure 9.9 *This patient is being fed a saline solution intravenously. Explain the underlying principle.*

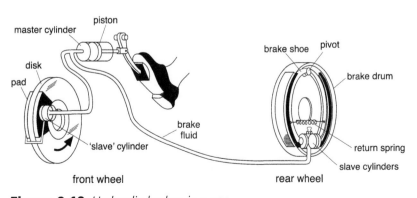

Figure 9.10 *Hydraulic brakes in a car.*

Note

The atmosphere extends for 100 km or so above the Earth's surface. Atmospheric pressure decreases with height.

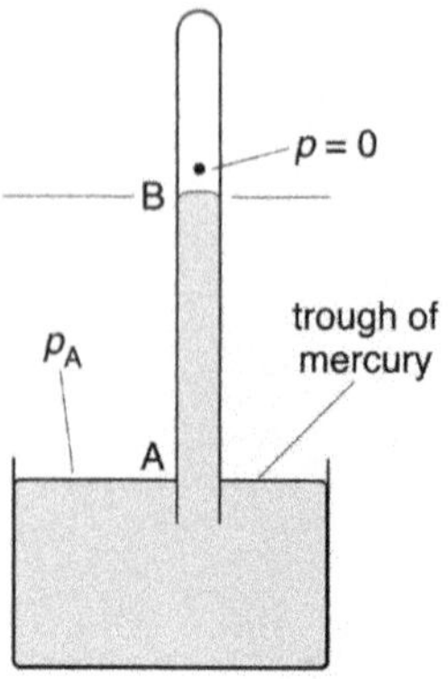

p = 0
B
trough of mercury
p_A
A

Figure 9.11 *A simple barometer. The atmosphere is able to hold up a column of mercury AB. The pressure at B = 0 (why?). The pressure at the surface = atmospheric pressure, p_A.*

Thinking it through

Why is mercury more suitable than water for use in the simple barometer?

9.4 Atmospheric pressure

Air particles move to occupy any available space. Gravity pulls air towards the Earth's surface. Particles of air strike the Earth's surface all the time, creating a pressure known as atmospheric pressure.

In Figure 9.11, atmospheric pressure acting on the surface of the mercury holds up the column of mercury in the tube. The atmosphere normally supports a column of mercury (density 13 600 kg m^{-3}) of height 0.76 m. Using the relationship:

$$p = h \times g \times \rho$$

$$\text{atmospheric pressure} = 0.76\ (\text{m}) \times 9.81\ (\text{N kg}^{-1}) \times 13\,600\ (\text{kg m}^{-3})$$

$$= 1.01 \times 10^5\ \text{N m}^{-2} \text{ or } 1.01 \times 10^5\ \text{Pa}$$

Atmospheric pressure varies with height and locality.

The aneroid barometer

Barometers are used to measure atmospheric pressure. The aneroid barometer (Figure 9.12) contains a sealed metal vacuum chamber (capsule) with a flexible cover. When pressure changes, the cover flexes. A system of levers magnifies the movement. An attached pointer moves over a scale.

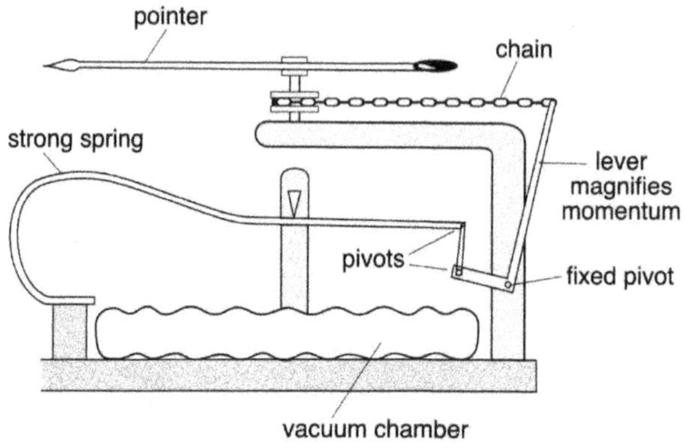

pointer
chain
strong spring
lever magnifies momentum
pivots
fixed pivot
vacuum chamber

Figure 9.12 *The aneroid barometer measures atmospheric pressure and height above sea level.*

The aneroid barometer can be used to measure not only atmospheric pressure but also height above sea level (atmospheric pressure decreases with height). Aneroid barometers, then, are used as altimeters in aeroplanes.

The aneroid barometer is widely used in weather forecasting. For example, there is rapid drop in barometer pressure if a storm is approaching. A steady barometer reading indicates stable weather conditions.

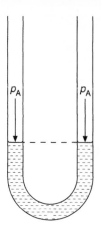

Figure 9.13 *This manometer measures pressure difference by the difference in the heights of two liquid columns.*

9.5 Measuring excess gas pressure

The **manometer** is a device which is used to measure excess gas pressure. A simple manometer (Figure 9.13) consists of a U-tube which contains a liquid of suitable density. Both surfaces of the liquid are exposed to atmospheric pressure.

To measure the pressure of a gas supply, for example, one arm of the U-tube is attached to the supply. The excess pressure supports a column of the liquid of height h, say, as shown in Figure 9.14.

$$(\text{excess}) \text{ gas pressure} = h \times g \times \rho$$

where ρ is the density of the liquid.

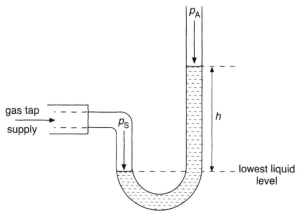

Figure 9.14 *Gas supply attached to manometer. The pressure of the supply is indicated by the difference in the heights of the liquid in the two arms of the manometer.*

9.6 Some effects of atmospheric pressure

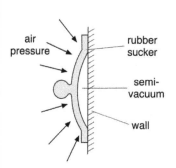

Figure 9.15 *When the sucker is pressed air is squeezed out. A near vacuum is created. The sucker clings to the wall.*

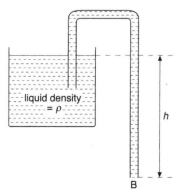

Figure 9.16 *A siphon is used to transfer liquids from one container to another. Atmospheric pressure acts on the surface of the liquid. What is the pressure at point B?*

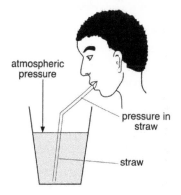

Figure 9.17 *The pressure on the surface of the liquid (atmospheric) is greater than that inside the straw where there is a partial vacuum (why?).*

Things to do

Practise using a siphon by transferring water from a fish tank or similar. Explain fully how the siphon works.

Thinking it through

- Reynold estimated the surface area of his chest was $520\,cm^2$. Given that atmospheric pressure $= 1 \times 10^5\,Pa$, calculate the force on his chest. Explain why the chest does not collapse.
- Account for the following:
  a Reservoirs are built on high ground.
  b Water flows more powerfully through taps on the first floor of a high rise building than through taps on the twelfth floor.
- a Why does your ear pop when the plane in which you are travelling changes altitude?
  b How easy would it be to use a straw to sip orange juice on the Moon?
- An oil tanker takes on $100\,000\,kg$ of crude oil at a port. The tanker sinks deeper into the sea by $3.8\,cm$ as a result. What is the cross-sectional area of the tanker at water level?

Figure 9.18 *Archimedes proved that a crown was made of pure gold. The idea of submerging the crown and comparing its water displacement to that created by an equal weight of pure gold came to him in the bath.*

9.7 Archimedes' principle

A boat floats, an elephant sinks, a helium-filled balloon rises. If you are stretched out you float easily in a pool whereas if you wrap your arms around your legs you sink. Objects appear to lose weight underwater. A boat carrying tons of stuff floats. However, the stuff, when thrown overboard, sinks. A piece of clay in the shape of a sphere sinks when placed in water whereas the same mass of clay in the shape of a boat floats . . . How do we explain these and similar observations?

Archimedes (287–212 BC)

Archimedes – the greatest mathematician of ancient times – was also an excellent engineer. He formulated the 'law of buoyancy' – which is the subject of this section – and the law of levers.

Archimedes was probably the first 'streaker'. Legend has it that he ran naked through the streets of Syracuse shouting 'Eureka, eureka' (I have found it, I have found it), when he solved the problem that led to his law of buoyancy.

1 Attach an object to a spring balance and record its weight in air.
2 Now, completely submerge the object in water in a eureka can.

(a)

(b)

displaced water

3 Record the weight of the object in water.
4 Measure the volume of the water displaced.

Note that the object appears to weigh less in water. Is gravity still pulling on the object? Has the object really lost weight? If not, how can the apparent loss in weight be explained?

The object pushes down on the water, moving liquid particles out of the way. In accordance with the action–reaction principle, the water pushes back on the object. This pushing back or upward force is called an **upthrust**. The object appears to weigh less in water because of this upthrust.

You should find that the upthrust (the apparent loss in weight of the object) equals the weight of the water displaced.

Archimedes' principle states that:

When a body is fully or partly submerged in a fluid the upthrust equals the weight of fluid displaced.

Thinking it through

A ship which is floating is displacing a weight of sea water equal to the ship's weight plus that of its cargo. Why do ships sink deeper in fresh (river) water than in the open sea? (*Hint:* the density of river water is less than that of sea water.)

Floating and sinking in water

If a cork is held under water and then released, it rises to the surface. The cork rises to the surface because the upthrust is greater than the weight of the cork (which acts downwards). A piece of wood, on the other hand, remains partly submerged in water because the weight of the wood equals the upthrust.

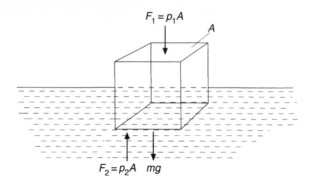

$$F_1 = p_1 A$$

$$F_2 = p_2 A \quad mg$$

Figure 9.19 *Force diagram for a cube in water.*

Consider the force diagram for a cube which is in water.

$$\text{The net force upwards} = \sum F = F_2 - F_1 - mg$$
$$= p_2 A - p_1 A - mg$$

(A = area of cross-section of the cube, and p_1 and p_2 are the pressures on the upper and lower surfaces respectively.)

$$\text{upthrust } F_u = p_2 A - p_1 A = \rho g\, \Delta h\, A = \rho g V = m_w g$$

$F_u = m_w g$ where m_w = mass of water displaced. If $F_u > mg$ the object rises to the surface; if $F_u < mg$ the object sinks.

Submarines sink when they take water into their ballast or buoyancy tanks. When sufficient water is pumped into the tanks the weight of the submarine becomes greater than the upthrust and the submarine sinks. When the tanks are emptied, a point is reached where the upthrust equals the weight of the submarine. The submarine then floats.

The hydrometer

The hydrometer (Figure 9.20) is an instrument used to determine the density of liquids such as beers, wines and other alcoholic drinks. The hydrometer has a long stem to which the density scale is attached and a large weighted bulb. The bulb (which is full of air) provides the upthrust which makes the hydrometer float. The hydrometer sinks deeper in liquids of low density. Density is read directly from the scale of the hydrometer.

Floating and moving in air

The gravitational field acting on your mass pulls you down towards the Earth. When you stand on the ground your weight is balanced by an equal and opposite reaction force.

For birds, hot air balloons and planes the air provides the upward push. The airship of Figure 9.21 floats because its weight (acting downwards) is balanced by a buoyant force

Describe the behaviour of the object if $F_u = mg$.

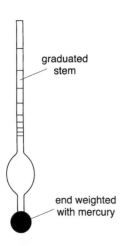

graduated stem

end weighted with mercury

Figure 9.20 *A hydrometer.*

Thinking it through

What would happen if the end of the hydrometer were not weighted?

Figure 9.21 *Airships are used in advertising and aerial surveys.*

Why must the gas used in balloons be less dense than air? Why is hydrogen no longer used?

Figure 9.22 *A hot-air balloon.*

provided by the air. The buoyant force acts up. Airship balloons are filled with helium, which is less dense than air.

The hot-air balloon of Figure 9.22 rises if the gas is heated (why?) and falls to lower levels if gas is let out or if the balloon is cooled.

Upthrust, wings and flight

Figure 9.23 shows the four forces acting on an aeroplane in level flight. The weight (W) acts downwards. Lift (L) is provided as follows. The wing has a curved upper surface. Air moving over the top of the wing travels faster and covers a bigger distance. The pressure below the wing is, therefore, greater than above it. This results in the upwards lift force. (This is called the Bernoulli effect.) The thrust (T) is provided by propeller or jet engines. As the plane moves forwards it experiences a drag force (D) which opposes its forward motion.

net force forwards = thrust − drag force

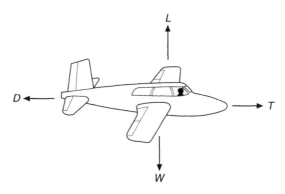

Figure 9.23 *Forces acting on an aeroplane in flight.*

Things to do

Using suitable resource materials, find out and present a written report on:

- the basic controls in an aeroplane
- the principle of operation of propeller and jet engines.

Support your report with annotated diagrams.

Checklist

After studying Chapter 9 you should be able to:

- define pressure
- recall the units of pressure
- understand and use the relationship between pressure, area and force
- recall that pressure can produce a force
- recall the variables on which pressure depends
- recall and use the equations: $p_{ex} = hg\rho$ and $p = p_{at} + hg\rho$
- recall the three basic pressure principles
- understand that fluid pressure can be used to magnify a force
- recall some practical examples of force magnification using fluid pressure
- understand how the simple barometer, aneroid barometer, manometer and hydrometer function
- recall some effects of atmospheric pressure
- define upthrust
- recall Archimedes' principle (the law of buoyancy)
- explain floating, sinking and rising to the surface in terms of the buoyancy law
- recall and explain practical applications of Archimedes' principle

Questions

1. What is the magnitude and the direction of the force exerted by the atmosphere on the top of a table $1\,m \times 1\,m$? [atmospheric pressure $= 1 \times 10^5\,Pa$]

2. The acceleration due to gravity on a distant planet is $2.0\,N\,kg^{-1}$. Would you float more easily in water on this planet than on Earth? Justify your answer.

3. A fish tank of dimensions $40\,cm \times 40\,cm \times 50\,cm$ (depth) is filled with water to a depth of $40\,cm$. What is the total pressure at the bottom of the tank? What is the force on an object of surface area $4\,cm^2$ which is resting on the bottom of the tank?

4. The input piston of a hydraulic jack has cross-sectional area $5.0\,cm^2$. The larger piston has cross-sectional area $500\,cm^2$.

a What force, applied to the input piston, will raise a car of mass 2500 kg?

b A student argues that the application of a small force to the hydraulic jack to lift a much larger load is a violation of the principle of 'conservation of energy'. Explain why this student's argument could not be valid.

An object weighs 250 N in air and 100 N when fully submerged in water.

a What is the weight of the water displaced?

b What is the mass of the water displaced?

c What is the volume of water displaced? [density of water $= 1000 \, \text{kg m}^{-3}$]

A basketball of mass 0.60 kg floats in a pool. Given that the ball has a diameter of 22 cm, calculate: **a** the volume of water displaced, **b** the buoyant force, **c** the mean density of the basketball.

A student decided to make a simple barometer. He used water because he did not have access to mercury. What is the minimum length of his water barometer? Based on your answer, discuss the advantages and disadvantages of using water as the liquid in a simple barometer.

8 An airship contains $6 \times 10^3 \, \text{m}^3$ of helium. Helium has a density of $0.180 \, \text{kg m}^{-3}$. Given that air has a density of $1.20 \, \text{kg m}^{-3}$ at the height at which the airship is floating in equilibrium, what load is the airship carrying?

9 Water is coming through a hose. Explain why, when the end of the hose is partially closed off (leading to a reduction in cross-sectional area), the fluid velocity is increased.

10 A block of metal of mass 1.5 kg and density $4.0 \times 10^3 \, \text{kg m}^{-3}$ is suspended from a string and immersed in water. What is the tension in the string before and after the block is immersed in water?

B Revision questions for Section B

Multiple-choice questions

1 Which of the double-decker buses A–D is least likely to topple?

Number of persons on upper deck	30	35	25	10
Number of persons on lower deck	30	25	35	50
	A	B	C	D

2 The see-saw is balanced at its midpoint, with forces of 600 N and F_x as shown. What is the value of F_x?

4 m	2 m
600 N	F_x

A 1800 N B 1200 N
C 600 N   D 300 N

3 Which of the following does *not* require the application of a net force?
A to change an object from a state of rest to a state of motion
B to maintain an object in motion with constant velocity in a straight line
C to change the speed of an object without changing its direction of motion
D to change an object's direction of motion without changing its speed

4 The velocity of a car is doubled. Which of A–D, concerning both the kinetic energy and the momentum of the car, is true?

	Kinetic energy is	Momentum is
A	doubled	doubled
B	quadrupled	quadrupled
C	quadrupled	doubled
D	doubled	quadrupled

5 A car of mass 1200 kg travelling at 20 m s^{-1} brakes and comes to rest in 4 seconds. What is the average force exerted by the brakes on the car?
A 80 N   B 240 N
C 300 N   D 6000 N

6 The joule is the same as one
A kilogram.metre   B kilowatt.hour
C newton.metre   D ampere.hour

Questions 7 and 8

A car is brought to rest uniformly in 6 seconds. The initial velocity of the car was 24 m s^{-1}.

7 What is the deceleration of the car in m s^{-2}?
A 0.25   B 4.0   C 96   D 144

8 How far does the car travel while decelerating?
A 24 m   B 72 m   C 144 m   D 576 m

Questions 9 and 10

An object of mass 60 kg, travelling at 8 m s^{-1} collides with a stationary object of mass 20 kg. The objects stick together and move off in the same direction with a common velocity.

9 The initial momentum of the moving object, in kg m s^{-1}, is:
A 480   B 900   C 1920   D 3840

10 The common velocity of the objects after collision is:
A 6 m s^{-1} B 7.5 m s^{-1}
C 24 m s^{-1} D 60 m s^{-1}

Questions 11 and 12

A J   B kg m^{-3}
C N m^{-2} D kg m s^{-1}
Which of the above is the unit of

11 kinetic energy?

12 momentum?

13 What is the power output of a boy of 65 kg who climbs a coconut tree 20 m high in 50 seconds?

A 32.5 W   B 260 W

C 162.5 W   D 1300 W

14 Which of the following statements is true of uniform circular motion?

During uniform circular motion

A no unbalanced force exists

B speed changes

C velocity remains constant

D direction constantly changes.

15 A rock of mass 2 kg falls 125 m from a cliff into the sea below. Its speed just before it hits the water is:

A $10\,\mathrm{m\,s^{-1}}$ B $25\,\mathrm{m\,s^{-1}}$

C $50\,\mathrm{m\,s^{-1}}$ D $100\,\mathrm{m\,s^{-1}}$

16 Which of the following is the same as power?

A mass × acceleration

B mass × acceleration ÷ time

C force × distance

D force × distance ÷ time

Questions 17 and 18

The diagrams A–D show different arrangements of two coplanar forces of value 24 N and 10 N.

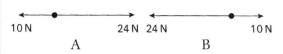

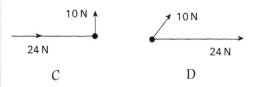

For which arrangement will the value of the resultant be:

17 34 N?

18 26 N?

19 Which of the following statements is correct?

A Momentum is a scalar quantity.

B Momentum is conserved in all collisions.

C An object could have kinetic energy but no momentum.

D The unit of momentum is the same as that for force.

20

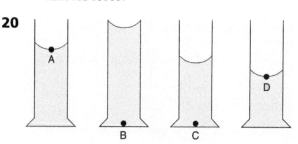

Each of the vessels above contains water. At which of the locations A–D is the pressure greatest?

21 A rectangular box of mass 50 kg measures 0.5 m by 0.4 m by 0.1 m. What is the least pressure that this box can exert?

$[g = 10\,\mathrm{N\,kg^{-1}}]$

A 500 Pa   B 2500 Pa

C 10 000 Pa   D 12 500 Pa

22 Which of the objects A–D, dropped from the same height and at the same place, experiences the greatest air resistance?

A cricket ball

B basketball

C crumpled sheet of paper

D uncrumpled sheet of paper

Questions 23–25

A–D below are velocity–time graphs for the motions of different objects.

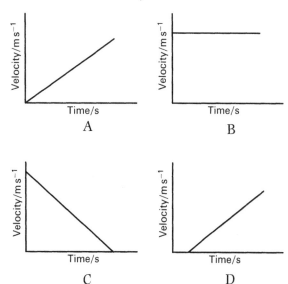

Which graph represents the motion of an object travelling with

23 zero acceleration?

24 uniform deceleration?

25 uniform acceleration, starting from rest?

Questions 26 and 27

Different weights were attached to a spiral spring of natural length 6 cm and the new lengths measured. Results from a typical experiment are shown in the table.

Load/N	0	50	100	150	200
Length of spring/cm	6.0	8.0	10.0	12.0	14.0

26 What would be the length of the spring when the load is 75 N?
  A 8.25 cm   B 8.50 cm
  C 9.00 cm   D 9.25 cm

27 From the results it may be deduced that the extension of the spring is:
  A directly proportional to the load
  B inversely proportional to the load
  C directly proportional to the square of the load
  D inversely proportional to the square of the load.

28 Which of the following energy resources is non-renewable?
  A solar energy   B natural gas
  C biogas   D wind energy

29 Which of the following statements is correct?
  A Friction improves the efficiency of machines.
  B Machines multiply forces but not distances.
  C The hydraulic press magnifies the effect of forces.
  D Bodies with the narrowest bases are the most stable.

30 What frictional force is being overcome by a vehicle which is generating 5000 W while travelling along a horizontal road at a constant velocity of $10\,\mathrm{m\,s^{-1}}$?
  A 50 000 N   B 5000 N
  C 500 N   D 50 N

31 A particle moving with constant speed in a uniform circular path has a:
  A constant acceleration along the radius of the path
  B constant acceleration at right angles to the radius of the path
  C constant velocity
  D variable acceleration.

32 Which of the following *best* expresses Aristotle's views of force? Force is:
  A directly proportional to acceleration
  B directly proportional to velocity
  C directly proportional to velocity squared
  D directly proportional to distance squared.

Questions 33 and 34

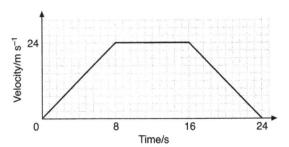

The graph above shows the motion of a vehicle which starts from rest and accelerates uniformly for 8 seconds until its speed is $24\,\mathrm{m\,s^{-1}}$. It travels at this speed for another 8 seconds and finally decelerates to rest again in a further 8 seconds.

33 What is the acceleration of the car during the first 8 seconds?
  A $0.33\,\mathrm{m\,s^{-2}}$ B $1.5\,\mathrm{m\,s^{-2}}$
  C $3.0\,\mathrm{m\,s^{-2}}$ D $72\,\mathrm{m\,s^{-2}}$

34 The total distance travelled by the vehicle in the 24 second period is:
  A 96 m   B 192 m
  C 288 m   D 384 m

35 Two toy trains are moving towards each other at the same speed on the same track. Upon colliding, the trains lock together. If one train has twice the mass of the other, what will be the ratio of the speed of the trains after the collision to their speed before the collision?

A 1:3   B 1:2   C 2:1   D 3:1

36 Which of A–D is correct concerning the kinetic energy and gravitational potential energy of an object which is falling freely under gravity?

	Kinetic energy	Gravitational potential energy
A	increases	remains constant
B	increases	decreases
C	remains constant	increases
D	decreases	increases

Structured and free-response questions

1 **a** Describe **two** situations which show that a force may cause a change of (i) shape of a body and (ii) motion of a body.
   b Distinguish between a scalar and a vector quantity. Place the following quantities into two sets (scalars or vectors): mass, weight, momentum, kinetic energy, work, velocity, density, displacement, power and acceleration.
   c Two forces F_w and F_e act on a body O as shown.

F_w ← ———O——— → F_e
180°

What happens to the size of the resultant as the angle between the two forces decreases from 180° to 0°?
   d 'The order in which displacements are combined changes the magnitude of the resultant.' Is this statement true or false? Explain, using a suitable example.

2 A famous scientist is reputed to have dropped balls of different masses from the top of the tower of Pisa and timed their fall.
   a Who was the scientist? List two important contributions that this person made to science.
   b Why do balls of different masses, dropped simultaneously from the same height, reach the ground in the same time?
   c When a hammer and a feather were dropped simultaneously from the same height above the Earth's surface, the hammer struck the ground first. When the astronaut David Scott dropped a hammer and a feather simultaneously from the same height on the Moon, the two objects struck the surface of the Moon at the same time. Account fully for these observations.
   d Discuss the features that allow parachutists to reduce their rate of fall to safe speeds.

3 **a** Define acceleration and average velocity. Show that the average velocity of a body undergoing constant acceleration, and starting from rest, is half of its final velocity.
   b State Newton's second law and use it to define the newton, the unit of force.
   c A crate of 35 kg is pulled by a steady horizontal force of 50 N across a table. The crate moves with a uniform acceleration of $1.2 \, \text{m s}^{-2}$. Calculate
   (i) the net force causing motion;
   (ii) the frictional force between the crate and the table.
   d Why is friction considered to be a nuisance?
   e Discuss two ways by which friction between the crate and the table in **c** could be reduced.

4 A spacecraft of mass 6000 kg lifts off the surface of Jupiter, where the acceleration due to gravity is $25 \, \text{N kg}^{-1}$ ($25 \, \text{m s}^{-2}$).
   a What is the weight of the spacecraft on the surface of Jupiter?
   b What will be the mass of the same spacecraft on Earth?

c What force must the rocket engines exert for the spacecraft to lift off the surface of Jupiter with an acceleration of $15\,\mathrm{m\,s^{-2}}$?

d Explain how a jet engine or rocket gets its propelling force. How can the set-up shown be used to model how a jet engine functions?

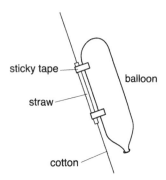

5 A car and drive have a combined mass of $1575\,\mathrm{kg}$. The car was travelling at $28\,\mathrm{m\,s^{-1}}$ when the driver, sensing danger, managed to reduce speed to $18\,\mathrm{m\,s^{-1}}$ before the car crashed into a barrier. The car came to rest in 1 second during the crash.

a Determine the change in kinetic energy when the car reduced speed from $28\,\mathrm{m\,s^{-1}}$ to $18\,\mathrm{m\,s^{-1}}$.

b What became of this change ('loss') of kinetic energy?

c Determine (i) the momentum of the car just before the crash; (ii) the change of momentum during the crash; (iii) the decelerating force on the car during the 1 second of the crash; (iv) the average deceleration of the car during the 1 second of the crash.

d During the crash the driver, whose mass was $75\,\mathrm{kg}$, moved forward a distance of $0.2\,\mathrm{m}$ against the force exerted by his seat belt. (i) What was the kinetic energy of the driver when the car was moving at $18\,\mathrm{m\,s^{-1}}$? (ii) Calculate the average force exerted by the seat belt on the driver during the crash. (iii) Discuss the advantage, in situations like this one, of using a broad seat belt rather than a narrow one.

e What features (other than seat belts) could be built into the construction of cars to protect passengers from serious injury during crashes?

6 a Which of the following statements concerning vectors is false? (i) A vector has magnitude and direction. (ii) The resultant is the arithmetic sum of two vectors acting at the same point. (iii) The resultant of two vectors acting in opposite directions is their sum. (iv) Two vectors in equilibrium neutralize each other.

b The diagram below shows two forces acting at right angles on a small object O.

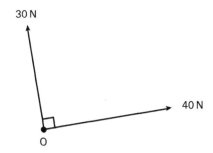

Using a scale drawing, find the magnitude and the direction of the resultant force acting on O.

c If the object O above accelerates uniformly at $12\,\mathrm{m\,s^{-2}}$ what is its mass?

d What is the speed of O 6.0 seconds after forces began acting on it?

7 a Distinguish between an elastic collision and an inelastic collision. Give a suitable example of each type of collision.

b How is the momentum of a body affected if (i) its speed is doubled; (ii) there is a change in the direction of motion of the body.

c Comment on the statement 'momentum is a measure of how difficult it is to stop a moving body'.

d Explain the following as fully as possible: 'It is dangerous to jump from a height of 2 metres onto a concrete floor, but it is safe for a high jumper to jump the same height onto a foam mattress'.

e A bullet of mass $0.05\,\mathrm{kg}$ leaves the muzzle of a gun of mass $5\,\mathrm{kg}$ with a velocity of $375\,\mathrm{m\,s^{-1}}$. What is the velocity of recoil of the gun?

8
a State Newton's third law.
b Use Newton's third law to explain the following: (i) The recoil of a gun which is firing bullets. (ii) The difficulty of getting a car to move off a road surface on which lubricating oil has been accidentally spilled.
c Two teams are involved in a tug-of-war. When the teams are momentarily in equilibrium, team GG pulls on the rope with a force of 500 N. (i) What is meant by the phrase 'in equilibrium'? (ii) What is the force exerted by the second team, JP?
d If it takes a force of 550 N to break the rope, will the rope break under the conditions described in **c** above?

9
a Explain each of the following: (i) Objects in a satellite orbiting the Earth appear weightless. (ii) Water drops move off at a tangent from the wheels of a car which is moving at high speed down a wet roadway.
b A boy stands on a bathroom scale which is placed on the floor of an elevator. He records his weight when the elevator (i) is stationary, (ii) has just started to move, (iii) reaches a normal steady speed and (iv) begins to decelerate. [If possible, carry out this experiment.] Describe and explain the changes in the boy's weight during the above sequence.

10
a What is meant by the term 'a centripetal force'?
b State three factors upon which the centripetal force keeping a body in circular motion depends.
c In what direction does a centripetal force accelerate a body upon which it acts?
d Explain why a body which is moving at constant speed is still accelerating.
e A glass is three-fifths filled with water and stirred with a glass rod. Describe the expected appearance of the water in the glass. Also, describe and account for the observations expected if the water is stirred at a faster rate. [Why not try this out experimentally.]

11 Explain each of the following:
a The Moon is in free fall towards Earth, yet it is not getting any closer.
b Satellites in orbits close to the Earth remain in their orbits for a limited time only.
c

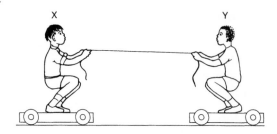

If boy X pulls boy Y, the two trolleys move towards each other.
d A special fluid – brake fluid – is used in the braking and transmission systems of cars. These systems do not work well if air gets into the fluid.
e It is easier to slice a loaf of bread using a sawing motion of the knife than by trying to cut directly through the loaf.

12 In an experiment to investigate the relationship between the load attached to a spiral spring and its extension, a fifth-former obtained the following results.

Load/N	0.5	0.75	1.0	1.25	1.50	1.75	2.00
Extension/ mm	17.5	27.5	37.5	47.5	54.5	67.4	77.5

a Plot a graph of load (y-axis) against extension (x-axis).
b The student incorrectly recorded one of the readings. Identify this reading and write the corrected value(s).
c State Hooke's law.
d Comment on the shape of the graph. Is Hooke's law obeyed for the spiral spring under the conditions of this experiment?
e Find the gradient of the graph and use the value of the gradient to obtain a value for the force constant of the spiral spring, given that the force constant, k, is equal to 1000 times the gradient. State the unit of the force constant.

13 Copy and complete the table below which links type of force with situation. An example has been given.

Situation	Type(s) of force involved
Dust collecting on a TV screen	
The interaction of protons in the nucleus	
The Moon in orbit around the Earth	
The heat produced by contact between tyres of a car and road surface	
The needle of a compass always pointing in a northerly direction	magnetic

14 What do you understand by 'the principle of conservation of energy'? A bank of rain-bearing clouds exists at a height of 400 metres. Raindrops falling from these clouds hit the ground with a velocity of $25\,\mathrm{m\,s^{-1}}$.
   a Calculate the gravitational potential energy associated with 50 kg of rain, before falling.
   b What is the kinetic energy of 50 kg of rain when it arrives at ground level?
   c Account fully for the difference between the answers to **a** and **b**.

15 A simple pendulum, consisting of a length of string and a bob, was used to determine a value for the acceleration due to gravity. The pendulum was displaced sideways through a small angle and set in oscillation. The time for 15 to and fro swings was determined. The procedure was repeated for different lengths of string. A fifth-former, carrying out this experiment, obtained the following results:

Length, l, of the pendulum/ m	Time, t, for 15 oscillations/ s	Time, T, for one oscillation/ s	T^2/s^2
0.25	14.85		
0.50	21.30		
0.75	26.25		
1.00	30.60		
1.25	33.90		
1.50	36.70		
2.00	42.40		

   a Copy and complete the table by filling in the values of (i) T and (ii) T^2.
   b Plot a graph of T^2 (y-axis) against l (x-axis).
   c Comment on the relationship between T^2 and l.
   d Determine the gradient (S) of the graph, stating its units.
   e According to theory, the gradient S of the graph is equal to $\dfrac{4\pi^2}{g}$. Use the value of S obtained in **d** above to find a value for the acceleration due to gravity.
   f Using the same apparatus, design an experiment to test the hypothesis: 'the time for one complete oscillation of a simple pendulum depends on the angle through which it is displaced'.

16 A rubber ball falls from a table on to the floor and finally comes to rest after rebounding many times.
   a What changes in energy were taking place when the ball was falling?
   b What happened to the initial energy after the ball came to rest?
   c State the principle of conservation of energy. Does it apply to this situation?
   d After being dropped from a height of 1 m the ball rebounds (the first time) to a height of 0.8 m. Neglecting air resistance, explain what happens to the energy 'lost'.
   e Is the collision between the ball and the floor elastic? Explain.

17 Write down the equation which links atmospheric pressure p_{at} with the pressure p_h at a depth h metres below the surface of a liquid. Explain the meaning of all terms used in the equation.

a A school pond is 2 metres deep. What is the pressure on a 5 kg ball which is at the bottom of the pond?

b A bubble forms at the bottom of the pond. Describe and explain the changes in the volume of the bubble as it rises to the surface of the pond.

c Study the diagram of the braking system shown.

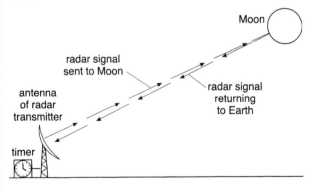

cross-sectional area of A $= 8$ cm^2
cross-sectional area of B $= 5.5$ cm^2
(i) Explain what happens when the driver presses down the brake pedal with a force of 240 N. (ii) Calculate the pressure in the master cylinder. (iii) Calculate the force between a brake pad and a disc brake.

18 a

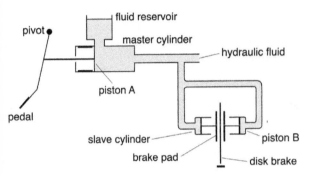

The Earth–Moon distance is determined using this arrangement. Fully explain the basis of this method, stating how the distance is obtained.

b

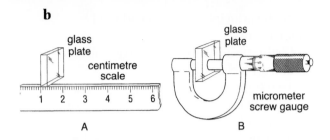

Which method, A or B, gives a more precise measurement for the thickness of the block? Explain.

c Describe how you would attempt to obtain a velocity–time graph for a runner in a 200 metre race, given that 10 persons are available with stop watches. Give a sketch of what the expected graph would look like.

d A ball is thrown vertically upwards. It rises to its highest point, then returns to its starting point. Sketch how (i) distance travelled changes with time, (ii) velocity changes with time, (iii) acceleration changes with time.

19 The diagram (below) shows a speed–time graph for a car.

a What was the speed of the car (i) at A, (ii) after 20 seconds (iii) after 80 seconds, (iv) after 100 seconds?

b During which time (intervals) was the car (i) increasing speed, (ii) decreasing speed, (iii) moving at constant speed?

c What was the average speed over (i) the first 10 seconds, (ii) the next 30 seconds (iii) the entire journey?

d Calculate the total distance covered.

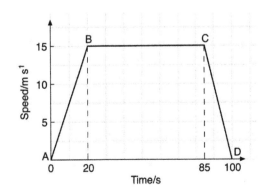

20 State the principle of conservation of momentum.

 a Distinguish between an elastic collision and an inelastic collision.

 b Two toy cars are moving towards each other at the same speed. After collision, the cars lock together. One car has a mass twice that of the other. (i) Is the collision of the cars elastic or inelastic? Explain. (ii) Calculate the combined speed of the cars, after collision, in terms of their speed, v, before collision.

 c One of Walsh's express deliveries to a New Zealand batsman was timed at $144 \, km \, h^{-1}$. The batsman hit back the ball at $20 \, m \, s^{-1}$. The bat and the ball were in contact for 0.025 seconds. The mass of the ball was 210 g. (i) Convert $144 \, km \, h^{-1}$ to $m \, s^{-1}$ and 210 g to kilograms. (ii) Calculate the momentum of the ball just before the batsman struck it. (iii) Calculate the momentum of the ball just after it was struck by the batsman. (iv) Calculate the momentum change and hence the force with which the bat hit the ball. (v) Discuss the energy changes from the moment the batsman struck the ball until the ball eventually comes to rest, after rebounding off the boundary wall.

21 a State Archimedes' principle. How does the principle apply to the buoyancy of (i) submarines, (ii) hydrogen-filled balloons?

 b A wooden cube of volume $0.25 \, m^3$ and density $750 \, kg \, m^{-3}$ is placed in a liquid of density $1000 \, kg \, m^{-3}$. (i) What fraction of the cube floats above the surface of the liquid? (ii) What force must be applied to the surface of the cube to bring it level with the surface of the liquid?

 c You are provided with a known volume of water in a container, a salt solution of known concentration and a raw egg. The egg sinks in the water. (i) Describe steps that can be taken to get the egg to float. (ii) Describe how you would attempt to determine the density of the egg. [No additional apparatus or material will be issued.]

22 a What is meant by the term 'atmospheric pressure'?

 b Discuss the principle of operation of a simple device for measuring atmospheric pressure.

 c Describe three different situations in which atmospheric pressure plays a part.

 d How does atmospheric pressure vary with altitude?

 e Discuss (briefly) the influence of atmospheric pressure on weather patterns.

10.1 The nature of heat – a historical perspective

It seems natural to think of heat as 'something that flows'. Well into the nineteenth century, heat was thought to be an invisible, weightless, self-repellent fluid – called caloric – which flowed easily from a hot body to a cold one. A hot body was thought to have more caloric than a cold one and caloric flowed from the hot body to the cold one until each had the same amount.

According to the caloric theory:

- caloric flows more easily through metals than non-metals;
- since caloric repels itself, a body with a lot of caloric expands, i.e. a heated body expands;
- when a nail is heated, caloric squeezes to the surface and the nail becomes hotter;
- the smaller an object the less caloric it can hold.

Scientists of the time clung to the caloric theory because it seemed to explain what they were interested in. However, it gradually became obvious that the caloric theory could not explain such thermal effects as:

- melting and evaporation;
- why the same quantity of heat given to different objects of the same mass results in different temperature changes;
- the heat produced by friction.

Rumford's cannon-boring experiments described below threw serious doubts on the caloric theory, which was finally abandoned when Joule showed that heat was a form of energy.

Rumford's experiments

As Minister of War in Bavaria, Count Rumford supervised the making of cannons. These were made by boring large holes in brass using a blunt boring tool which was driven by horses (see Figure 10.1). In one experiment, Rumford found that sufficient heat was produced in 2.5 hours to raise the temperature of 12 kg of water to 100 °C.

From his experiments, Rumford concluded that the heat produced depended on the amount of work done by the horses. Rumford discredited the caloric theory and established a link between heat and mechanical work.

Figure 10.1 *A model of Count Rumford's cannon-boring apparatus. The results of experiments with this apparatus in 1798 caused Rumford to doubt that a material substance (caloric) was flowing into the cannon, causing its temperature to rise.*

The important contribution of James Joule

In the 1840s Joule carried out experiments to determine the amount of work required to raise the temperature of a fixed mass of water by 1 °C. Using apparatus similar to that of Figure 10.2, he determined that:

- the gravitational potential energy of falling weights was converted into thermal energy (E_H);
- E_H was related to the change in the temperature of the water ($\Delta\theta$) by the relationship:

$$E_H = m \times c \times \Delta\theta$$

where m is the mass of the water used and c is a constant known as the **specific heat capacity** of the water. Joule found that 4.2 kJ of work raised the temperature of 1 kg of water by 1 °C.

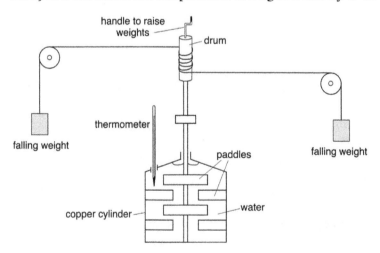

Figure 10.2 *An illustration of the apparatus used by James Joule in his famous experiment for determining the 'mechanical equivalent of heat'.*

Note

In the late 18th century two opposing views of heat were held. Some saw heat as the consequence of the *ceaseless motion of particles*. Others thought of heat as *a fluid which moved from a hot object to a colder one*.

Joule also measured, experimentally, the quantity of heat produced when electric currents flowed through conductors (resistors).

The total energy was constant in all systems that Joule investigated. The quantity of energy 'lost' as work or electricity 'reappeared' as heat.

Joule's experiment established:

- heat as a form of energy;
- the important principle of the **conservation of energy**.

With the fall of the caloric theory, scientists sought alternative explanations for thermal effects. The 'kinetic theory' – discussed later – provides a satisfactory explanation for many physical phenomena.

10.2 Heat and its effects

When a substance is heated, the particles within it gain energy and vibrate more rapidly, i.e. the internal energy of the substance increases. Increases in internal energy may lead to:

- changes in chemical composition;
- changes in shape, resulting from expansion;
- changes of state (e.g. from solid to liquid or liquid to vapour);
- a rise in temperature.

We use heat in a number of ways:

- to cook our food;
- to keep us warm;
- to raise steam to run engines and drive turbines.

Heat, however, often appears when we do not want it: for example, in the moving parts of machinery or when we turn on an electric light.

Note

The internal energy of a system is the sum of the kinetic and potential energies of all the particles in the system.

10.3 Temperature is not the same as heat

In earlier times, the distinction between heat and temperature was not clear; the two were often used interchangeably. A burning splint (temperature relatively high) is not likely to make the chemistry laboratory appreciably warmer, but constant temperature water baths, which are at a much lower temperature, could make the room uncomfortably warm. Whereas temperature is independent of the amount of substance, the quantity of heat is dependent on the amount of substance and also on the nature of the substance. We shall

Note

Before the 19th century it was believed that the hotness or coldness of an object depended solely on the amount of heat it contained. Joseph Black (1728–1799) was the first person to make a clear distinction between heat and temperature.

return to discuss quantity of heat later. Let us, for now, focus on the concept of temperature.

You already know that temperature is a sense impression, and temperature is one of the base SI quantities. Many physical properties vary with temperature and can, therefore, be used as the basis for making themometers and for establishing temperature scales. Among these 'thermometric' properties are:

- the expansion of mercury and alcohol;
- the resistance of wires;
- the pressure of a gas, held at constant volume;
- the volume of a gas, held at constant pressure;
- the radiation emitted by hot objects;
- the expansion of bimetallic strips, and so on.

10.4 Thermometers

A thermometer quantitatively measures the temperature of an object or system. Common laboratory thermometers are mercury (or alcohol)-in-glass thermometers. These work on the principle that liquids expand slightly when heated.

The **mercury-in-glass thermometer**:

- may be used in the range $-39\,°C$ to $400\,°C$, but is not suitable for use at very low temperatures;
- is easy to use;
- responds relatively quickly.

Alcohol-in-glass thermometers can be used at lower temperatures than mercury-in-glass thermometers (alcohol freezes at $-115\,°C$). Alcohol, however, wets the sides of the capillary tube and the thread often breaks. Alcohol-in-glass thermometers are not used above $110\,°C$. Why?

How to calibrate a mercury-in-glass thermometer

As with all measuring instruments, readings must be made with reference to a standard. Two standard temperatures are needed to define a temperature scale. These references are:

- the **lower fixed point** – the temperature of pure melting ice (0°C on the Celsius scale). To determine this point see diagram (a) overleaf.
- the **upper fixed point** – the temperature of steam at 1 atmosphere pressure (100°C on the Celsius scale). To determine this point see diagram (b) overleaf.

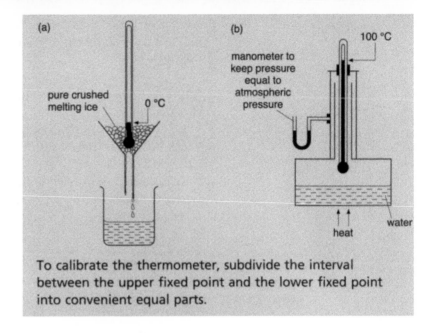

To calibrate the thermometer, subdivide the interval between the upper fixed point and the lower fixed point into convenient equal parts.

The clinical thermometer

This is a special type of mercury-in-glass thermometer. It is used to measure the temperature of the human body. It covers a narrow range of temperature on either side of 37 °C (why 37 °C?).

Observe the constriction of the bore of the clinical thermometer (Figure 10.3). This prevents the mercury thread from running back into the reservoir, once the body temperature has been taken. The clinical thermometer is reset by shaking it.

Figure 10.3 *A clinical thermometer.*

The thermocouple

The thermocouple consists of two wires of different metals joined together as part of an electric circuit (see Figure 10.4) When one junction is cold and the other hot, a small current

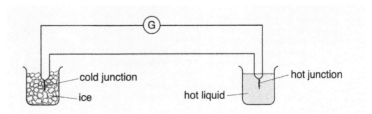

Figure 10.4 *A thermocouple.*

flows. The greater the difference in the temperature of the junctions, the greater the current which flows. Thermocouples:

- may be used to measure temperatures in the range $-200\,°C$ to $1600\,°C$;
- respond rapidly to changing temperatures;
- take little heat from the object whose temperature is being measured;
- are used to measure temperatures in small enclosures or remote places.

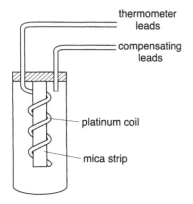

Figure 10.5 *A resistance thermometer.*

Resistance thermometers

Resistance thermometers work on the principle that the resistance of a conductor increases with increasing temperature. The thermometer usually consists of a length of platinum wire wound onto a mica strip, which is connected to a special electrical circuit (Figure 10.5). Resistance thermometers are used to measure steady temperatures in the range $-200\,°C$ to $1200\,°C$.

Many different thermometric properties are used in the making of thermometers. The thermometric property selected should ideally:

- vary regularly (preferably linearly) with temperature;
- give reproducible readings;
- respond readily to rapidly changing temperatures;
- give accurate readings;
- give readings over a wide range of temperatures.

Things to do

1 Using suitable resource materials, find out about the temperature scales proposed or used by the following people: Galen (AD 170), Roemer (1708–1709), Fahrenheit (1724) and Celsius (1742). Discuss ONE important advantage of the Celsius scale compared with the Fahrenheit scale.

2 Find out how the infrared pyrometer in Figure 10.6 works.

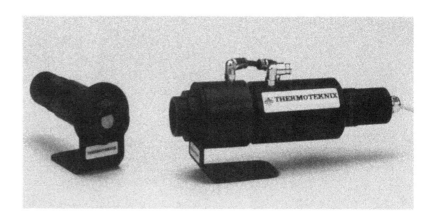

Figure 10.6 *An infrared pyrometer.*

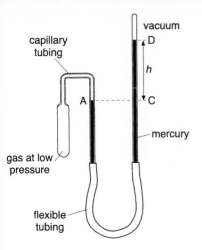

capillary tubing

vacuum

D

h

A - - - - - - C

mercury

gas at low pressure

flexible tubing

Figure 10.7 *A constant volume gas thermometer. Find out how it works.*

10.5 The absolute (Kelvin) temperature scale

In 1780, the Frenchman Jacques Charles showed that all gases expanded by very nearly the same amount for the same increase in temperature. His experiments showed that a gas could be used as a thermometric medium. Two types of gas thermometers are in use: the **constant volume gas thermometer** and the **constant pressure gas thermometer**.

Experiments show that, for gases at low pressure, a simple relationship exists between pressure, volume and temperature. This relationship may be expressed as follows:

$$pV = \text{constant} \times T$$

where T is the *thermodynamic* (or *absolute*) temperature.

The **absolute temperature scale** (related to the Celsius scale) has been in use since 1954. The unit of temperature on the absolute scale is the **kelvin (K)**, after Lord Kelvin. The kelvin is denoted by the letter K (*please do not refer to it as °K.*)

Although the kelvin is the same size as one degree on the Celsius scale, the zero point on the kelvin scale is 0 K – commonly called **absolute zero**. Absolute zero is equivalent to −273 °C. It follows, then, that:

temperature in °C = temperature in K − 273

The thermometers in use today measure a very wide range of temperatures: from about one-millionth of a kelvin to about 10 000 kelvin.

10.6 Some important temperatures

−273 °C	0 K	absolute zero: molecules have zero energy at this temperature
−37 °C	234 K	the freezing point of mercury
0 °C	273 K	the ice point: the melting point of ice
30 °C	303 K	average air temperature in the Caribbean
37 °C	310 K	healthy body temperature
79 °C	352 K	the boiling point of ethanol
100 °C	373 K	the steam point: the boiling point of water
357 °C	630 K	the boiling point of mercury
827 °C	1100 K	the approximate temperature of red-hot objects
1227 °C	1500 K	the approximate temperature of white-hot objects
5727 °C	6000 K	the approximate temperature of the surface of the Sun

Checklist

After studying Chapter 10 you should be able to:

- recall the caloric theory
- differentiate between the caloric and kinetic theories of heat
- discuss Rumford's experiment and its conclusion
- discuss Joule's experiment and its results
- recall and use the formula $E = mc\Delta\theta$
- define internal energy
- define heat and discuss its effects
- understand the difference between heat and temperature
- define a thermometric medium
- discuss the characteristics of a suitable thermometric medium
- discuss some commonly used thermometers
- define upper and lower fixed points
- calibrate any thermometer
- convert °C to kelvin and vice versa
- recall that kelvin is the unit of the absolute temperature scale

Questions

1 What do you understand by these terms: calibrating a thermometer, the upper fixed point and the lower fixed point of a thermometer? How are these established?

2 The interval between the lower and upper fixed points on a thermometer is 20 cm.
   a What is the temperature when the mercury column is (i) 10 cm and (ii) 12 cm long?
   b What is the length of the mercury column when the thermometer is placed in a liquid whose temperature is 68 °C?

3 The freezing point of gold is 1337 K. What is the corresponding temperature in degrees Celsius?

4 An ungraduated thermometer attached to a centimetre scale reads 13 cm in pure melting ice and 38 cm in steam at 1 atmosphere. When the thermometer is placed in a certain freezing mixture it reads 11.5 cm. When the thermometer is placed in pure boiling ethanoic acid it reads 42.5 cm. What is:
   a the temperature of the freezing mixture?
   b the boiling temperature of pure ethanoic acid?

Questions 5–9 concern the different types of thermometers labelled A–D.

A the thermocouple
B the mercury-in-glass thermometer
C the alcohol-in-glass thermometer
D the clinical thermometer

Which of the thermometers

5 has a constriction in its bore?

6 is best suited for measuring rapidly changing temperatures?

7 is used in the temperature range 37 °C to 42 °C?

8 is most convenient for measuring laboratory temperatures in the range $-20\,°C$ to $150\,°C$?

9 contains two dissimilar wires?

10 A student suspects that her alcohol-in-glass thermometer has an incorrectly calibrated lower fixed point.
  a What are the fixed points on a thermometer?
  b Why is a mercury-in-glass thermometer not suitable for measuring a temperature of $-80\,°C$?
  c Give the name of a thermometer suitable for measuring temperatures in the range $-80\,°C$ to $1200\,°C$.
  d Describe how the thermometer mentioned in **c** works. Support your description with a suitable diagram.
  e Design an experiment by which the student could determine if the alcohol-in-glass thermometer has an incorrectly calibrated lower fixed point.

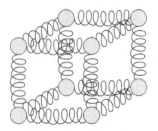

Figure 11.1 *You can imagine the particles in a solid being joined by 'invisible springs'.*

11.1 How solids expand

Most solids expand when heated and contract when cooled. Crystalline solids can be thought of as consisting of cubes (see Figure 11.1). Particles (atoms or molecules) are situated at the corners of the cubes; you can imagine the particles being joined by 'invisible springs'. At room temperature, the particles vibrate about the springs with definite *frequencies* and definite *amplitudes*. When temperature increases, the amplitude and frequency of the vibrations increase and the distance between particles increases, i.e. the material expands. When temperature decreases, the amplitude and frequency of the vibrations decrease and the material contracts.

The expansion and contraction of solids can be troublesome in some situations.

Making allowance for expansion and contraction

Figure 11.3 *Expansion joints in railway lines prevent buckling of the rails when the metal expands.*

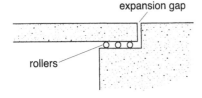

Figure 11.2 *This bridge is supported on rollers. It moves without damage during contraction and expansion. The gap compensates for both expansion and contraction.*

11.2 Linear expansion

All materials expand to different extents. How much a material expands depends on a characteristic of the material called the **coefficient of linear expansion** (symbol α). As heat is applied to the material, the expansion of the material ΔL can be found as

$$\Delta L = L \times \alpha \times \Delta\theta$$

where L = original length (in metres) and $\Delta\theta$ = change in temperature (in °C). Hence the total length of the material after expansion = $L + \Delta L$.

Example

12 metre lengths of metal were laid down at 0 °C with a gap of 1.2 mm between them. At what temperature will the gap just close? (The coefficient of linear expansion of the metal is $1.0 \times 10^{-5}\,°C^{-1}$.)

Solution

$L = 12\,m$, $\Delta L = 1.2 \times 10^{-3}\,m$, $\alpha = 1.0 \times 10^{-5}\,°C^{-1}$

Substituting in the equation above gives a values of 10 °C for $\Delta\theta$.

Thinking it through

Use any available resources to find the values of the coefficient of linear expansion for the materials mentioned in these questions.

- Why is Pyrex less likely to shatter than ordinary glass when heated?
- Discuss why concrete is reinforced with steel and not copper.
- It is observed that when a mercury-in-glass thermometer is placed in contact with a hot object, the mercury column first drops before it begins to rise. How can you explain this observation?

11.3 Making use of expansion

Expansion can be useful. We make use of expansion in four major ways:

- in thermometry;
- in thermostats for temperature control;
- for binding metal plates together;
- in shrink fitting.

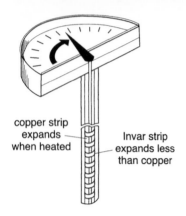

copper strip expands when heated

Invar strip expands less than copper

Figure 11.5 *This bimetallic strip functions as a thermometer. The pointer moves as the temperature changes. Name one household appliance that has a thermometer like this.*

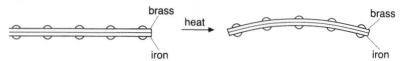

brass

heat

brass

iron

iron

Figure 11.4 *A bimetallic strip.*

In Chapter 10 you saw that liquid-in-glass thermometers depend for their operation on the expansion of mercury and alcohol. Bimetallic strips may also be used as thermometers. A bimetallic strip consists of two different metals fused together (Figure 11.4). The strip shown is made of brass and iron – two metals which have different coefficients of linear expansion. When heated, the strip bends with brass on the outside as shown, because brass has the greater coefficient of linear expansion.

Many thermostats depend on the ability of bimetallic strips to bend when heated. Thermostats are fitted to electric irons, refrigerators, ovens and a range of other electrical equipment. Some examples are illustrated in Figures 11.6–11.8.

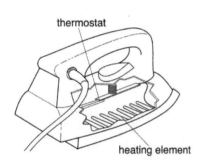

thermostat

heating element

Figure 11.6 *This electric iron contains a bimetallic strip which functions as a thermostat. The iron never becomes too hot. When the temperature rises the strip bends downwards and the circuit is broken. When the iron cools down contact is made again and current flows. In this way a steady temperature is maintained.*

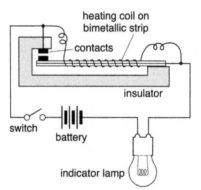

heating coil on bimetallic strip

contacts

insulator

switch

battery

indicator lamp

Figure 11.7 *This is a circuit for a flashing light. Why is the coil wrapped around the bimetallic strip? What happens when the current flows through the coil? Explain how this circuit produces a flashing light.*

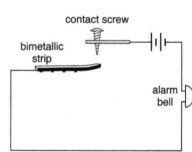

contact screw

bimetallic strip

alarm bell

Figure 11.8 *In this alarm circuit the bimetallic strip bends towards the contact screw when temperature rises. The alarm is set off when the strip touches the screw.*

11.4 The expansion of liquids

Most liquids expand when heated. This expansion leads to an increase in volume.

The mercury or alcohol-in-glass thermometer works on this principle.

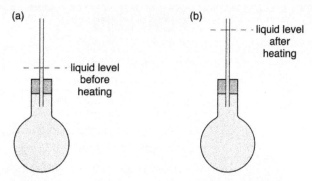

Figure 11.9 *(a) System before it is heated. (b) System after it has been heated. Expansion has occurred and the liquid level rose in the capillary tube.*

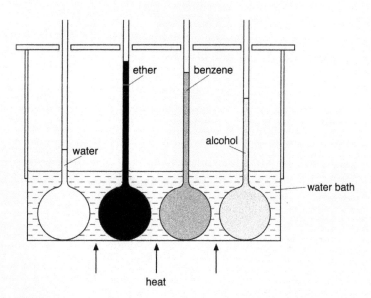

Figure 11.10 *This apparatus shows that different liquids expand by different amounts when heated to the same temperature.*

Note

The volume of freezing water increases by about 9%. This often leads to the bursting of pipes in cold countries.

Thinking it through

Water behaves unusually. It has its greatest density at 4 °C. Water at 4 °C sinks through the colder or warmer water around it.

Use this fact to discuss how water in a lake or pond freezes. Why do fish move to the bottom of a lake that is freezing?

Checklist

After studying Chapter 11 you should be able to:

- recall that solids expand when heated and contract when cooled
- discuss the useful and troublesome situations resulting from expansion and contraction
- discuss ways of allowing for expansion and contraction
- discuss ways in which the expansion of solids can be used
- discuss the expansion of liquids, especially water

Questions

1 When an ungraduated thermometer was placed in pure melting ice, the height of the mercury column was 15 cm. When the same thermometer was placed in steam, the height of the mercury column rose to 40 cm.

  a What will be the height of the mercury column when the thermometer is placed in a liquid that is boiling at 120 °C?

  b When the thermometer was placed in another boiling liquid the height of the mercury column was 27 cm. What is the boiling point of the liquid?

2 Explain why railway lines are laid down in short lengths, with gaps in between.

3 Explain how a steel rim is fitted onto a wooden wheel in a traditional blacksmith's shop.

4 Explain each of the following.

  a Floorboards creak at the end of a hot day.

  b A drinking glass shatters if boiling water is poured into it.

  c Power lines may snap if the temperature drops significantly.

12 Particulate nature of matter

The work of Rumford and Joule (pp. 131–2) led to the fall of the caloric theory and the establishment of the kinetic theory. The kinetic theory 'explains' the behaviour of bulk matter in terms of particles, which may be atoms, molecules or ions.

The kinetic theory had a humble beginning some two thousand years ago when Democritus suggested that all matter consisted of tiny particles. Some time later, Lucretius suggested that these particles moved all the time, colliding with and rebounding from one another. However, little further progress was made in this direction until the 17th century when Gassendi and Hooke, between them, provided explanations for phenomena such as the existence of the three states of matter and the transitions between states.

12.1 The solid, liquid and gaseous states

Table 12.1 **A comparison of the three states of matter**

Property	Solids	Liquids	Gases
Arrangement of particles	Closely packed; arranged in orderly fashion	Randomly distributed clusters of particles	Particles further apart than in either the solid or liquid states
Relative strength of forces between particles	Very strong	Moderate	Very weak
Shape/ structure	Regular firm lattice; many have three-dimensional structures, crystalline, definite shape	No fixed lattice; variable shape; take the shape of the container	No definite shape; take the shape of the container
Volume	Constant (definite) volume	Definite volume	Fill any available space

Continued on next page

Property	Solids	Liquids	Gases
Motion of particles	Restricted motion; particles vibrate about their mean positions, but no translational motion	Freer to move than in solids; motion is random; strength of the forces between particles determines the extent of motion; liquid, as a whole, flows because of the movement of clusters of particles	Random motion; each particle moves freely
Compressibility	Not easily compressed; large repulsive forces operate when particles are pushed closer than their average positions; can be stretched if large forces are applied; pressure has no effect on volume	Pressure has a small effect on volume; they are compressible to a small extent	Easily compressed; volume highly sensitive to changes in pressure and temperature
Kinetic energy	Low	Higher than for solids	Higher than for solids or liquids

12.2 Some evidence for the existence of particles

Brownian motion is the continuous, random movement of microscopic solid particles when these are suspended in a fluid (liquid or gas) medium.

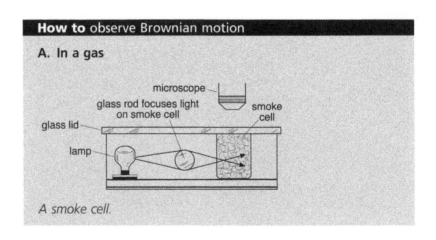

How to observe Brownian motion

A. In a gas

microscope
glass rod focuses light on smoke cell
smoke cell
glass lid
lamp

A smoke cell.

1 Fill the smoke cell with smoke.
2 Shine a bright light through the cell.
3 View through a microscope placed at right angles to the beam of light.

4 Observe the random, 'jiggling' motion of the smoke particles.

B. In a liquid
1 Mix very dilute solutions of sodium carbonate and lead ethanoate. Very fine particles of lead carbonate are produced.

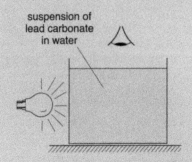

suspension of
lead carbonate
in water

2 Shine a bright light through the suspension of lead carbonate and view from above.
3 Observe the twinkling of lead carbonate particles as they move about.

The random movements of the smoke specks or the lead carbonate particles result from the uneven bombardments of these particles (see Figures 12.1, 12.2) by invisible particles of air in the case of the smoke cell experiment, and by water molecules in the case of lead carbonate particles.

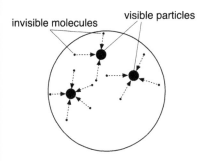

Figure 12.1 *The random motion of the particles.*

invisible molecules   visible particles

Figure 12.2 *The uneven bombardment of the particles creates an unbalanced force which causes the irregular motion of the particles.*

12.3 Evidence for the existence of interparticle forces

Many phenomena, including surface tension, adhesion and cohesion, and capillary action provide evidence for the existence of interparticle (molecular) forces.

Surface tension

Surface tension is that property of a liquid which makes it behave as if its surface is enclosed in an elastic skin. This property results from interparticle forces.

A particle in the interior of the liquid (A in Figure 12.3) is surrounded on all sides by other particles. There is no net force on such a particle (why?). However, a particle on the surface (B) is affected only by those particles below it in the liquid and

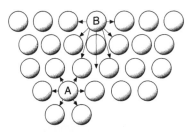

Figure 12.3 *Forces on a particle at the surface of a liquid are unbalanced.*

Figure 12.4 *The drops from this dripping tap are spherical because of the strong interparticle forces.*

experiences a net or overall downward force. The surface acts as if there is a 'skin' holding it there.

Capillary action

If one end of a capillary tube is placed in water, the water rises up the tube. The water rises to a higher level in narrower tubes. The force which causes the water to rise is called **capillarity**. It acts between the water molecules and the molecules of the material of the tube.

Paper towels and blotting paper 'take up' liquids by capillary action. Capillary action is also responsible for 'rising damp', when the floors and walls of concrete buildings become moist. Suggest one way of preventing this.

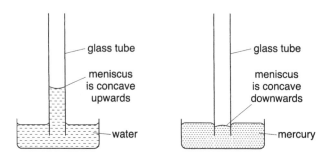

Figure 12.5 *Capillary action: water rises, mercury falls. The difference in the shape of the meniscus can be explained in terms of the attractions of the particles concerned, i.e. the water–water, water–glass, mercury–mercury and mercury–glass attractions.*

12.4 Diffusion

Diffusion is the movement of particles in a fluid from a region of high concentration to a region of low concentration.
Figures 12.6–12.8 show some experiments that demonstrate diffusion.

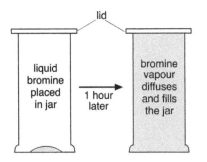

Figure 12.6 *Bromine vapour diffuses through air.*

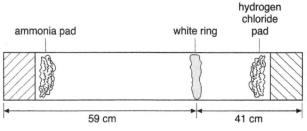

Figure 12.7 *Small cotton-wool pads soaked with aqueous ammonia and hydrogen chloride respectively are placed in a glass tube 100 cm long. The gases diffuse. After about 15 minutes a white ring of ammonium chloride appears where they meet.*

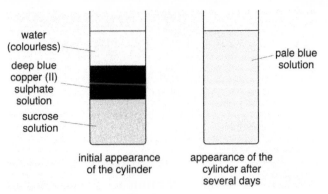

Note

Diffusion is the process by which different substances mix because of the random motion of atoms, molecules or ions.

water (colourless)

deep blue copper (II) sulphate solution

sucrose solution

pale blue solution

initial appearance of the cylinder

appearance of the cylinder after several days

Figure 12.8 *An example of diffusion in a solution. 50 cm³ of a sucrose solution is placed in a tall measuring cylinder. 50 cm³ of a copper sulphate solution is then carefully added, followed by 50 cm³ of water.*

12.5 The kinetic theory applied to gases

The kinetic theory offers a satisfactory explanation for the behaviour of gases. The important assumptions of this theory as it applies to gases are:

- gases are made up of very small particles which are very far apart and which occupy very little volume compared with the bulk of the gas – that is, most of the given volume of gas is empty space;
- the particles of a gas are in rapid, random motion, colliding with each other and with the sides of the container. The pressure exerted by the gas results from the collisions of the particles with the sides of the container;
- collisions of the particles are elastic – that is, no kinetic energy is lost on collision;
- there is very little attraction between the particles of the gas;
- the average kinetic energy of the particles is proportional to the absolute temperature;
- as temperature increases the particles move faster.

From the above you can see that:

- the volume of a gas is sensitive to changes in pressure and temperature;
- as temperature is increased, at constant volume, the particles hit the sides of the container harder and more often: pressure, therefore, increases.

This model describes an 'ideal' gas. However, real gases do behave quite similarly so it is a good model.

12.6 Spotlighting the behaviour of gases

Gaseous systems are completely described if any three of the following are accurately known:

p – the pressure of the gas

V – the volume of the gas

T – the absolute temperature

n – the number of particles of gas under study. This is sometimes referred to as 'a fixed mass of gas'.

Boyle's law

Boyle's law states that the volume of a gas is inversely proportional to pressure, provided that temperature and the number of particles remain constant.

$$p \propto \frac{1}{V} \quad \text{(if } n \text{ and } T \text{ are kept constant)}$$

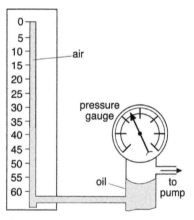

Figure 12.9 *Boyle's law apparatus.*

How to verify Boyle's law

1 Set up the apparatus as shown in Figure 12.9.
2 Connect a pressure pump to the inlet tube.
3 Alter the pressure inside the apparatus using the pump.
4 Record the volume of trapped air for each pressure.
5 Obtain about ten pairs of readings.
6 Tabulate the results using the headings Volume, Reciprocal volume (1/V), Pressure, and Product of pressure and volume (pV).

Figure 12.10 shows graphs drawn from the results of an experiment to verify Boyle's law.

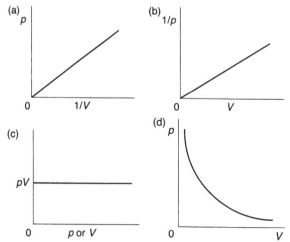

Figure 12.10 *Graphs of (a) p vs 1/V, (b) 1/p vs V, (c) pV vs p or V, (d) p vs V.*

You can deduce from graphs (a) to (d) that for the same system, i.e. for n and T constant:

- Pressure is directly proportional to the reciprocal of volume.

$$p \propto \frac{1}{V}$$

Therefore

$$p = \frac{k}{V}$$

and so

$$pV = k$$

- The product (pressure × volume) is a constant for all values of pressure.

$$p_1 V_1 = p_2 V_2$$

- Pressure is inversely proportional to volume.

Why are helium balloons under-inflated when released from the ground?

Example

In the apparatus of Figure 12.9, when atmospheric pressure acts on the piston the volume of gas enclosed is $1.0 \, dm^3$. What pressure is required to reduce the volume of gas to $0.25 \, dm^3$?

Solution

$$p_1 = 1 \, atm \qquad p_2 = ?$$
$$V_1 = 1.0 \, dm^3 \qquad V_2 = 0.25 \, dm^3$$
$$p_2 = \frac{p_1 \times V_1}{V_2}$$
$$= \frac{1 \times 1}{0.25}$$
$$= 4 \, atmospheres$$

Problem

The following data were obtained from an experiment to verify Boyle's law.

Volume/$m^3 \times 10^3$	20	10	5	4	2	1
Pressure/kPa	120	240	480	600	?	?

Using the data, plot a suitable linear (straight line) graph and indicate whether the data fit Boyle's law.

Use your graph to determine

a the volume when the pressure is 750 kPa;
b the pressure when the volume is $3.5 \times 10^{-3} \, m^3$.

Which variables were controlled in this experiment?

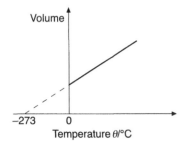

Figure 12.11 *A graph of volume against absolute temperature.*

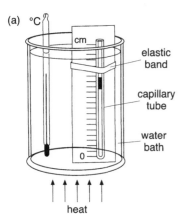

Figure 12.12 *A graph of volume against temperature in °C.*

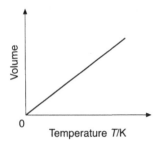

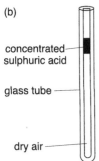

Figure 12.13 *(a) Charles' law apparatus. (b) Details of the capillary tube.*

Charles' law

Charles' law states that the volume occupied by a gas is directly proportional to the absolute temperature, provided that the pressure and number of particles remain constant.

$$V \propto T$$

$$V = \text{constant} \times T = kT$$

$$\frac{V}{T} = \text{constant}$$

$$\frac{V_1}{T_1} = \frac{V_2}{T_2}$$

for the same system, i.e. for n and p constant.

From the graphs in Figures 12.11 and 12.12) you can see that $0\,°C = 273\,K$.

$$T/K = \theta/°C + 273$$

The temperature 0 K is known as absolute zero. All molecular motion ceases at this temperature.

How to demonstrate Charles' law

1 Set up the apparatus shown in Figure 12.13. About 20 cm of capillary tubing is sealed at one end. It contains an index (thread) of concentrated sulphuric acid, which keeps the trapped air dry. The tube is of uniform cross-section, so the length of the air column is a measure of its volume.
2 Note (a) the steady temperature of the water bath and (b) the volume of trapped air.
3 Vary the temperature of the bath and note the corresponding volume of trapped air.
4 Plot a graph of volume versus temperature, drawing the best straight line through the points.

Example

A 100 cm³ sample of a gas is kept in a syringe at 400 K. The syringe is then heated to 500 K. What volume will the gas now occupy?

Solution

$$V_1 = kT_1 \qquad\qquad V_2 = kT_2$$

$$\frac{V_1}{T_1} = \frac{V_2}{T_2} = k$$

$$V_2 = \frac{V_1 T_2}{T_1} = \frac{100 \times 500}{400} = 125\,cm^3$$

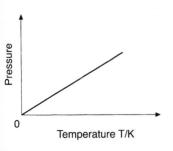

Figure 12.14 *A graph of pressure against absolute temperature.*

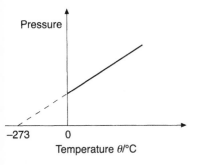

Figure 12.15 *A graph of pressure against temperature in °C.*

The pressure law

The pressure law states that the pressure of a gas is directly proportional to the (absolute) temperature, provided that the volume and number of particles (n) are held constant.

The combined gas equation

In everyday situations each of the quantities – pressure, volume and temperature – changes. The combined gas equation can be used to solve problems in these situations.

The combined gas equation can be stated as follows:

$$\frac{p_1 V_1}{T_1} = \frac{p_2 V_2}{T_2} = \frac{pV}{T} = \text{constant}$$

If $T_1 = T_2$, then $p_1 V_1 = p_2 V_2$ which is a statement of Boyle's law.

If $p_1 = p_2$, then $\dfrac{V_1}{T_1} = \dfrac{V_2}{T_2}$ which is a statement of Charles' law.

If $V_1 = V_2$, then $\dfrac{p_1}{T_1} = \dfrac{p_2}{T_2}$ which is a statement of the pressure law.

Example

A sample of gas has a volume of $0.2\,\text{m}^3$ at $27\,°\text{C}$ and $1.2 \times 10^5\,\text{Pa}$. What volume will this gas occupy if the temperature is raised to $127\,°\text{C}$ and the pressure lowered to $0.6 \times 10^5\,\text{Pa}$?

Solution

$$p_1 = 1.2 \times 10^5\,\text{Pa} \quad V_1 = 0.2\,\text{m}^3 \quad T_1 = 273 + 27 = 300\,\text{K}$$

$$p_2 = 0.6 \times 10^5\,\text{Pa} \quad V_2 = ? \quad\quad\quad T_2 = 273 + 127 = 400\,\text{K}$$

$$\frac{p_1 V_1 T_2}{p_2 T_1} = V_2 = 0.53\,\text{m}^3$$

Checklist

After studying Chapter 12 you should be able to:

- compare and contrast solids, liquids and gases
- discuss Brownian motion
- describe an experiment which demonstrates Brownian motion
- recall and discuss phenomena which prove the existence of interparticle forces
- define and discuss diffusion

- discuss the assumptions of the kinetic theory of gases and the resulting conclusions
- use the theory to explain the difference in macroscopic properties of gases
- recall Boyle's law and its equation $p_1 V_1 = p_2 V_2$, if $T_1 = T_2$
- recall Charles' law and its equation $\dfrac{V_1}{T_1} = \dfrac{V_2}{T_2}$, if $p_1 = p_2$
- recall the pressure law and its equation $\dfrac{p_1}{T_1} = \dfrac{p_2}{T_2}$, if $V_1 = V_2$
- recall and use the combined gas equation
  $$\frac{p_1 V_1}{T_1} = \frac{p_2 V_2}{T_2}$$

Questions

1 At constant pressure, a fixed mass of gas occupied a volume of $20\,m^3$ at $27\,^\circ C$. The temperature of the system was increased to $177\,^\circ C$. What is the new volume occupied by the gas?

2 In an experiment to verify Boyle's law, a student obtained the results below.

Pressure/kPa	100	80	50	40	25	10
Volume/mm^3	20	25	40	50	80	200

   a What is meant by Boyle's law?
   b What factors should the student control in this experiment?
   c State the manipulated and responding variables in the experiment.
   d What is the mathematical relationship between the responding and the manipulated variable? [*Hint:* draw a suitable graph.]
   e Draw a diagram of the apparatus the student could have used.

3 In an experiment to verify a gas law, the following data on volume of gas and corresponding Celsius temperature were obtained.

Volume/m^3	208	238	328	388	418	448
Temperature/$^\circ$C	−101	−75	−50	0	75	100
Temperature/K						

   a Copy the table and complete it by filling in the Kelvin temperatures.
   b Plot a graph of volume versus Kelvin temperature.
   c What is the mathematical relationship between volume and Kelvin temperature?
   d Identify the gas law being investigated.
   e What factors should be controlled in this experiment?
   f What is the significance of the temperature $0\,K$?

13 Measuring heat

13.1 Specific heat capacity

The rise in the temperature of a body which is heated, without a change of state, depends on:

- the mass of the body;
- the nature of the substance of which the body is made.

The **heat capacity** (C) of an object is defined as the heat required to raise its temperature by one degree. Heat capacity has units $J\,K^{-1}$.

The **specific heat capacity** (c) of a substance is the amount of heat required to raise the temperature of 1 kg of the substance by one degree. Specific heat capacity has units $J\,kg^{-1}\,K^{-1}$.

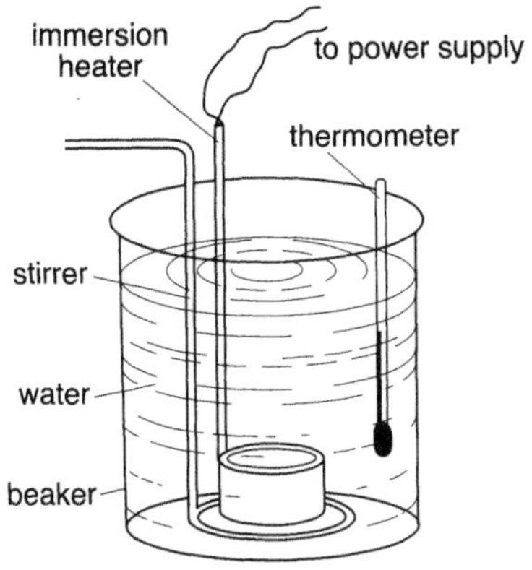

Figure 13.1 *This apparatus may be used to investigate the relationship between heat and temperature.*

> **How to** investigate the relationship between quantity of heat supplied and temperature change
>
> 1 Place a 50 W immersion heater in 250 cm³ of water in a beaker. Make sure that the heater is completely covered by the water (Figure 13.1).
> 2 Record the steady initial temperature of the water.
> 3 Turn on the heater and, stirring the water continually, note the temperature rise $\Delta\theta$ at 1-minute intervals for 5 minutes.
>
> The following results were obtained by a fourth-former, using 250 cm³ (250 g) of water:
>
> | Time/min | 1 | 2 | 3 | 4 | 5 |
> |---|---|---|---|---|---|
> | Temperature rise/°C | 2.9 | 2.8 | 2.9 | 2.9 | 2.9 |
>
> What can you deduce from these results?

If you assume that the heater is supplying heat at a constant rate and that heat exchanges with the surroundings are negligible, then you can deduce that:

Equal quantities of heat supplied to the same mass of a given substance (in this case, water) produce the same temperature rise.

When the same student heated 125 cm³ (125 g) of water under the same conditions as above, the average temperature rise per minute was 5.8 °C. From this you can deduce the following.

The temperature rise for a given quantity of heat supplied depends on the mass of substance involved.

When 250 g each of water and oil were heated for the same time on the same hotplate, the temperature rise of the oil was 20 °C while that of water was 11 °C. From this you can deduce:

The same quantity of heat supplied to equal masses of different substances results in different temperature changes, depending on the different specific heat capacities of the materials.

Putting it all together, quantitatively:

$$\begin{array}{ccccc}
\text{heat supplied} \\ \text{or given out} & = & \text{mass of} \\ \text{substance} & \times & \text{specific heat} \\ \text{capacity} & \times & \text{temperature} \\ \text{rise}
\end{array}$$

$$E_H = \quad m\,(\text{kg}) \quad \times c\,(\text{J}\,\text{kg}^{-1}\,\text{K}^{-1}) \times \quad \Delta\theta\,(\text{K})$$

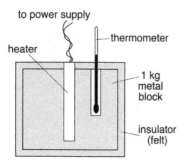

to power supply

—thermometer

heater

1 kg metal block

insulator (felt)

Figure 13.2 *An electrical method for finding the specific heat capacity of a solid.*

How to determine the specific heat capacity of an aluminium block by an electrical method

1 Insert the immersion heater (Figure 13.2) in one hole in the block as shown and the thermometer in the other.
2 Record the initial steady temperature of the block.
3 Turn on the heater for a fixed time (say, 5 minutes).
4 Record the steady final temperature.

Typical results for such an experiment are:

$$\text{Time} = 5\,\text{min} = 300\,\text{s}$$

$$\text{Initial temperature of the block } (\theta_i) = 30.3\,°C$$

$$\text{Final temperature of the block } (\theta_f) = 46.5\,°C$$

$$\text{Power rating of the immersion heater} = 50\,\text{W}$$

$$\text{Temperature change} = (46.5 - 30.3)\,°C = 16.2\,\text{K}$$

$$\text{Mass of aluminium block} = 1\,\text{kg}$$

To calculate the specific heat capacity:

$$\text{heat supplied } (E_H) = 50\,(\text{W}) \times 300\,(\text{s}) = 15\,000\,\text{J}$$

$$E_H = \text{mass of block} \times \begin{array}{c}\text{specific heat capacity} \\ \text{of block}\end{array} \times \begin{array}{c}\text{temperature} \\ \text{rise}\end{array}$$

$$= \quad 1\,(\text{kg}) \quad \times \quad c_{Al}\,(\text{J}\,\text{kg}^{-1}\,\text{K}^{-1}) \times \; 16.2\,\text{K}$$

$$\therefore \quad c_{Al} = \frac{15\,000}{1 \times 16.2}\,\text{J}\,\text{kg}^{-1}\,\text{K}^{-1}$$

$$= 925\,\text{J}\,\text{kg}^{-1}\,\text{K}^{-1}$$

This method may be modified to determine the specific heat capacities of liquids.

Note

Water, because of its high specific heat capacity, is a useful material (substance) for storing and transferring energy.

Problem

The following results were obtained in an experiment to determine the specific heat capacity of water.

Mass of water $= 0.5\,kg$, temperature rise $= 8.2\,K$, time $= 360\,s$, power rating of heater $= 50\,W$.

Use these results to obtain a value for the specific heat capacity of water. Is this value higher or lower than that quoted in Table 13.1? Suggest possible reason(s) for the difference, if any.

Thinking it through

- Why is the temperature of land higher than that of the surrounding body of water during the day?
- Why does land cool more rapidly than a surrounding body of water at night?

Table 13.1	The specific heat capacities of selected materials	

Material	Specific heat capacity ($J\,kg^{-1}\,K^{-1}$)
Water	4200
Alcohol	2290
Paraffin	2000
Aluminium	900
Iron	430
Copper	380
Lead	128
Soil	840

Note

The method used for finding the specific heat capacity here ignores the heat capacity of the container and assumes that no heat is lost to or gained from the surroundings.

The method of mixtures

When two objects at different temperatures are placed in thermal contact the hot object loses heat and the cold gains heat until both reach the same temperature. In the process,

heat lost by the hot object $=$ heat gained by the cold object

This method, known as the **method of mixtures**, may be used to determine the specific heat capacity of a material.

Example 1

A block of metal of mass 0.8 kg at 100 °C was placed in 0.4 kg of water at 30 °C. The final steady temperature of the mixture was 43.5 °C. What is the specific heat capacity of the metal?

Solution

$$\text{Temperature rise of the water} = (43.5 - 30)\,°C = 13.5\,°C$$
$$\text{Temperature drop of the block} = (100 - 43.5) = 56.5\,°C$$
$$\text{Heat lost by block} = \text{heat gained by water}$$
$$0.8 \times 56.5 \times c_{block} = 0.4 \times 4200 \times 13.5$$
$$\Rightarrow \quad c_{block} = 502\,J\,kg^{-1}\,K^{-1}$$

13.2 Heat and the transitions between states

Heat is supplied to change ice to water and to change water to water vapour (gas). The reverse processes occur if heat is extracted or removed. See Figure 13.3.

These changes of state, sometimes called 'phase changes', take place at constant temperature. A thermometer placed in a beaker of crushed ice reads 0 °C until the ice has completely melted. A thermometer placed in boiling water registers a constant temperature of 100 °C.

Figure 13.4 opposite shows how temperature varies with time when a solid is heated until it melts.

- From O to A, the heat supplied causes an increase in temperature. There is no change of state.
- From A to B, liquid and solid exist together. The temperature remains constant although heat is being supplied.
- From B to C, there is no change of state. The heat supplied causes an increase in temperature.

Figure 13.5 shows how temperature varies with time when a liquid is cooled.

- From O to A, the substance exists in the liquid state only, heat is given up and the temperature falls.

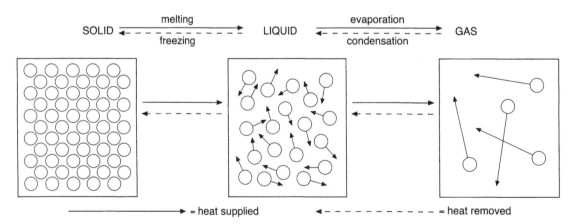

Figure 13.3 *The transitions between the states of matter (phase changes).*

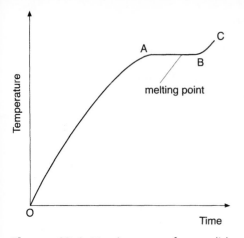

Figure 13.4 *Heating curve for a solid.*

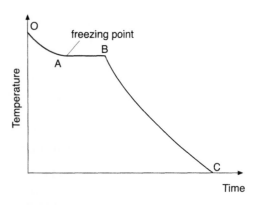

Figure 13.5 *Cooling curve for a liquid.*

- From A to B, the heat released as the liquid freezes keeps the temperature constant; liquid and solid exist together. The temperature remains constant throughout the solidification process. Solidification starts at A and is completed at B.

The heat supplied during melting represents the work done to overcome the attractive forces between the particles in the solid state. This work is 'used' to separate the particles. As more heat is absorbed the particles move further apart. During freezing (or solidification) heat is released as the particles come closer together to occupy the positions in the lattice structure of the solid.

Factors affecting melting and boiling points

The melting temperature of a solid is lowered by:

- the presence of impurities;
- increasing external pressure.

The boiling temperature of a liquid is raised by:

- increasing external pressure;
- the presence of impurities.

The pressure cooker

A pressure cooker (Figure 13.6) is a cooking utensil which has a tight-fitting lid. The lid has a safety valve and a pressure regulator. When the liquid in the cooker boils the steam produced cannot escape. The trapped steam increases the pressure inside the cooker. Boiling temperature increases with increasing pressure.

The temperature of the water inside the cooker can get as high as 120 °C. Foods cook faster at the higher temperatures.

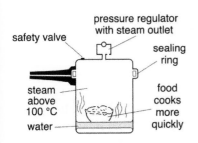

Figure 13.6 *A pressure cooker.*

Boiling and evaporation

A liquid changes into a gas by boiling or evaporation.

Boiling takes place when a liquid changes to vapour at a constant temperature. When a liquid is heated it gives off a vapour which exerts a pressure on the liquid surface. This pressure is the **saturated vapour pressure** at that given temperature. As the liquid is heated, its saturated vapour pressure increases. The liquid boils when this vapour pressure is equal to atmospheric pressure.

As a liquid is heated, the kinetic energy of the particles increases until clusters of particles (liquid state) break down into individual particles (gaseous state), with very weak forces between the particles.

Evaporation takes place when a liquid changes into a vapour at a temperature lower than its boiling point. Rapidly moving particles at the surface of the liquid have sufficient kinetic energy to escape the attractive forces of the particles within the bulk of the liquid and become a gas or vapour.

The escape of these high-energy particles lowers the average kinetic energy of the particles remaining in the liquid state. Since temperature is directly proportional to the average kinetic energy of the particles, the temperature of the liquid drops, unless heat is taken in from the surroundings.

Table 13.2	A comparison of evaporation and boiling
Evaporation	**Boiling**
• takes place only from the surface of the liquid	• takes place throughout the bulk of the liquid
• takes place at all temperatures	• takes place at one fixed temperature (constant external pressure)
• rate increases if surface area increases	• boiling temperature is not affected by change of surface area
• rate decreases if external (atmospheric) pressure increases	• temperature is increased by increases in external pressure
• rate is increased by increased movement of air over the surface of the liquid	

Evaporation at work

Why do we feel uncomfortable on a hot, humid day? Why is it not advisable to wear wet clothing? You may feel cold when you have come out of the sea because heat is taken from your body to evaporate the water on your skin.

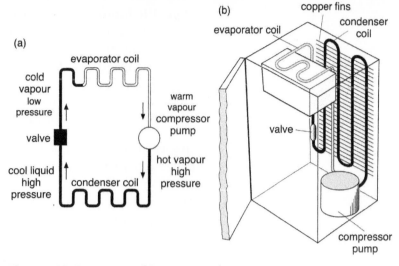

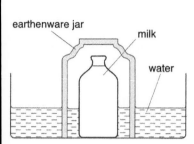

Figure 13.7 *Using evaporation to keep milk cool. Water moves up through the earthenware by capillary action and evaporates from the surface of the jar.*

Figure 13.8 *How a refrigerator works.*

Water stored in an unglazed earthenware jug is cooler than water stored in other containers under identical conditions, because water evaporates through the small pore spaces of the jug. The heat needed for evaporation is taken from the liquid itself. The liquid, therefore, becomes cooler. The same principle is at work in Figure 13.7.

Evaporation plays an important role in cooling the contents of a refrigerator (Figure 13.8). The cooling system contains a refrigerant substance which alternately evaporates and condenses.

- Low pressure vapour enters the compressor pump. Its temperature and pressure rise.
- The vapour flows through the condenser coil, where it loses heat to the surroundings. The refrigerant cools and becomes a liquid.
- Cooled liquid refrigerant, still at high pressure, passes through a valve into the evaporator coil. The pressure drops and the liquid takes in heat from the food compartment and evaporates.
- The cycle is repeated.

Thinking it through

Why are the fins of the condenser made of copper?

13.3 Latent heat

The heat absorbed or released when a substance changes state or phase, e.g. melting or freezing, is known as *latent* or *hidden* heat.

- The heat absorbed during the melting process is called **latent heat of fusion** (L_f). An equivalent amount of heat is released when the liquid freezes.
- The heat needed to convert a liquid into a gas is called the **latent heat of vaporization** (L_v).

- **Specific latent heat of fusion** (l_f) is the heat required to convert 1 kg of a substance at a fixed temperature (usually its melting point) from solid to liquid state.
- **Specific latent heat of vaporization** (l_v) is the heat required to convert 1 kg of a substance at a fixed temperature (usually its boiling point at 1 atmosphere pressure) from liquid to gaseous (vapour) state.

Both specific latent heat of fusion and specific latent heat of vaporization are measured in J kg^{-1}.

$$\frac{\text{Heat supplied or released}}{\text{during a change of state}} = \text{mass (kg)} \times \frac{\text{specific}}{\text{latent heat (J kg}^{-1})}$$

$$E_H = m \times l_f \quad \text{or} \quad m \times l_v$$

where l_f = specific latent heat of fusion
l_v = specific latent heat of vaporization

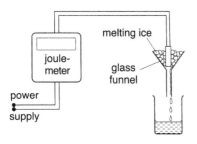

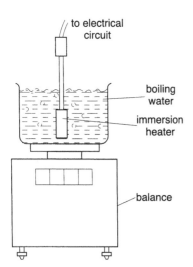

Figure 13.9 *Determining the specific latent heat of fusion of ice.*

Figure 13.10 *Determining the specific latent heat of vaporization of water.*

How to determine the specific latent heat of fusion of ice

1 Set up the apparatus shown in Figure 13.9.
2 Weigh the beaker empty (M_a kg).
3 Place the 50 W immersion heater in pure ice in the funnel.
4 Turn on the heater for a fixed time (t s).
5 Collect the water produced and re-weigh beaker with water (M_b kg).
6 Mass of ice which melted $= (M_b - M_a)$ kg

Heat supplied $= (50 \times t)$ J

$$50 \times t = (M_b - M_a) \times l_f$$

from which you can calculate a value for the specific latent heat of ice.

How to determine the specific latent heat of vaporization of water

1 Place a beaker on a top-pan balance as shown in Figure 13.10.
2 Switch on the immersion heater and bring the water to the boil.
3 Record the reading on the balance.
4 Continue heating the water for a fixed time (several minutes).
5 Turn off the heater and, at the same time, take the reading on the balance.
6 Use the power rating of the heater and the time for which the water was boiling to obtain a value for the heat supplied.

7 The difference of the two readings of the top-pan balance gives the mass of water which turned to vapour.
8 Calculate a value for the specific latent heat of vaporization of water.
9 List possible sources of errors in the experiment as carried out.

Thinking it through

- Orchard owners spray their fruit trees with water when temperatures are expected to fall to −4 °C. Explain how this procedure protects blossom from destruction.
- In general, more energy is required to vaporize than to liquefy a substance. Why is this?

Example

The following results were obtained using the apparatus of Figure 13.10.

$$\text{Mass of water evaporated} = 19.5\,\text{g}$$
$$\text{Power rating of the immersion heater} = 50\,\text{W}$$
$$\text{Time for which the boiling water was heated} = 900\,\text{s}$$

Find a value for the specific latent heat of vaporisation of water.

Solution

$$\text{Heat supplied} = \text{power rating (W)} \times \text{time (s)}$$
$$= 50 \times 900\,\text{J}$$
$$= 45\,000\,\text{J}$$
$$45\,000\,(\text{J}) = \text{mass of water evaporated (kg)} \times l_v$$
$$l_v = \frac{45\,000}{1.95} \times 10^2\,\text{J}\,\text{kg}^{-1}$$
$$= 2.3 \times 10^6\,\text{J}\,\text{kg}^{-1}$$

Checklist

After studying Chapter 13 you should be able to:

- define specific heat capacity and heat capacity and recall the units
- describe an experiment used to investigate the relationship between heat and temperature
- discuss the properties on which temperature rise depends
- recall and use the formula $E = mc\Delta\theta$
- describe an experiment used to determine the specific heat capacity of a block of metal using an electrical method
- describe an experiment used to determine the specific heat capacity of a block of metal using the method of mixtures
- solve method of mixtures mathematical problems
- describe phase changes
- recall that changes of phase occur at constant temperatures
- graphically describe phase changes
- compare and contrast boiling and evaporation

- recall the factors that affect melting and boiling points
- discuss practical applications of varying melting and boiling points
- discuss practical applications of evaporation
- define specific latent heat of fusion
- define specific latent heat of vaporization
- recall and use the formula $E_H = m \times l_f$ or $m \times l_v$
- describe an experiment to determine the specific latent heat of fusion of ice
- describe an experiment to determine the specific latent heat of vaporization of water

Questions

1 Air passes over an electrical heater at the steady rate of 2500 cm^3 per second. The steady inlet temperature of the air is $20\,°C$ and the steady outlet temperature is $40\,°C$.
  a What heat is absorbed by air passing over the heater in 2 hours?
  b Obtain an estimate for the power rating of the heater. Is this estimate too high or too low? Explain.

$$\text{Density of air} = 1.2 \text{ kg cm}^{-3}$$

$$\begin{array}{l}\text{Specific heat} \\ \text{capacity of air}\end{array} = 1000 \text{ J kg}^{-1}\text{K}^{-1}$$

2 A piece of metal weighing 0.5 kg, at $100\,°C$, was transferred to 0.25 kg of a liquid at $20\,°C$.

$$\begin{array}{l}\text{Specific heat capacity} \\ \text{of the metal}\end{array} = 400 \text{ J kg}^{-1}\text{K}^{-1}$$

$$\begin{array}{l}\text{Specific heat capacity} \\ \text{of the liquid}\end{array} = 2000 \text{ J kg}^{-1}\text{K}^{-1}$$

What was the final temperature of the mixture?

3 How much heat energy is required to convert 10 g of ice at $0\,°C$ to water at $0\,°C$?

$$\begin{array}{l}\text{Specific latent heat} \\ \text{of fusion of ice}\end{array} = 3.36 \times 10^5 \text{ J kg}^{-1}$$

4 5×10^4 J of heat were required to raise the temperature of 500 g of a liquid from $31\,°C$ to $56\,°C$. What is the specific heat capacity of the liquid?

5 100 g of a liquid metal at $375\,°C$ cools as shown below:

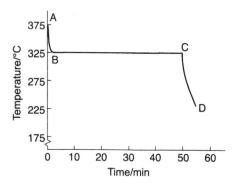

a What is the melting temperature of the metal?
b Account for the stages AB, BC and CD.
c Heat is lost at the average rate of 540 J min^{-1} during the segment BC.
(i) What is meant by the 'specific latent heat of the metal'? (ii) Calculate a value for specific latent heat of the metal.
d What changes would be expected in the cooling curve above, if an impurity is added to the molten metal?
e Calculate the heat lost to the surroundings as the metal cools from C to D ($327\,°C$ to $227\,°C$). (Assume that the specific heat capacity of the metal is $130 \text{ J kg}^{-1}\text{K}^{-1}$.)
f If 1.33 kg of solid metal (at its melting point) is placed in 0.2 kg of water at $30\,°C$, calculate the final temperature of the mixture. (The specific heat capacity of water is $4200 \text{ J kg}^{-1}\text{K}^{-1}$.)

14 Methods of heat transfer

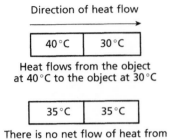

Direction of heat flow

| 40 °C | 30 °C |

Heat flows from the object
at 40 °C to the object at 30 °C

| 35 °C | 35 °C |

There is no net flow of heat from
one object to the other

Figure 14.1 *Heat flow between objects.*

We want to keep some things hot and others cold. We, in the Caribbean, need to keep ourselves and our homes cool throughout the year. People in colder climates need to keep heat out of their homes in summer and to keep heat in, in winter. We need to design better radiators so that car engines heat up less. To meet these and a whole range of other challenges, we need to understand how heat is transferred from one place to another.

Heat flows from one place to another, from one object to another, only if a temperature difference exists between them. Heat flows from the hotter to the colder object until both are at the same temperature (Figure 14.1). There are three ways in which heat is transferred: convection, conduction and radiation (Figure 14.2).

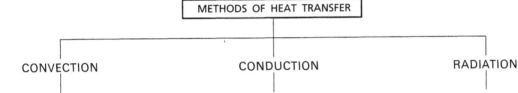

METHODS OF HEAT TRANSFER

CONVECTION

The process of heat transfer through a fluid (liquid or gas). Convection is heat transfer by bulk movement in fluids as a result of differences in density

CONDUCTION

The process of heat transfer through a solid medium which is not free to move. Conduction may be viewed, at the atomic level, as the exchange of kinetic energy between colliding particles

RADIATION

The process of heat transfer which does not require a medium. All objects radiate energy continuously in the form of electromagnetic waves. Objects also absorb electromagnetic radiation from their surroundings. An object whose temperature is steady is radiating and absorbing radiation at the same time

Figure 14.2 *Methods of heat transfer.*

14.1 Convection at work

Convection takes place in fluids (liquids and gases) but not in solids. Convection currents are more easily set up in gases than in liquids.

Fluid in contact with a hot surface acquires heat energy from that surface. The particles in contact with the hot surface spread out to occupy a greater volume, and that part of the fluid, therefore, becomes less dense (recall that density is mass divided by volume). This part of the fluid, being less dense than surrounding fluid, rises, and fluid in contact with a colder surface moves in to take the place of the rising fluid. This gives rise to a circulatory motion which is called a **convection current**.

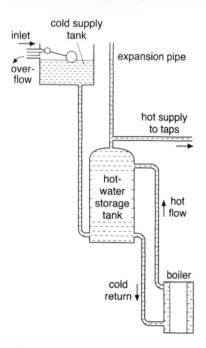

Figure 14.3 *Convection currents play an important role in the working of hot-water systems.*

1 Place some aluminium powder at the bottom of a beaker.
2 Add water, carefully, so as not to disturb the aluminium powder.
3 Using a gentle flame, heat the beaker.

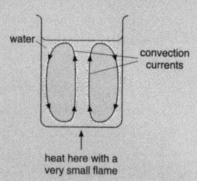

4 Observe the path of disturbed aluminium particles. Particles rise up the middle of the water in the beaker and come down near the sides, which are cooler. Movement of liquid particles is responsible for the rise and fall of the aluminium powder.

Breezes blow from sea to land during the day and from land to sea during the night. How can we explain this phenomenon?

The specific heat capacity of water is about five times that of soil (land). This means that the temperature over the land is higher than that over the sea during the day (the sea absorbs more of the Sun's heat). As a consequence, the warmer, less dense air over the land rises and cooler, more dense, air from over the sea rushes in – giving rise to 'sea breezes'.

Water cools more slowly than land. At night the air over the sea is therefore warmer than the air over the land. The warmer air over the sea rises and the cooler air from over the land rushes in to take the place of the rising air, giving rise to 'land breezes'.

Thinking it through

Which is the better place to install an air conditioning unit, near the roof or near the floor? Give a reason for your answer.

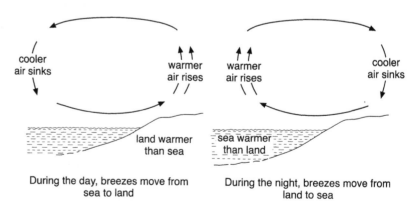

Figure 14.4 *Land and sea breezes.*

Note

Materials that conduct heat poorly are called insulators, materials that totally prevent heat flow are called perfect insulators.

14.2 Conduction at work

If one end of a metal rod is in a fire and you are holding the other end in your hand, heat passes from the fire to your hand via the metal rod. A metal consists of a lattice of positive ions, with free electrons that can move through the lattice. Before the metal rod is placed in the fire, the metal ions vibrate about their 'equilibrium' positions. When the fire heats the rod, the metal ions nearest the end being heated gain energy and vibrate with bigger and bigger amplitudes. These rapidly vibrating ions transfer some of their energy to neighbouring ions, and also to the free electrons. The free electrons rapidly transfer energy to other ions and electrons along the rod.

Compared with convection, conduction is a slow process. The rate of conduction depends on:

- the nature of the material;
- the thickness of the material [Δx];
- the area of cross-section.

Thinking it through

Why is a vacuum a perfect insulator?

Whereas most metals are good conductors, non-metals are generally poor conductors. There are no 'free electrons' in non-metals. Conduction, here, is due to the vibrations of atoms only. Good conductors of electricity are, in general, good conductors of heat. The mechanism of thermal conduction is essentially the same as for electrical conduction.

How to compare thermal conductivities

1 Obtain rods of different material but of the same length.

2 By means of a bit of wax, attach a thumb tack to one end of each rod.

3 Place the other end of the rods, at the same time, into a beaker of boiling water.

4 From which material did the thumb tack first fall off? Comment on the ability of these materials to conduct heat.

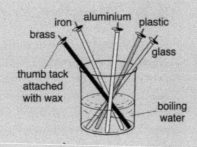

Thinking it through

Discuss how you should present the results of this experiment to compare the conductivities of the materials under test.

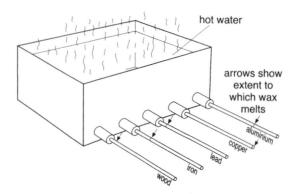

Figure 14.5 *An alternative way of comparing the thermal conductivities of metals. The rods were coated with wax before hot water was placed in the tank.*

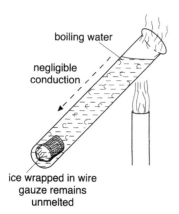

boiling water

negligible conduction

ice wrapped in wire gauze remains unmelted

Figure 14.6 *Water is a poor conductor of heat.*

The experiment in Figure 14.6 shows that water is a poor conductor of heat. Water at the top boils, but it takes a long time for the block of ice to melt.

Conduction is essentially a two-way process. Those materials which prevent heat from leaving a system will also prevent heat from entering the system. Air is a good insulator but hot objects which are surrounded by air cool quickly because convection currents develop readily in air. The insulating properties of air can be enhanced by trapping the air in such materials as fibreglass, styrofoam and wool.

14.3 Radiant energy

Radiant energy consists of electromagnetic waves which are emitted by all hot bodies. Whereas convection and conduction require a material medium for their transmission, **radiation** (radiant energy) does *not* require a material medium. It is transmitted in a vacuum. Energy from the Sun reaches the Earth 150 million kilometres away through the 'vacuum' of space.

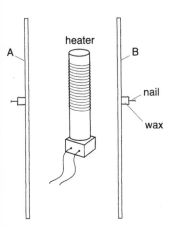

Figure 14.7 *Which surface is the best absorber of radiant energy?*

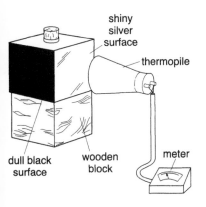

Figure 14.8 *Leslie's cube.*

The radiation falling on an object may be partly reflected and partly absorbed. The absorbed radiation is converted to heat.

Radiant energy may be detected by the skin, by a thermometer with a blackened bulb, by a thermopile or by a phototransistor.

Every object emits some radiant energy. The human body has a temperature of about 37°C and radiates mainly in the far infrared region of the electromagnetic spectrum. Thermal pictures of the body or body parts can be taken with cameras which are sensitive to wavelengths in this region.

How to show that black surfaces are good absorbers of radiant energy

1 Set up the apparatus shown in Figure 14.7. A is a thin sheet of polished metal. B is made of the same metal (and of the same thickness) but is blackened by lampblack. A nail is attached by wax to each metal sheet.
2 Place a lighted Bunsen burner or radiant heater equidistant from A and B.
3 The nail attached to the blackened metal sheet falls off first. This indicates that a blackened surface absorbs heat better than a polished one.

How to show that black surfaces are good emitters of radiant energy

Figure 14.8 shows a hollow copper cube (called Leslie's cube) which has four different side surfaces – one is blackened, one roughened, a third white and the fourth highly polished.

Boiling water is placed in the cube and a thermopile or other detector is then placed, in turn, equal distances from each surface.

The greatest radiation is received from the black surface and least radiation from the highly polished surface.

From the above we can deduce that:

good absorbers of radiant energy (i.e. black, rough surfaces) are also good emitters of radiant energy.

The greenhouse effect

The increasing use of fossil fuels has caused a significant rise in the amount of carbon dioxide in the atmosphere. Neither carbon dioxide nor water vapour prevents the energy from the Sun from reaching the surface of the Earth. However, the radiant energy given off from the Earth has a different

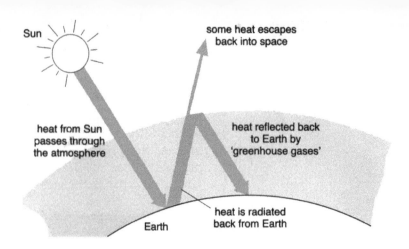

Figure 14.9 *The greenhouse effect.*

wavelength and *is* absorbed by both the carbon dioxide and the water vapour (as it is by the glass or plastic in greenhouses).

This absorption of radiant energy causes the carbon dioxide molecules to vibrate and heat up. The heat so produced is reflected back to Earth which then heats up as shown in Figure 14.9. This phenomenon is known as the **greenhouse effect**.

The greenhouse effect has already caused an increase in average global temperatures. This may, in turn, lead to:

- changes in global climate;
- melting of the polar ice caps which may cause a rise in the sea level.

14.4 Using heat transfer

Cooking with direct sunlight is one of the oldest applications of solar energy. The **solar cooker** is a curved reflector with a highly polished surface: it collects the Sun's rays and focuses them at a point (or over a small area). The pot or dish should be placed here.

Radiant energy is also used in solar water heaters, as you saw in Chapter 8 (p. 103).

A vacuum (thermos) flask is designed to reduce heat losses by convection, conduction and radiation. The flask contains a double-walled glass vessel with a vacuum between the walls. The vacuum (no particles present) prevents loss of heat by conduction or convection.

- The silvering on the inner surfaces of the double wall reduces loss of heat by radiation.
- The top and the support at the base are made of insulating materials. These reduce, but do not entirely cut out, loss of heat by conduction. The flask keeps hot liquids hot and cold liquids cold.

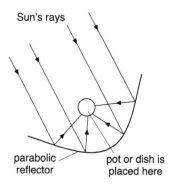

Figure 14.10 *A solar cooker.*

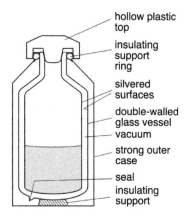

Figure 14.11 *A vacuum flask.*

Using materials readily available around the home, including aluminium foil, construct a custom-made solar cooker. Use it to cook a simple meal.

Use your knowledge of the methods of heat transfer to account for the following features in the design of solar water heaters:

- copper pipes;
- blackened metal base;
- tilting the assembly so that the Sun's rays fall on it at right angles;
- the insulated box;
- transparent tops which sometimes consist of two sheets of glass with air sandwiched between them.

Checklist

After studying Chapter 14 you should be able to:

- define and discuss convection
- describe an experiment which demonstrates convection
- discuss practical applications of convection
- define and discuss conduction
- describe an experiment which demonstrates conduction
- discuss practical applications of conduction
- describe an experiment to compare the thermal conductivities of different materials
- describe an experiment to demonstrate that water is a poor conductor of heat
- recall that air is a good insulator, hence its use in insulating applications
- define and discuss radiation
- discuss practical applications of radiation
- describe experiments to determine the factors on which the absorption and emission of radiation depend
- recall that good absorbers of radiation are also good emitters
- discuss thermal energy transfer
- discuss the greenhouse effect

Questions

1 Why does the water in a beaker which is being heated eventually reach a uniform temperature?

2 Select five conductors which are used in everyday situations. For each conductor, state two uses and say why the conductor is suitable for these uses. Present your answer in tabular form. Repeat the above for five selected non-conductors.

3 A space shuttle returning from a mission is prevented from getting too hot and burning up by covering it with a suitable material.
   a What causes heat to be produced when the space shuttle returns?
   b What type of material might be suitable for use as covering for the space shuttle?

4 Explain why convection losses are usually high in solar cookers.

C Revision questions for Section C

Multiple-choice questions

Questions 1 and 2

A Lord Kelvin   B Count Rumford
C Joule   D Watt

Which of the above scientists first carried out experiment(s) which:

1   showed that heat is a form of energy?

2   provided evidence against the 'caloric theory'?

Questions 3 and 4

Select from the temperatures A–D the one which best matches the descriptions of questions 3 and 4.

A 273 K   B 310 K
C 373 K   D 6273 K

3   The boiling point of pure water at a pressure of 1 atmosphere.

4   The normal body temperature.

Questions 5 and 6

An immersion heater is rated at 12 V, 50 W.

5   What current does the heater draw from the power supply?
(power = current × voltage)
A 38 A   B 4.2 A   C 2.9 A   D 0.35 A

6   How much energy does this heater use when it is functioning normally for 5 minutes?
A 60 J   B 250 J
C 3600 J   D 15 000 J

7   Which of the following is best explained by convection?
A the flow of heat along a metal rod
B the movement of air masses in the atmosphere
C heat transfer from the Sun to the Earth
D heat sinks are painted matt black

8   Which of the following statements is true of conduction? Conduction:
A is heat transfer by rapidly vibrating particles
B is heat transfer in a vacuum
C results from changes in density in a fluid
D causes coastal winds and breezes.

9   Which of the following statements is correct?
A Boiling takes place at all temperatures.
B Evaporation leads to the cooling of liquids.
C Fusion is the change from the liquid to the solid state.
D Condensation is the change from liquid to the gaseous state.

10 Equal masses of water at 95 °C and cool water were poured into a vessel. The steady final temperature of the mixture was 35 °C. What was the most likely temperature of the cold water?
A 45 °C   B 25 °C
C −25 °C   D −45 °C

11 Which of the following phenomena is NOT explained by the kinetic theory?
A conduction   B convection
C radiation   D evaporation

12 Which of A–D is the correct unit for specific heat capacity?
A J kg K   B J kg
C $J K^{-1}$ D $J kg^{-1} K^{-1}$

13 Heat energy reaches someone standing in front of an open fire mainly by:
A diffusion   B radiation
C convection   D conduction.

14 The transfer of heat in solids occurs mainly by:
A conduction   B convection
C radiation   D diffusion.

15 Which of the following is the correct sequence of energy changes which takes place in an oil-fired electricity generating plant?

A chemical → mechanical → heat → electrical

B heat → chemical → mechanical → electrical

C mechanical → chemical → heat → electrical

D chemical → heat → mechanical → electrical

Questions 16–18

Select from the options A–D the one that best matches each of the descriptions in questions 16–18.

A condensation   B evaporation
C diffusion   D conduction

16 the rapid spread of alcohol vapour throughout a room

17 the change from the vapour to the liquid state

18 particles escaping from the surface of a liquid

19 A cylinder of metal of mass 0.2 kg was heated to 200 °C, and then transferred to 0.4 kg of water at 30 °C. The final temperature of the mixture was 40 °C. What is the specific heat capacity of the metal? (The specific heat capacity of water = 4000 J kg^{-1} K^{-1})

A 500 J kg^{-1} K^{-1} B 400 J kg^{-1} K^{-1}
C 160 J kg^{-1} K^{-1} D 50 J kg^{-1} K^{-1}

Questions 20 and 21

These concern gaseous systems. Select the option A–D which answers questions 20 and 21.

A At a constant volume and a given number of particles, pressure is proportional to the absolute temperature.

B At a constant volume and temperature, pressure is directly proportional to the number of particles present.

C At a constant pressure and a given number of particles, volume is directly proportional to the absolute temperature.

D At a constant temperature and for a fixed number of particles, the product of pressure and volume is a constant.

Which of A–D is a statement of:

20 Boyle's law?

21 Charles' law?

22 I   Radiation occurs in a vacuum.
II Heated objects emit radiant energy.
III Silvered surfaces are good absorbers of radiant energy.

Which of the above statements is/are correct?

A I and III only   B II and III only
C III only   D I and II only

23 At constant volume, a fixed mass of gas had a pressure of 50 kPa when the temperature was 360 K. At what temperature will the pressure be 37.5 kPa?

A 267 °C   B 130 °C
C 58 °C   D −3 °C

24 In falling through a height of 250 metres, the temperature of water increases by 0.6 °C. Ignoring heat losses, what value does this information give for the specific heat capacity of water in J kg^{-1} K^{-1}?

A 4.17 × 10^3 B 3.6 × 10^3
C 5.2 × 10^3 D 2.4 × 10^3

Questions 25 and 26

I It is more difficult to unscrew a bottle top after it has been held in hot water.

II Bubbles usually burst on reaching the surface of a pond.

III The freezing compartment is placed at the top of a refrigerator.

IV Thermal energy must be supplied to turn water into ice.

25 Which of the above statements is/are correct?

A I only   B IV only
C III and IV only D II and III only

26 Which of the above statements is an application of Boyle's law?

   A I   B II   C III   D IV

Questions 27 and 28 concern the cooling of a liquid. The variation in the temperature of the substance with time is shown on the graph.

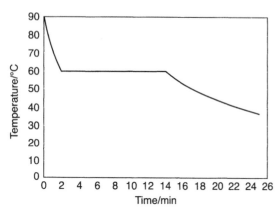

27 How long did liquid and solid co-exist in equilibrium?

   A 2 minutes   B 12 minutes
   C 14 minutes   D 25 minutes

28 What is the melting temperature of the substance?

   A 90 °C   B 75 °C
   C 60 °C   D 40 °C

Structured and free-response questions

1 Use your knowledge of the kinetic (molecular) theory to explain the following.
   a When a car is driven at high speeds over a road the tyres become firmer.
   b Solids and liquids expand when they are heated.
   c Gases exert pressure on the walls of containers.
   d The smell of perfume, accidentally spilled on the floor, gradually spreads to all parts of the room.
   e The haphazard motion of smoke particles seen in a smoke cell which is viewed through a microscope.

2 **a** What is meant by the terms 'boiling' and 'evaporation'?
   b State three differences between boiling and evaporation.
   c Explain, in terms of molecular theory, why a liquid evaporates more rapidly when it is heated.
   d Explain, as fully as you can, the statement: 'Ice, water and steam – the same but different'.
   e Explain why evaporation leads to cooling.

3 A thermos flask is so designed as to reduce heat exchange with the surroundings. It tends to keep hot liquids hot and cold liquids cold.
   a Is heat exchanged between an object and its surroundings when both are at the same temperature? Explain.
   b How does the 'caloric theory' account for heat exchanges between an object and its surroundings?
   c Briefly discuss **three** ways by which heat may be transferred from one place to another.
   d With the aid of a labelled diagram, discuss features of the thermos flask which allow it to reduce heat losses by each of the methods described in **c** above.

4 **a** What is meant by the efficiency of a device?
   b Design an experiment to test the efficiency of an electric coffee maker when it heats water to the boil. Pay attention to (i) the principles involved, (ii) the variables, if any, that should be controlled, (iii) the readings that should be taken and how you would use these readings to determine the efficiency of the coffee maker, (iv) possible sources of error.

5 A student finds an immersion heater in a laboratory drawer. The name plate is missing. He suspects, but is not sure, that the 'power rating' is 50 W.
   a What is meant by the 'power rating' of the immersion heater?

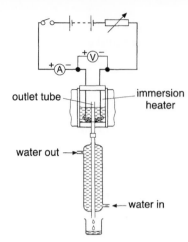

He sets up the apparatus shown. After the liquid has been brought to its boiling point, he finds that 3.5 g (0.0035 kg) of vapour escaped each minute through the outlet tube.

b What is meant by the term 'specific latent heat of vaporization of the liquid'?

c Why did the temperature of the liquid remain constant, once the liquid had been brought to the boil?

d State **one** way of getting the liquid to boil below 80 °C.

e Calculate the heat required to convert 3.5 g of liquid L, at its boiling point, to 3.5 g of vapour at the same temperature.

f Use your answer to **e** above to determine the quantity of heat supplied by the heater (i) each minute (ii) each second.

g Does the anser to **f** (ii) confirm the suspicion that the immersion heater had a power rating of 50 W? Explain.

6 A student used the data in I and II below to obtain a value for the specific latent heat of fusion of ice.

I She placed ice in an insulated but uncovered container. After 1.25×10^4 s, she found that 0.5 kg of ice was converted to water at 0 °C.

II In a separate experiment, she found that heat was gained from the surroundings at the average rate of 14.4 joules per second.

a What is meant by the term 'an insulated container'?

b Give **two** examples of good insulators.

c Use the data from I and II above to calculate the quantity of heat used to convert 0.5 kg of ice at 0 °C to water at 0 °C.

d Use your answer to **c** above to obtain a value for the specific latent heat of fusion, l_f, for ice.

e Compare the value you obtained in **c** above with the theoretical value of $3.4 \times 10^5 \, \mathrm{J\,kg^{-1}}$, suggesting reason(s) for the difference, if any.

f Suggest **one** way of lowering the freezing point of ice, and discuss **one** practical application of this lowering.

g Draw a fully labelled sketch to show how temperature varies with time when heat is supplied to ice cubes until steam is obtained.

7 You are required to determine the specific heat capacity of copper, using an electrical heating method (see diagram). The immersion heater has a known power rating and is connected to a power supply.

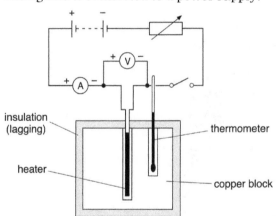

a What additional item(s) of apparatus is (are) needed?

b Outline the measurements you would take to obtain a value for the specific heat capacity of copper.

c What is the purpose of the lagging? Suggest a suitable material for the lagging.

d The experiment was repeated without lagging. Would you expect the experimental value of the specific heat capacity to be (i) greater (ii) smaller (iii) the same? Justify your answer.

e In one such determination using a 50 W immersion heater, the temperature rose by 20 K when 7800 J of energy was supplied to a 1 kg copper block. (i) For how long was the heater on? (ii) Use the data to obtain a value for the specific heat capacity of copper.

8 a What is meant by the phrase 'an ideal gas'?

b What assumptions are made in the 'kinetic model' of an ideal gas concerning (i) the size of the molecules, (ii) the type of collision between molecules, (iii) the force of attraction between molecules?

c Compare the kinetic energies of two gases which are at the same temperature, pressure and volume but which have different masses (relative molecular masses).

d A rigid vessel contains x particles of gas A; x particles of gas B are then added. Assuming that the particles of gas A and gas B have the same average kinetic energy before mixing, discuss how the mixing of the two gases affects (i) their temperature, (ii) the pressure exerted on the walls of the vessel.

e What would be the effect on pressure in **d** (ii) above if the temperature is suddenly increased, while the volume of the vessel remains constant?

9 a Explain the significance of each of the following features of solar cooking devices: (i) they have parabolic reflectors, (ii) the pot or cooking utensil is located (placed) at the focus of the reflector, (iii) the pot is made of good conducting materials, (iv) the pot is blackened.

b The diagram below shows a cross-section

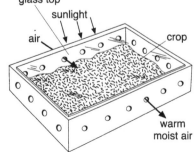

of a solar device. Explain how the device works. In your answer make reference to all features shown.

10 The apparatus shown below, with a suitable addition, may be used to determine the temperature of a liquid, i.e. as a thermometer.

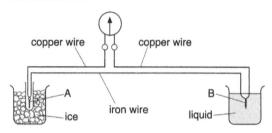

a Explain how the thermometer works.

b When junction A is placed in contact with pure melting ice, the galvanometer deflects in one direction. When junction A is placed in steam the galvanometer deflects in the opposite direction. Account for these observations.

c Fully explain how you would convert the galvanometer to a thermometer which reads temperature directly.

11 A thermopile is a series arrangement of thermocouples, the junctions of which are blackened. When a 48 W lamp, a convex lens and a special prism are arranged in the diagram, the light splits into colours. A thermopile moved in the plane AC shows a significant increase in reading when in region B.

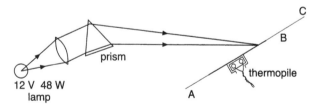

a Suggest **one** reason for each of the following: (i) connecting the thermocouples in series, (ii) blackening the junctions of the thermocouples.

b What is the name given to the splitting of white light into component colours?

c Which are the violet and red ends of the spectrum?

d Suggest **one** other method by which white light may be split into its component colours.

e Account for the significant increase in the reading of the thermopile when it is in the region B.

12 a Comment on the following features of a hot-water storage tank.

   I   The storage tank is double-walled, with fibreglass filling the space between the inner and outer walls.

   II The cold water inlet is at the base of the tank, while the hot water outlet is at the top.

   III The electrical heater is placed near the base of the tank.

   Draw a diagram of a hot-water system which incorporates the above-mentioned features.

b A hot-water storage tank has a capacity of 40 dm^3. The tank is filled with water at 25 °C. (i) Calculate the heat that must be supplied to bring the water to 55 °C. (Assume that 1 dm^3 of water has a mass of 1 kg and that the specific heat capacity of water is 4000 J kg^{-1} K^{-1}.) (ii) State the assumption(s) made in your calculation in **b** (i) above.

c If the electrical heater has a power rating of 5 kW, what is the time required to bring the water from 25 °C to 55 °C?

13 The diagram below shows a solar water heating system.

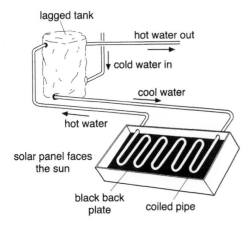

lagged tank

hot water out

cold water in

cool water

hot water

solar panel faces the sun

black back plate

coiled pipe

a Account for the following features of the solar panel: (i) a clear transparent top, (ii) a base (floor) made of copper which is blackened, (iii) the slope of the panel is such that the Sun's rays strike at right angles.

b Why is the coiled tube made of copper?

c Why is the copper tube bent as shown rather than straight?

d Discuss how the system exchanges heat with the surroundings.

e What steps could you take to minimise these heat exchanges?

14 Discuss the physics of the following:

a The freezing compartment is positioned at the top, not at the bottom, of a tall deep-freeze unit. Opening the door of such a unit for a short time has little effect on the temperature inside the unit.

b Small engines may be air-cooled, but larger engines, e.g. in motor cars, are usually water-cooled. Electrical equipment, e.g. transformers, is oil-cooled.

c Air is cooled when it is blown past coils in which a liquid is continuously being evaporated.

d Ice cubes dropped into a cup of hot water float on the surface of the water. When the ice cubes melt, there is no change in the water level.

e On a sunny day at the seaside, the water is colder than the sand.

f The base of a saucepan (pot) is made of stainless steel but the handle is made of hardened plastic.

g A burn resulting from contact with steam at 100 °C may cause greater injury than one from contact with the same mass of water at 100 °C.

15 a Describe the process(es) by which (i) energy is produced in the Sun (ii) energy is transferred from the Sun to Earth.

b Compare the heat radiated by a hot body with the nature of the light emitted from a lamp under the following headings: (i) speed of travel (ii) ability to pass through a vacuum (iii) ease with which they are reflected (iv) ease with which they are absorbed.

c An ice tray containing 0.30 kg of water at 27 °C is placed in the freezing compartment of a refrigerator. How much heat must the refrigerator remove to turn the water into ice at 0 °C? (specific heat capacity of water $= 4200 \, \text{J kg}^{-1} \, \text{K}^{-1}$; specific latent heat of ice $= 3.3 \times 10^5 \, \text{J kg}^{-1}$)

15 General properties of waves

15.1 Waves transfer energy

Waves are produced whenever vibrations (or oscillations) disturb a medium. A wave is a means of transferring energy (and information) between two points. Perhaps you are a little surprised at this. But just think of the rumble and rattle of seismic (earthquake) waves and their potential for destruction, or the poor surfer who is trying to negotiate a sea that has suddenly become unfriendly, or the heavy vibrations from the speakers at the disco, or the fact that solar cells convert light directly into electrical energy . . .

15.2 Types of wave

Waves may be mechanical or electromagnetic. Waves may also be transverse or longitudinal.

Mechanical waves, e.g. waves on slinkies, on stretched strings, water waves and seismic waves, are produced whenever particles in an elastic medium are displaced. Since the medium is elastic, these particles, in turn, cause their neighbours to be displaced from their rest positions. A single disturbance produces a **pulse**. If, however, the disturbance is repetitive a wave is generated.

Electromagnetic waves surround us. Light, radiant heat, radio signals, X-rays, microwaves . . . are all examples of electromagnetic waves. Electromagnetic waves consist of oscillating electric and magnetic fields which transfer energy through the vacuum of space at the speed of $3 \times 10^8 \, \text{m s}^{-1}$.

For **transverse waves** the direction of movement of the particles or fields is at right angles to the direction of wave travel (Figure 15.1). Transverse waves include water waves, light and other electromagnetic waves. Transverse waves may be demonstrated using a slinky (Figure 15.2).

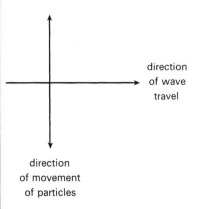

direction of wave travel

direction of movement of particles

Figure 15.1 *Transverse wave.*

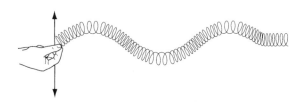

Figure 15.2 *Transverse wave on a slinky spring.*

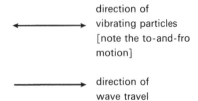

direction of
vibrating particles
[note the to-and-fro
motion]

direction of
wave travel

Figure 15.3 *Longitudinal wave.*

In **longitudinal waves**, particles vibrate about a rest (mean) position in the direction of wave travel (Figure 15.3). If a slinky is given a pulse as shown in Figure 15.4, the coils become compressed in some regions and spaced out in others. Longitudinal waves, for example sound waves, travel through the air, through liquids and through solids by a series of compressions and rarefactions. Longitudinal waves are not transmitted in (do not travel through) a vacuum. In a longitudinal wave, the wave energy is transmitted by physical contact between particles or layers of the transmitting medium.

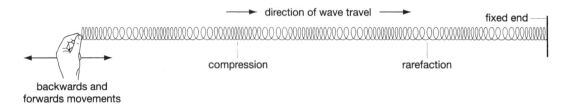

direction of wave travel

fixed end

compression

rarefaction

backwards and
forwards movements

Figure 15.4 *Longitudinal wave on a slinky spring.*

15.3 Wave properties

Amplitude and power

Figure 15.5 shows how the displacement of particles from their rest position, at a given time, changes with distance from the source.

The **amplitude** (A) of a wave is the maximum displacement of any particle on the wave from its rest position or equilibrium position. Displacement is a vector quantity, measured in metres.

It can be shown that a wave travelling through a medium corresponds to energy transport through the medium. However, there is no net transfer of *matter*. The **power** transmitted by any sinusoidal wave (e.g. that of Figure 15.5) is proportional to the square of the amplitude and to the square of the frequency. Frequency is defined on page 181.

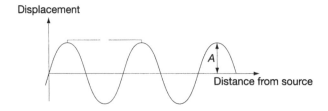

Displacement

A

Distance from source

Figure 15.5 *Graph of displacement against distance for particles in a wave at an instant of time.*

Wavelength

The **wavelength** (λ) can be taken as the distance between the midpoints of two adjacent crests or two adjacent troughs (Figure 15.5). More generally, the wavelength is the distance between one particle and *the nearest one* which is at the same stage of its motion. (The two particles are said to be **in phase** – they have the same displacement and move in the same direction with the same speed.) Wavelength is measured in metres.

Period and frequency

Figure 15.6 is a plot of displacement against time for a given particle.

The **period** (T) is the time taken for any particle to make one complete oscillation (one complete to-and-fro movement). The period is also equal to the time for the wave to travel one wavelength. Period is measured in seconds.

The **frequency** (f) is equal to the number of oscillations per second or the number of wavelengths that pass a reference point in one second. Frequency is measured in **hertz** (Hz).

$$\text{frequency } (f) = \frac{1}{T}$$

The frequency of a wave depends on the source of the disturbance.

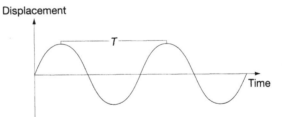

Figure 15.6 *Graph of displacement against time for a particle in a wave.*

Wave speed

It follows from our definitions of period and wavelength that the wave speed is related to wavelength and period:

$$v = \frac{\lambda}{T} \quad \begin{array}{l} \longleftarrow \text{wavelength (distance)} \\ \longleftarrow \text{period (time)} \end{array}$$

wave speed

Since $T = \dfrac{1}{f}$ then

$$v = f \times \lambda$$

The speed of a wave depends on the medium through which it is travelling. For example, light has a speed of $3 \times 10^8\,\mathrm{m\,s^{-1}}$ in a vacuum but a speed of $2 \times 10^8\,\mathrm{m\,s^{-1}}$ in glass.

Example

What is the frequency of ultraviolet light with a wavelength of $2 \times 10^{-7}\,\mathrm{m}$?

Solution

$$v = f \times \lambda$$

$$3 \times 10^8 = f \times 2 \times 10^{-7}$$

$$f = 1.5 \times 10^{15}\,\mathrm{Hz}$$

15.4 Studying wave phenomena

Waves can:

- be reflected;
- be refracted;
- be diffracted;
- interfere.

These properties are shown by longitudinal and transverse waves and by mechanical and electromagnetic waves.

The general properties of waves can be examined using water waves. A convenient way to study waves is to use a ripple tank. This is a shallow container with straight edges, filled with water.

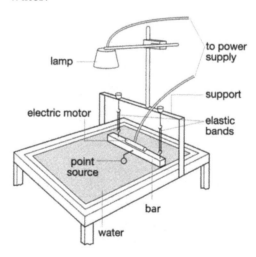

Figure 15.7 *A ripple tank.*

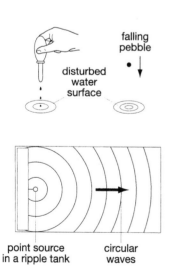

Figure 15.8 *Point sources cause circular waves.*

Point and extended sources

When you drop a pebble into a pond it disturbs the surface of the water and serves as a **point source**. Point sources, such as those illustrated in Figure 15.8, generate circular waves.

Circular ripples spread out from the point of disturbance. All the points on a particular ripple have the same phase. These points of constant phase define a surface known as a **wavefront**.

Thinking it through

- 'Long waves sometimes travel great distances over the ocean, yet the water does not flow with the wave.' Explain this statement as fully as you can.
- Longitudinal waves move along a slinky at $2.5\,\mathrm{m\,s^{-1}}$. Do individual coils of the slinky move 2.5 metres each second? Justify your answer.
- Observe, if possible, the motion of a buoy at sea or of a float on the surface of a lake or pond. Based on your observations, are water waves fully transverse waves, fully longitudinal waves or a mixture of both?
- The speed, v, of a wave on a taut string is given by the relationship:

$$v = \frac{\sqrt{T}}{\mu}$$

  where T is the tension in the string and μ is the mass per unit length.
  a By what factor should the tension in the string be increased for the speed of the wave to double?
  b A rubber hose is stretched and plucked. What happens to the speed of the pulse if the hose is stretched tighter, the mass and the length of the hose remaining constant?
  c What happens to the speed of the pulse if the hose is filled with water, tension remaining constant?

Extended sources, such as the flat bar in Figure 15.9, generate plane waves. If the bar is vibrated at low frequency the waves produced have long wavelength. On the other hand, if the bar is vibrated at higher frequencies the waves have shorter wavelength (Figure 15.10).

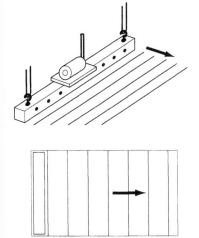

Figure 15.9 *The generation of plane waves.*

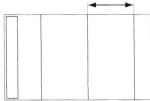

Low frequency, long wavelength

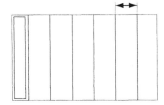

High frequency, short wavelength

Figure 15.10 *The frequency at which the bar vibrates determines the wavelength of the waves.*

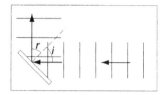

Figure 15.11 *The reflection of plane waves from a barrier.*

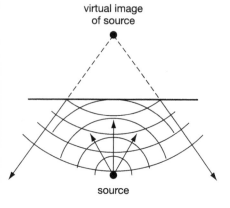

Figure 15.12 *The reflection of circular waves from a plane barrier.*

Note

You will learn in Chapter 19 that:

$$\frac{v_1}{v_2} = \frac{\lambda_1}{\lambda_2} = \frac{\sin \theta_1}{\sin \theta_2}$$

where v is the speed of the wave, λ the wavelength, θ the angle with the normal, and 1 and 2 are the two media the wave travels through.

15.5 Reflection

The reflection of water waves from plane surfaces

Using plane waves

If a plane barrier is placed at an angle i to the path of plane waves, the waves are reflected at an angle r, where $i = r$.

Using circular waves

Here the reflected waves behave as if they originated at a point behind the plane surface (Figure 15.12).

15.6 Refraction

How to study the refraction of water waves

1 Place a few washers on the bottom of the ripple tank.
2 Rest a sheet of glass on the washers. (With the washers, the position of the glass sheet can be easily adjusted.)
3 Generate plane waves so that they approach the sheet of glass at an angle, as shown in diagram (a).
4 Copy diagram (b) and complete it to show the path of the water waves as they emerge from the shallow water.
5 What is the ratio of the wavelength of the wave in deep water to that in shallower water? (Find it by actual measurement.)

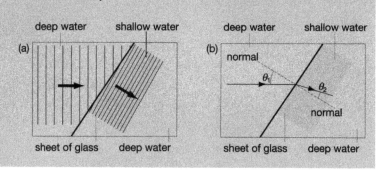

In the set-up above, the depth of water over the glass is less than at other places in the ripple tank. The shallow water constitutes a change of medium.

You should notice the following:

- The direction of the wave changes (i.e. refraction takes place) in the shallower water.
- The wavelength decreases in the shallower water.
- The speed of the wave decreases in the shallower water. (How can you check this experimentally?)

Refraction takes place when there is a change of medium (in this case, from deep water to shallower water).

Refraction is accompanied by changes of wavelength and wave speed. Frequency (which depends on the source) remains unaffected.

Things to do

Look at a distant street lamp through a pinhole. Observe rings of light around the pinhole. The rings are formed by the diffraction of light.

15.7 Diffraction

Diffraction is the spreading of waves when they encounter obstacles or when they pass through openings which have widths close to the wavelength of the wave. The longer the wavelength of the wave or the narrower the opening, the greater the extent of diffraction.

How to demonstrate diffraction of water waves

1 Set up the ripple tank for plane waves and keep the frequency constant.
2 Arrange barriers with an opening as shown, starting with a wide opening.
3 Observe the ripple patterns for different gap widths.

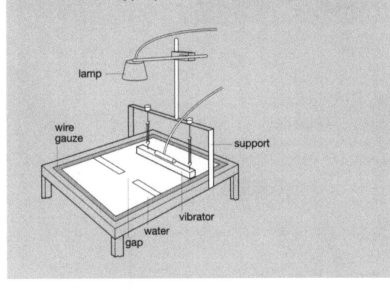

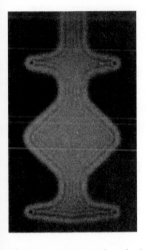

Figure 15.13 *The dark and bright bands outlining the shadow of the razor blade result from diffraction of light.*

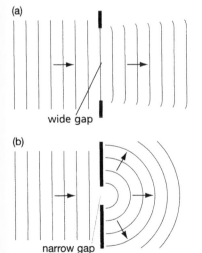

Figure 15.14 *Diffraction pattern with (a) wide opening, (b) narrow opening.*

You should see patterns of ripples like those in Figure 15.14. In effect, the narrow opening serves as a point source, producing circular waves to the right of the opening.

There is no change of wavelength as a result of diffraction.

15.8 The superposition of waves

The pattern you see is called an **interference pattern**. The two sets of waves combine (interfere) to give:

- maxima – regions of large disturbance, where crests overlap with crests and troughs with troughs. These are indicated on Figure 15.15 by full lines.
- minima – regions of near zero disturbance, where crests of one wave overlap with troughs of another. These are indicated on Figure 15.15 by broken lines.

Consider a point P along one of the full lines. The distance PS_2 minus the distance PS_1 is a whole number of wavelengths.

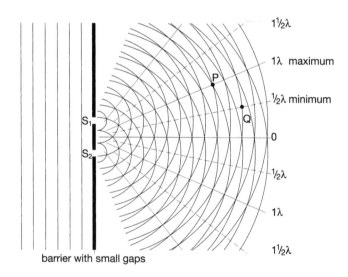

Figure 15.15 *Interference between two sets of circular waves.*

Waves meet at P **in phase** and give rise to **constructive interference**. The same is true for all points along the full lines.

On the other hand, QS_2 minus QS_1 equals $\frac{n}{2}$ times λ, i.e. it is not a whole number of wavelengths. Waves meet at Q (or at any other point along the broken lines) **out of phase**. They cancel each other's effect. This is known as **destructive interference**.

Note

Constructive interference occurs when waves superpose to give a resultant amplitude which is greater than either of the individual amplitudes. Constructive interference occurs when two wave trains arrive at a certain point *in phase.*

In destructive interference the resultant amplitude is less than either of the individual amplitudes.

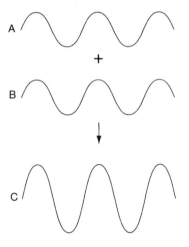

Figure 15.16 *Constructive interference (amplitude of C = amplitude of A + amplitude of B.*

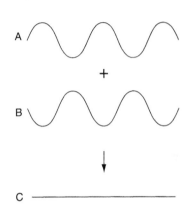

Figure 15.17 *Destructive interference.*

Things to do

1 Predict how the interference pattern would change if you altered

- the space between the gaps;
- the wavelength of the waves.

2 Carry out an experiment to test your prediction.

Checklist

After studying Chapter 15 you should be able to:

- recall that waves may be longitudinal or transverse and mechanical or electromagnetic
- recall that mechanical waves are produced whenever particles in an elastic medium are displaced
- recall that electromagnetic waves consist of transversely oscillating electric and magnetic fields which transfer energy through a vacuum at the speed of $3 \times 10^8 \, \mathrm{m\,s^{-1}}$

- recall that a transverse wave has the direction of vibration perpendicular to the direction of transfer of energy; light waves and water waves are examples of transverse waves
- recall that a longitudinal wave vibrates in the same plane as the transfer of energy; sound waves are an example
- use cylindrical coiled springs (slinkies) to model both transverse and longitudinal waves
- describe a wave in terms of its amplitude, wavelength, speed, frequency and period
- recall that the period of a wave is the reciprocal of frequency
- recall that the product of frequency and wavelength gives the speed of the wave
- recall that a wavefront is a line or surface, in the path of a wave motion, on which the disturbances at all points are in phase
- recall that point sources generate circular waves, whereas extended sources generate plane waves
- use the ripple tank to demonstrate reflection, refraction and diffraction of water waves
- recall that refraction takes place at a boundary because speed changes across the boundary
- recall that speed and wavelength change, but frequency remains unchanged after refraction
- recall that diffraction of plane wavefronts occur at edges and at slits of different widths
- recall that the longer the wavelength of the wave, the greater the extent to which it is diffracted
- recall that diffraction is greatest when the slit width is approximately equal to the wavelength of the wave
- recall that diffraction effects are not normally observed with light, because light has a very short wavelength
- recall that waves can interfere constructively and destructively

---

Questions

1 The speed of sound in air is $340\,\mathrm{m\,s^{-1}}$. What is the wavelength of a sound of frequency $1.7\,\mathrm{kHz}$?

2 Plane waves are generated on the surface of water in a ripple tank.
The distance between ripple I and ripple VI is $10\,\mathrm{cm}$.
   a What is the wavelength?
   b If ripple VI now occupies the position occupied by ripple I 10 seconds ago, what is the speed of the ripples?
   c What is the frequency of the ripples?

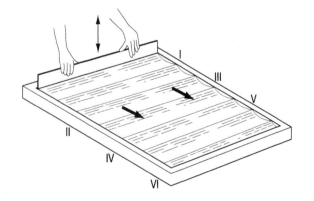

3 You are provided with a 2.8 cm microwave transmitter, metal plates which may be used to make slits, a microwave detector and a meter to measure the strength of the detector signal. Describe how you would set up this apparatus to illustrate that microwaves undergo interference. Draw a diagram of the interference pattern that you would expect.

4 Two loudspeakers X and Y are set up as shown. The speakers are fed a single note of frequency 256 Hz, the sound having a speed of $338\,\mathrm{m\,s^{-1}}$. Determine whether an observer stationed at Z will hear a loud sound or no sound at all. Explain fully how you arrive at your answer.

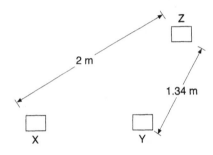

5 Radio waves of wavelength 13 m transmitted by a radio station are reflected from the ionosphere back to Earth.

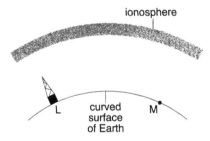

a Copy the diagram and draw three different rays to show how radio waves may get from the station L to a distant place M.
b Reception of the broadcast received by M from station L is often poor in quality because of interference from multiple reflection of the radiowaves. How is this interference related to the multiple reflections?
c On what frequency is the station broadcasting?

6 A fireworks display is taking place 5 km away from you. Calculate
a the time it takes the flash of light to reach you ($c = 3 \times 10^8\,\mathrm{m\,s^{-1}}$);
b the time the sound takes to reach you (speed of sound $= 340\,\mathrm{m\,s^{-1}}$).
c Based on your answers to **a** and **b** describe how you can estimate the distance between a thunderstorm and yourself.
d A US president is addressing a crowd on the lawn of the White House. The broadcast is carried live on TV. Explain why someone 10 000 km away receives the broadcast before someone 100 metres away from the podium gets the president's message.

7 Consider a particle on a string which is undergoing transverse wave motion.
a Does the speed of the particle remain constant? Justify your answer.
b Does the speed of the wave remain constant? Justify your answer.
c Is the speed of the particle the same as the speed of the wave? [*Hint:* The speed of a wave depends on the medium.]

8 A ripple tank is incorrectly levelled. The water in its trough is shallower at the left-hand end.
a Predict the shape of the wavefront produced if the surface of the water in the trough is touched with a pencil point.
b Draw the pattern of the wavefront if the water at the left-hand side of the trough is touched, sharply, with the edge of a 30 cm rule.

In 1860 Maxwell predicted the existence of electromagnetic waves and was the first to suggest that light was a form of electromagnetic radiation. In 1887, Hertz confirmed Maxwell's theory when he used a 'tuned' electrical circuit to generate electromagnetic waves and a similar circuit to detect them.

16.1 What are electromagnetic waves?

Figure 16.1 is a representation of an electromagnetic wave. It shows electric (E) and magnetic (B) fields at right angles to each other, each varying sinusoidally. Each field is perpendicular, i.e. at right angles, to the direction of wave travel: electromagnetic waves are transverse waves.

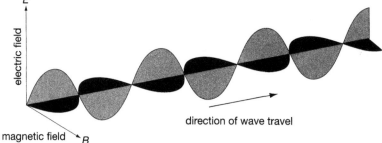

Figure 16.1 *An electromagnetic wave travelling in free space.*

We are surrounded by a 'family of electromagnetic waves', commonly referred to as the **electromagnetic spectrum**. Figure 16.2 shows that the electromagnetic spectrum ranges from the short wavelength (high frequency) gamma rays to the long wavelength (low frequency) radio waves.

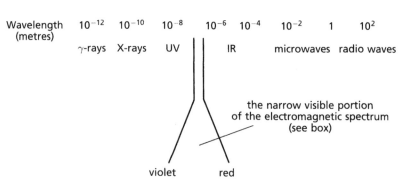

Figure 16.2 *The electromagnetic spectrum.*

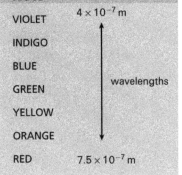
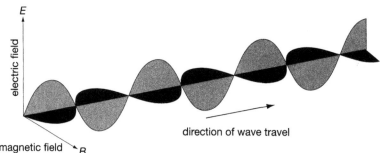

The properties of electromagnetic waves

All electromagnetic waves:

- travel at the speed (c) of $3 \times 10^8 \, \text{m s}^{-1}$, in free space;
- are transverse;
- obey the laws of reflection and refraction;
- show interference and diffraction effects;
- can be polarized;
- are unaffected by electric and magnetic fields.

Table 16.1 Electromagnetic waves

Waves	How produced	Major uses	How detected
Gamma-rays (ionizing radiation)	In nuclear reactions In cosmic radiation From radioactive elements	In medicine – to sterilize instruments, to kill cancerous growths In nuclear medicine and research – to trace the path of substances through the body	By the Geiger–Müller tube
X-rays (ionizing radiation)	By the sudden deceleration of high-speed electrons	To ionize gases To detect flaws in metals, forgeries, broken bones To determine crystal structure	By special photographic films
Ultraviolet (UV) radiation	In mercury vapour lamps	To kill bacteria In fluorescent lighting Causes human skin to tan (but causes sunburn and may cause skin cancer – recall the role of the ozone layer) Causes some metals to emit electrons	By photographic films By fluorescent materials
Infrared (IR) radiation	Emitted by all hot objects	In infrared-sensitive cameras to detect diseases In the heat drying of paints In thermal photocopiers	Detected through a rise in temperature when it falls on objects
Microwaves	In klystron tubes	Microwave cooking Telecommunication links Radar	By wave guides
Radio waves	By electrical transmitters By the motion of charged particles in fields	Communication Navigation	By aerials and diodes

16.2 A tale of two models

Newton (1642–1727) and Huygens (1629–1695) proposed opposite theories of light. Newton thought of light as 'swarms of particles wriggling through at great speeds'. He advanced evidence for the particle behaviour of light. Huygens, on the other hand, was convinced that light had a wave-like behaviour. Among other things, he observed 'that light rays crossed one another without hindrance'. Huygens was here referring to the principle of superposition.

So scientists had, at this time, two conflicting models of the nature of light to choose from. Not surprisingly, controversy was both lively and sustained. More scientists were inclined to accept the particle model of the more famous Newton.

In 1801, Thomas Young (1773–1829) performed a historic experiment in which two overlapping light 'waves' exhibited interference effects. From the results of this experiment (described on page 193), Young determined the wavelength of his light source. This was the first determination of this important property. The wave theory of light received further boosts from Maxwell and Hertz and by the late 19th century it was generally accepted that 'all radiation – light, heat ... – were waves of the electromagnetic field in space ...'

Intriguingly, this very absorbing story took yet another turn early in the 20th century when Max Planck, Albert Einstein and others breathed new life into the particle model with the quantum theory and experimental evidence which is satisfactorily explained by assuming that light behaves as particles.

So, the question remains: is light a particle or is it a wave? Scientists have decided to live with the **wave–particle duality of light**. This means that they accept the fact that sometimes light behaves as a wave and at other times as streams of particles.

In 1928, Davisson and Germer showed that electrons (small, negatively charged particles) were diffracted (i.e. they behave as waves under certain experimental conditions). G. P. Thomson (1928) showed that electrons displayed both diffraction and interference effects. But things did not stop there. Protons, neutrons, atoms (in fact all particles) also display wave properties under appropriate experimental conditions.

We have come to accept that the wave–particle duality applies to everything. But do we know when to use the wave part of the model and when to use the particle part? As a general guide:

- When light is interacting with matter it is displaying its particle characteristics.
- The wave model enables us to make predictions about the way in which light moves (i.e. how it *propagates*).

electric
field

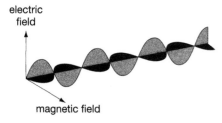

magnetic field

WAVE MODEL
Predicts well the way in which
light travels, diffraction, etc.

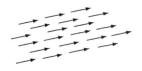

PARTICLE MODEL
Adequately explains the interaction
of light with matter

Figure 16.3 *The two models for light.*

16.3 Young's double-slit experiment

The results of this experiment, first carried out by Thomas Young in 1801, confirm that light has a wave-like nature.

The experiment may be carried out by setting up the apparatus of Figure 16.4, preferably in a darkened room. The interference pattern consists of alternate bright and dark bands (fringes) as shown in Figure 16.5.

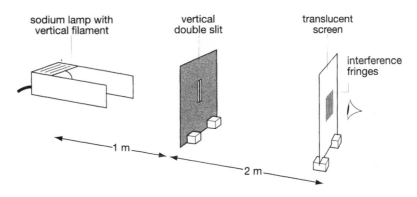

Figure 16.4 *An arrangement for carrying out Young's double-slit experiment.*

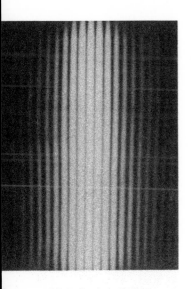

Figure 16.5 *An interference pattern.*

In Figure 16.6, light from the sodium lamp first falls on slit S_0. Light reaches slits S_1 and S_2 which act as **coherent** sources (see Note). Diffraction takes place at each of the slits and wavefronts reach the screen as a visible pattern of alternate bright and dark bands (fringes).

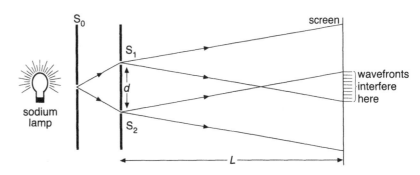

Figure 16.6 *Young's double-slit interference.*

Note

If two light sources are **coherent**, the light waves coming from one have a constant phase relationship with the waves coming from the other. They may be in phase, i.e. oscillating together, or out of phase, but the phase difference is constant.

When light from slits S_1 and S_2 reaches the screen and gives rise to constructive interference a bright band results. A bright band occurs if the path difference for wave trains from the two slits is $m\lambda$, where m is a whole number and λ is the wavelength of the light.

Dark bands correspond to regions of destructive interference. They occur where the path difference for wave trains from the two slits is $(m + \frac{1}{2})\lambda$.

An inspection of the interference pattern reveals:

- a central bright band fringed by two equally spaced dark bands;
- each dark band is fringed by two bright bands and so on . . .
- the intensity of the bands decreases from the centre outwards.

Determining wavelength

Let the distance between successive bright (or dark) bands be y.

Let the perpendicular distance from the double slit to the screen be L.

Let the slit separation be d.

Then, it can be shown that:

Thinking it through

Use this relationship to predict how increasing the source (double slit)-to-screen distance will affect the appearance of the fringes.

$$\text{wavelength } (\lambda) = \frac{\text{fringe separation} \times \text{slit separation}}{\text{source-to-screen distance}}$$

$$= \frac{yd}{L}$$

Checklist

After studying Chapter 16 you should be able to:

- recall that electromagnetic waves are transverse waves
- recall that all electromagnetic waves travel with the same speed in a vacuum
- recall that Newton first proposed the particle theory of light; Huygens first proposed the wave theory of light
- recall that it is now accepted that electromagnetic radiation can be modelled using both the wave theory and the particle model
- discuss how the wave theory is needed to explain diffraction and interference effects
- discuss how Young's double-slit experiment, first carried out in the early 19th century, provided crucial evidence that light is a wave
- define what is meant by coherent sources of light

Questions

1 Which part of the electromagnetic spectrum
   a is widely used in satellite
     communications?
   b has the highest frequency?
   c may be produced in mercury vapour
     lamps?
   d causes sunburn?

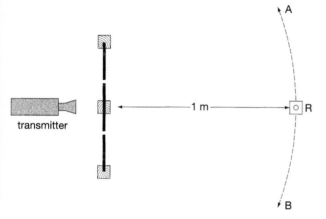

2 The transmitter shown in the left column
   emits 3 cm microwave pulses.
   a What is the frequency of the pulse?
     $c = 3 \times 10^8 \, \mathrm{m \, s^{-1}}$.
   b Describe expected changes in the
     intensity of the signal as the receiver R
     moves through the arc of the circle AB.
   c What width of gaps gives the best
     pattern?
   d What is meant by (i) a coherent source,
     (ii) monochromatic waves?

3 a Describe one phenomenon or one
     experiment which shows that light
     behaves as particles.
   b Describe the experiment carried out by
     Thomas Young which established the
     wave nature of light. Comment on the
     historical significance of this experiment.
   c Explain what is meant by the 'wave–
     particle' duality.

17 The physics of sound

Note

A sound wave is a series of alternating compressions and rarefactions.

Vibrating objects make sound by first pressing air molecules together and then letting them thin out.

Sound waves are longitudinal waves.

17.1 How are sounds made?

When objects vibrate they push on the air molecules nearest them, setting them in motion. These molecules, in turn, push on others. Since air is elastic, a series of compressions (air molecules close) and rarefactions (air molecules spread out) is produced (see Figure 17.1). The air molecules return to their original positions once the energy moves on.

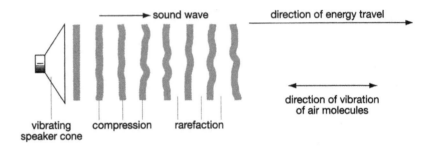

Figure 17.1 As the cone of the speaker vibrates, compressions and rarefactions are produced.

17.2 How are sounds transmitted?

Sound waves travel through solids, liquids and gases, i.e. through any medium which contains atoms or molecules. However, the sounds we hear are transmitted or sent mainly through the air.

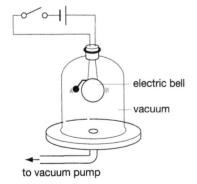

Figure 17.2 Can you hear the bell in a vacuum?

> **How to** find out if sound can travel in a vacuum
>
> 1 Set up the apparatus shown in Figure 17.2 and close the switch.
> 2 Observe:
>   a that you see the gong striking the bell;
>   b that you hear the bell ringing.
> 3 Slowly pump air out of the jar.
>
> Does the loudness of the sound you hear change? What happens if all the air is removed from the jar?

From this activity you can deduce that sound needs a medium in order to be transmitted. Sound does not travel in a vacuum.

Note

Waves of large amplitude produce loud sounds.

Waves of small amplitude produce soft sounds.

Sound levels are measured in **decibels**. The decibel scale is a *logarithmic* scale: an increase in the sound level of 10 dB means a two-fold increase in loudness.

17.3 Loudness, pitch and quality

Loudness

The loudness of a sound depends on:

- the amplitude of the vibrator producing it – the greater the amplitude of the vibration, the louder the sound;
- the 'amount of energy' reaching our ears. When a source (of sound) radiates in all directions, the sound intensity (I) falls off inversely as the square of the distance (d) from the source, since it is spread over a bigger area ($\propto d^2$).

$$I \propto \frac{1}{d^2}$$

Pitch

Pitch is a measure of a sound that depends on the frequency. The sound from a violin is of higher pitch than that from a double bass because the sound from the violin makes the eardrum vibrate at a faster rate. Sounds from the violin have higher frequencies.

Rapidly vibrating objects produce sounds of high frequency (and pitch). Slowly vibrating objects produce sounds of low frequency.

Note

Slowly vibrating objects produce low pitched sounds.

Rapidly vibrating objects produce high pitched sounds.

Quality

Because of differences in quality a note on the violin sounds different from the same note on the oboe. Notes played on each musical instrument, when displayed on an oscilloscope, show a regular pattern of frequencies (a waveform) – see Figure 17.3.

Each waveform has a fundamental or lowest frequency, but the pattern of harmonics (overtones), which are multiples of the

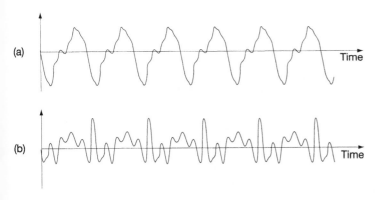

Figure 17.3 *(a) Waveform of a violin note displayed on a CRO; (b) CRO trace for the same note played on an oboe.*

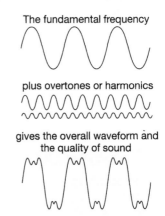

Figure 17.4 *Harmonics.*

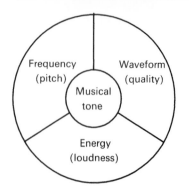

Figure 17.5 *Musical tones are made up of regular sound waves.*

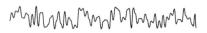

Figure 17.6 *The irregular vibrations of noise – no fixed pattern of frequencies.*

fundamental frequency, are different; the sounds are said to be of different quality (see Figure 17.4). The quality of a sound depends on the manner in which the sounding material vibrates. A musical tone is defined by its pitch, loudness and quality (Figure 17.5).

17.4 A question of noise

Whereas music is the regular repetition of vibrations, noise is caused by irregular vibrations (Figure 17.6).

Noise is unwanted or unpleasant sound. The evidence suggests that the present level of noise pollution is bad for our hearing and affects our physical and emotional health. Loud sounds lead to poor concentration, nervous tension, annoyance and anger.

17.5 Estimating the speed of sound

The principle of the method is to:

- determine the distance a sound (pulse) travels;
- measure the time taken to travel the distance.

$$\text{speed} = \frac{\text{distance sound travels}}{\text{time taken}}$$

Thinking it through

- Identify possible sources of error in this determination of the speed of sound in air.
- Discuss practical applications of echoes.

How to estimate the speed of sound in air

1 Select a hard, smooth surface (why?), e.g. a wall about 100 metres away. Let the distance between you and the wall be *d* metres.
2 Clap two wooden blocks together at such a rate that each clap coincides with the echo of the one before.
3 Count and time 25 claps and, hence, the time for one clap.

$$\text{speed of sound in air} = \frac{\text{twice the distance to the wall}}{\text{time for one echo}}$$

The speed of sound varies considerably with the material through which the waves are travelling.

Problem

Why is it necessary to state the temperature when giving the speed of sound in a given medium?

Table 17.1

Medium	Speed of sound at $0\,°C/m\,s^{-1}$
Air (dry)	330
Water	1400
Iron	2700
Concrete	5000
Carbon dioxide	265

Example

The time it takes a high frequency sound to travel from the surface of a lake to the bottom of the lake and back to the surface is 0.1 second. What is the depth of the lake?

Solution

$$\text{distance} = \text{speed} \times \text{time}$$

$$= 1400 \, \text{m s}^{-1} \times 0.05 \, \text{s}$$

$$= 70 \, \text{m}$$

Why was the time of 0.05 s used here, rather than 0.1 s??

17.6 The sound spectrum

Audio frequencies

The human ear is not sensitive to all frequencies (vibrations). We normally hear frequencies in the range 20–20 000 Hz. However, older folks have difficulties hearing notes of frequencies above 16 000 Hz.

Ultrasound

Any sound having a frequency greater than 20 000 Hz is described as **ultrasound**. Many animals, e.g. bats, cats, dogs and dolphins, hear frequencies that we cannot hear. Bats can hear frequencies as high as 100 000 Hz. By emitting high-frequency sounds and detecting the ultrasound reflected from objects, they can steer past obstacles and locate their food.

Ultrasound is used:

- for cleaning objects and materials. Optical companies clean lenses by placing them in a suitable liquid and shaking the dust and grit off with ultrasound. Surgical instruments are cleaned in this way too.
- for washing clothes and dishes. Lightweight machines use ultrasound generators to drive soap suds through clothes at such a speed that they are cleaned in a short time.
- in depth sounding and in finding submarines and shoals of fish by echo-location or 'sonar'. Pulses of high frequency sound are sent out from a generator on board the ship (Figure 17.7). The pulse travels down and is reflected from the sea floor or from a shoal of fish. A detector on board the ship detects the reflected signal. The return time for the signal is noted.

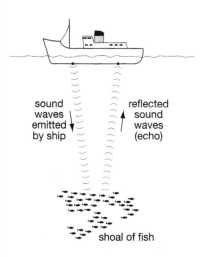

sound waves emitted by ship

reflected sound waves (echo)

shoal of fish

Figure 17.7 *Using a sonar detector to detect a shoal of fish.*

$$\begin{array}{c}\text{velocity of} \\ \text{sound in water}\end{array} \times \begin{array}{c}\text{return time} \\ \text{of signal}\end{array} \times \frac{1}{2} = \begin{array}{c}\text{depth of the} \\ \text{ocean floor}\end{array}$$

- for detecting flaws in metals and other materials.
- for mixing liquids at a molecular level.
- for scanning the human body to produce images of an unborn baby or a diseased organ.

17.7 Properties of sound waves

Reflecting sounds (echoes)

How to demonstrate reflection of sound

1 Obtain two long cardboard or PVC tubes.
2 Clamp one tube in position as shown and hold a watch or clock at the end furthest from a sheet of glass or hardboard.

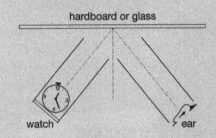

hardboard or glass

watch   ear

3 Listen to the watch or clock through the second tube. You will have to move this tube until you can hear the reflected sound (echo) clearly.

Things to do

How could you modify this experiment to compare the ability of different materials to absorb sound? Use the experiment to show that sound obeys the laws of reflection.

The echoes you hear in closed rooms, whispering galleries and caves all result from the reflection of sound from solid objects.

Refracting sounds

Consider the arrangement of Figure 17.8(a). The signal generator and loudspeaker unit send out pulses which are picked up by the microphone and displayed on the oscilloscope.

When a balloon filled with carbon dioxide is placed between the microphone and the loudspeaker the pattern on the screen of the oscilloscope changes (Figure 17.8(b)). This indicates that the sound has been refracted – sound travels more slowly in carbon dioxide than in air.

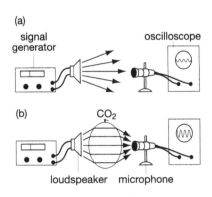

(a)
signal generator   oscilloscope

(b) CO_2

loudspeaker   microphone

Figure 17.8 *Refracting sound.*

Thinking it through

How can you tell that the sound waves have been refracted?

Note

The spreading of sound waves around the edges of a doorway is an example of diffraction. When a wave passes through a gap, the extent of diffraction is greater when $\lambda \approx d$, where λ is the wavelength and d is the width of the gap.

Diffracting sounds

Sound, because of its long wavelength, is easily diffracted. You can often hear the approach of a vehicle or a band before it comes around a corner. This diffraction or spreading of sound limits its range in air.

Human speech (average frequency 150 Hz and wavelength 2 metres) is easily diffracted. Doorways and windows (width 1–2 metres) provide ideal gaps for the diffraction of speech. We can hear street conversations etc. without going outside.

Sounds interfering

In Figure 17.9, L_1 and L_2 are matched speakers which are connected to a signal generator. The signal generator is emitting a note of constant frequency.

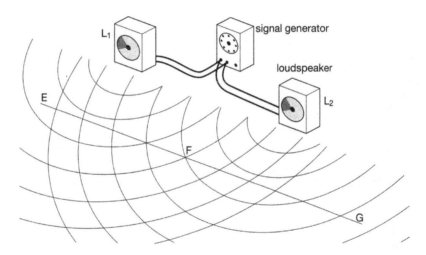

Figure 17.9 *Constructive interference of coherent sound waves.*

Someone walking along the line EG hears alternate loud sound and no sound at intervals. F is a position of maximum intensity because it is equidistant from L_1 and L_2.

L_1 and L_2 are **coherent** sources. They are emitting waves of the same frequency (phase difference is constant throughout). Sound waves from L_1 and L_2 superpose and interfere constructively and destructively at points along EG.

Checklist

After studying Chapter 17 you should be able to:

- recall that sounds are produced by vibrating systems
- recall that sound is transmitted as longitudinal waves

- recall that vibrating objects produce sound waves first by 'pressing' air molecules tightly together and then letting the air thin out (by a series of compressions and rarefactions)
- recall that sound does not travel through a vacuum
- recall that the pitch of a sound is related to its frequency; the higher the frequency the higher the pitch
- recall that loudness is related to the amplitude (energy) of the wave
- recall that musical instruments produce notes which have fundamental frequencies and harmonics (or overtones)
- describe how the speed of sound in air can be estimated by timing echoes
- recall that sound which has no regular pattern of frequencies is called noise
- recall that the human ear is normally sensitive to frequencies in the range 20 Hz to 20 000 Hz
- recall that sound with frequencies above 20 000 Hz is described as ultrasound
- recall that many animals use ultrasound to navigate and to find food
- recall that ultrasound (sonar) is used to locate underwater objects and to find the depths of ocean floors
- recall that ultrasonics is widely used in industry and medicine

Questions

1 A ship emits a pulse of sound and records the echo from the bottom of a lake 0.3 s later. If the speed of sound in water is 1500 m s^{-1}, how deep is the lake?

2 **a** What is the decibel a measure of?
   b Explain the purpose of the small holes found in some ceiling tiles.
   c Explain why a sports master uses a megaphone to make himself heard by all competitors.

3 Each musical instrument has a vibrating part and a means of changing pitch.
   a What is the purpose of the vibrating part?
   b What is meant by the term 'pitch'?
   c How is the pitch changed on a guitar?
   d Explain fully why musical instruments sounding the same note sound different.

4 **a** Distinguish between noise and musical sounds.

   b Identify four major sources of noise pollution.
   c Suggest a method for reducing the noise associated with each of the sources identified in **b**.
   d Should there be legislation to control noise pollution? Support the position you take.

5 Sound waves are incident on a water–air interface at 42° as shown. Sound waves travel about 4.3 times as fast in the water as in air. Copy and complete the diagram to show how the sound wave would behave at the water–air interface.

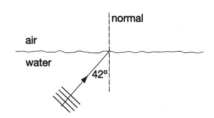

Note

Light is a form of energy.

Light is part of the electromagnetic spectrum.

Note

We can communicate using light.

Light travels at the speed of $3 \times 10^8 \, m \, s^{-1}$.

Light rays can be represented by straight lines.

light passes through
a **transparent** object
things clearly seen on this side

some light passes through
a **translucent** object
things not clearly seen on this side

*no light passes through
an **opaque** object*
nothing seen on this side

Figure 18.1 *Light travels through transparent and translucent materials, but not through opaque ones.*

18.1 How light travels

- Light travels through a vacuum or through transparent media.
- Light travels in straight lines, unless there is a change of medium.
- Light can be thought of either as waves or as a stream of particles.
- Light changes direction when it leaves one medium and enters another.
- Light travels through **translucent** and **transparent** materials (Figure 18.1).
- Light does not pass through **opaque** materials.
- Light affects some chemicals, e.g. those used in photography.
- Light plays an important role in **photosynthesis**.

Light sources

Light sources may be:

- **luminous**, e.g. a candle flame, a torch, a lighted match, the Sun. Luminous objects give off light of their own.
- **non-luminous**, e.g. the Moon. Non-luminous objects do not produce their own light. We see them because they reflect some or all of the light falling on them.
- **incandescent**, e.g. a Bunsen flame or a tungsten filament lamp. An object or material which gives off light when it is heated is described as incandescent.
- **point sources**: a point source is a source whose dimensions are small compared with other distances.
- **extended sources**, e.g. a fluorescent lighting strip. These have dimensions which are comparable with other distances.

How to show that light travels in straight lines

A Observing rays

1 Shine a light in a dust-filled, but darkened, room.
2 Observe the straight-line paths of the light rays.

Shadow formation and the nature of the images formed by pinhole cameras provide further evidence that light travels in straight lines.

B Shadow formation

1 Arrange a small light source about 60 cm in front of a screen.

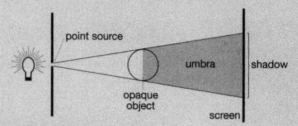

2 Place an opaque object midway between the source and the screen.
3 Move the opaque object until its shadow falls on the middle of the screen.
4 Note that the shadow has sharp edges. It is also equally dark throughout. The dark region is known as the **umbra** – meaning 'total shadow'.
5 Repeat the above steps using an extended source.

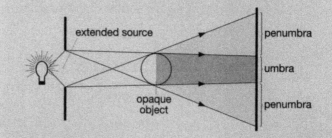

Thinking it through

Why are fluorescent lighting strips (extended sources) widely used in factories?

6 With this arrangement, note that there is a region of partial shadow (less dark) around the umbra. This region is called the **penumbra**. Only some of the light from the source falls in the penumbra.

Eclipses

Eclipses of the Sun and the Moon are 'spectacular' examples of shadow formation, involving an extended source.

In an eclipse of the Sun (Figure 18.2) the view of the Sun from Earth is partly or wholly blocked because the Moon moves between the Sun and Earth.

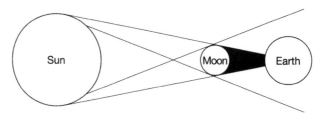

Figure 18.2 *An eclipse of the Sun.*

Note

In eclipses, the Sun is functioning as an extended source.

Recall that the Moon is a non-luminous object. We see the Moon because it reflects light from the Sun.

Figure 18.3 *An eclipse of the Moon.*

An eclipse of the Moon occurs when the Moon moves into the shadow of the Earth (Figure 18.3).

The pinhole camera

Thinking it through

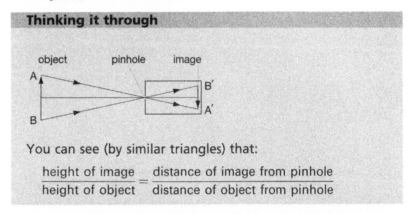

You can see (by similar triangles) that:

$$\frac{\text{height of image}}{\text{height of object}} = \frac{\text{distance of image from pinhole}}{\text{distance of object from pinhole}}$$

How to study the pinhole camera

1 Construct a pinhole camera like the one in Figure 18.4.
2 Hold the camera at arm's length and point it at brightly lit objects and dimly lit objects.
3 Comment on the size and other features of the image formed in each case.
4 Vary the distance between the camera and a chosen brightly lit object.
5 Comment on how the size of the image changes as you change the distance of the object from the camera.
6 Predict what would happen if:
   a multiple holes were used in place of a single hole;
   b a large hole were used instead of a pinhole. Try it and see!
7 In what way(s) is a pinhole camera similar to a lens camera?
8 State one disadvantage of the pinhole camera over the lens camera.

Figure 18.4 *Pinhole camera.*

Features of the image formed by the pinhole camera

- The pinhole camera forms an inverted (upside down), reversed, real image of the object.
- The pinhole camera forms multiple images if multiple holes are used.

- Bright but less sharp images are formed if a large hole is used.

You can take photographs with a pinhole camera, if you replace the screen with a piece of light-sensitive paper or film. However, you need a long exposure time. No focusing is necessary and clear pictures are obtained for all object distances.

18.2 Reflection of light

We see objects because light bounces off them. This 'bouncing off' is known as **reflection**.

Rough surfaces give irregular or scattered reflection (Figure 18.5).

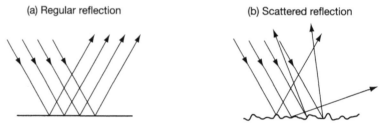

(a) Regular reflection **(b) Scattered reflection**

Figure 18.5 (a) A parallel beam of light is reflected as a parallel beam. (b) The parallel beam is scattered on reflection by a rough surface.

Mirrors

Mirrors and other highly polished surfaces change the direction in which light is travelling to produce images. Mirrors may be plane (flat) or curved. Common plane mirrors are made of thin sheets of glass which are silvered on the back. Curved mirrors may be convex or concave.

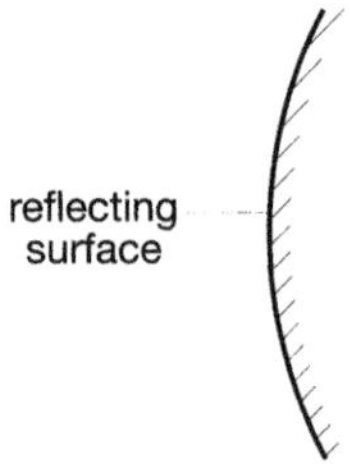

Convex mirrors (Figure 18.6) reflect a large part of the surroundings, i.e. they give a wide field of view. Convex mirrors are used as driving mirrors on cars or for surveillance, e.g. in large departmental stores.

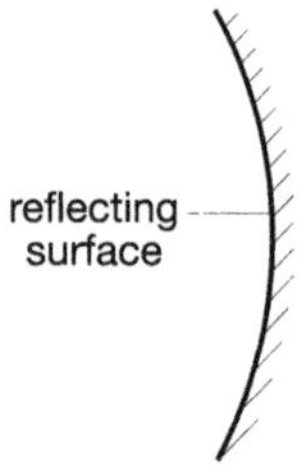

Concave mirrors (Figure 18.7) give images of close-up objects that are upright and enlarged. They are used as shaving mirrors, by dentists to examine the back of teeth and as reflectors in torchlights and vehicle headlamps. A special case is the parabolic mirror (see p. 170).

reflecting surface

Figure 18.6 A convex mirror.

reflecting surface

Figure 18.7 A concave mirror.

Things to do

A highly polished metal spoon can serve as both a concave and a convex mirror. Use a spoon to explore the properties of the images formed by these two types of mirror.

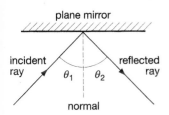

Figure 18.8 *Reflection in a plane mirror.*
$\theta_1 = $ *angle of incidence*
$\theta_2 = $ *angle of reflection*

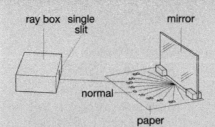

1. Draw a straight line on a sheet of paper. This line indicates where the back of the mirror should be placed.
2. Using a protractor map out angles as measured from the normal.
3. Allow a narrow pencil of light from a ray box or other suitable light source to fall along the 30° line.
4. Draw dots to show the direction of the incident and reflected rays.
5. Measure the angle the reflected ray makes with the normal.
6. Repeat for other angles of incidence.

The results of this experiment indicate that, for reflection:

- The angle of incidence equals the angle of reflection:

$$\theta_1 = \theta_2$$

- The incident ray, the normal and the reflected ray all lie in the same plane (Figure 18.8).

Features of the image formed by a plane mirror

- The image is **virtual**. Rays of light do not pass through the image. They only appear to do so.
- The image is the same size as the object.
- The image is the same distance behind the mirror as the object is in front.
- The image is upright.
- The image is **laterally inverted**. For example, your right hand, when viewed in the mirror, looks like a left hand.

Locating the image formed by plane mirrors

In Figure 18.9 two incident rays A and B from the lamp O strike the normals to the mirror at P_1 and P_2 respectively at angles i_1 and i_2. The rays are reflected as C and D so that

$$i_1 = r_1 \quad \text{and} \quad i_2 = r_2$$

The reflected rays C and D enter the eye, giving the impression that they came from point I.

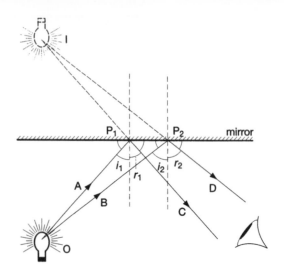

Figure 18.9 *How an image is formed in a plane mirror.*

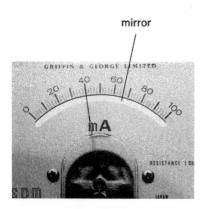

Figure 18.10 *Meters and other accurate measuring instruments often have plane mirrors mounted behind the scale to reduce parallax errors.*

How to locate an image by the no-parallax method

1 Place an object such as a pin (O in the diagram), 8–10 cm in front of a plane mirror.

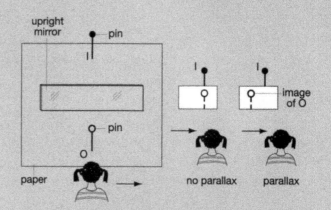

2 Looking into the mirror, place a second pin I behind the mirror, in such a way that the image of O (as seen in the mirror) is in line with I, as seen over the top of the mirror.

3 If the image and I remain lined up as you move your head from left to right, then there is *no parallax*. The image has been accurately located.

Under conditions of no parallax, the perpendicular distance from O to the mirror is equal to the perpendicular distance of I from the mirror.

Checklist

After studying Chapter 18 you should be able to:

- recall that light travels in straight lines if there is no change of optical medium
- recall that the working of the pinhole camera, eclipses and shadow-formation are all dependent on light travelling in straight lines
- recall that reflection occurs at plane and curved mirrors. For reflection: (i) the angle of incidence = the angle of reflection, (ii) the incident ray, the normal and the reflected ray all lie in the same plane
- recall that the image formed in a plane mirror is virtual, upright, laterally inverted and the same size as the object. The object to mirror distance equals the image to mirror distance
- recall that concave mirrors form upright, magnified images
- recall that convex mirrors give a wide field of view

Questions

1 A pinhole produces an image which is $\frac{1}{20}$ the size of the object. If the distance between the pinhole and the screen is 25 cm, how far away is the object in front of the pinhole?

2 Explain as fully as you can why the candle in the diagram appears to be burning in the glass of water.

3 Why does a solar cooker (see Chapter 14) have a parabolic reflector?

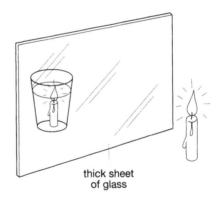

thick sheet
of glass

19 Refraction

19.1 Introduction

Light bends or changes direction if it leaves one medium and enters another of different optical density (see Figure 19.1). This bending or change in direction of light is known as **refraction**. The change in direction results from a change in the speed of light as it leaves one medium and enters another. (See Chapter 15.)

Rays of light travelling from one medium into an optically denser medium are bent *towards* the normal (Figure 19.2).

Rays travelling into an optically less dense medium are bent *away* from the normal (Figure 19.3).

Rays normal (at right angles) to the surface of separation between two media are not deviated (bent) (Figure 19.4).

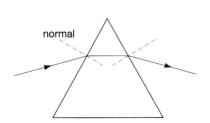

Figure 19.1 *Light is displaced sideways when it passes through a rectangular glass block.*

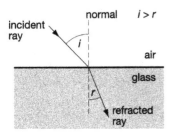

Figure 19.2 *The refracted ray is bent towards the normal when travelling into a denser medium.*

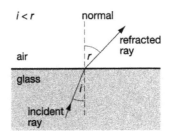

Figure 19.3 *The refracted ray is bent away from the normal when travelling into a less dense medium.*

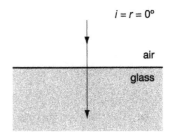

Figure 19.4 *Rays normal to the interface are not bent at all.*

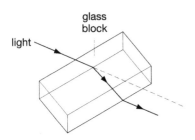

Figure 19.5 *Refraction of monochromatic light by a triangular prism.*

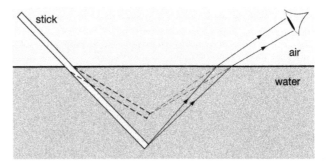

Figure 19.6 *A stick partially immersed in water appears to bend.*

19.2 The laws of refraction

1 The incident, the normal and the refracted rays all lie in the same plane.

2 The sine of the angle of incidence divided by the sine of the angle of refraction is a constant known as the **refractive index**.

This is known as **Snell's law**.

The refractive index is represented by the symbol $_1n_2$ where the 1 refers to the 'incident medium' and the 2 refers to the 'refracting medium'.

How to test Snell's law

1 Set up the apparatus below, allowing a thin pencil of light to fall on the parallel-sided glass block.

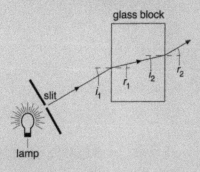

2 Observe the path of the light.
3 Make pencil marks (crosses) on the paper to indicate the direction of the ray into and out of the glass block.
4 Draw normals to the block at the points at which the light enters and leaves it.
5 Measure the angles of incidence and refraction.
6 Record the corresponding values of i, r, $\sin i$ and $\sin r$.
7 Plot a graph of $\sin i$ (y-axis) against $\sin r$ (x-axis). The slope of the graph gives a value for the refractive index of the glass.

Defining equations for refractive index

$$_1n_2 = \frac{\sin i}{\sin r}$$

$$= \frac{\text{real depth}}{\text{apparent depth}} \quad (\textit{especially applicable to liquids})$$

$$= \frac{\text{speed of light in incident medium}}{\text{speed of light in refracted medium}} = \frac{c_1}{c_2}$$

$$= \frac{\text{wavelength of light in incident medium}}{\text{wavelength of light in refracted medium}} = \frac{\lambda_1}{\lambda_2}$$

Example

Glass has a refractive index of 1.5.

Determine the angle of refraction for an angle of incidence of 45°.

Solution

$$n = \frac{\sin i}{\sin r}$$

$$1.5 = \frac{0.707}{\sin r}$$

$$r = 28° \, 7'$$

Problem

Copy the table and fill in the missing values. The table is for light travelling from a less dense to a more optically dense medium.

i	r	sin i	sin r	n
60°	40° 38′	–	–	–
22°	–	–	–	2.42
–	18°	–	–	1.5

19.3 Total internal reflection

Consider a semi-circular glass block (Figure 19.7) with ray A making an angle θ_A with the normal. The angle of refraction will be as shown by ray A_1.

An increase in the angle of incidence (ray B) leads to an increase in the angle of refraction. With a further increase in the angle of incidence (ray C), the angle of refraction is almost 90°.

Note

The conditions for total internal reflection to occur:

- light must be travelling to a less dense optical medium;
- the angle of incidence must be greater than the critical angle.

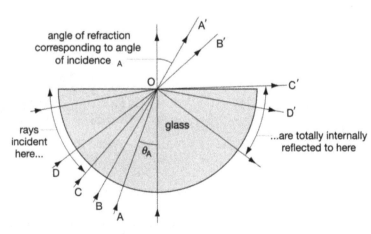

Figure 19.7 *Total internal reflection.*

For light travelling to a less dense medium, the angle of incidence for which the angle of refraction is 90° is known as the **critical angle**.

If the angle of incidence is increased beyond the critical angle **total internal reflection** takes place. This is what happens to ray D.

Let the critical angle be i_c. Then

$$\frac{\sin i_c}{\sin 90°} = {_g}n_a \quad \text{where g = glass, a = air}$$

Also,

$$\frac{\sin 90°}{\sin i_c} = {_a}n_g$$

$$\frac{1}{\sin i_c} = {_a}n_g$$

$$\sin i_c = \frac{1}{{_a}n_g}$$

Problem

The critical angle for a given medium is 48° 45′. What is the refractive index for the medium? Give your answer to two decimal places.

Example

Diamond has a refractive index of 2.42. What is the critical angle for diamond?

Solution

$$\sin i_c = \frac{1}{{_a}n_d} = \frac{1}{2.42} = 0.413$$

$$i_c = 24.4°$$

Some applications of total internal reflection

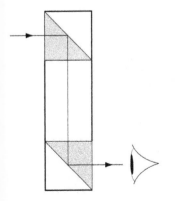

Figure 19.8 *Right-angled prisms are used in periscopes.*

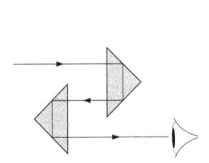

Figure 19.9 *Right-angled prisms are used in binoculars to give an image that is the right way up and not inverted.*

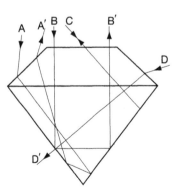

Figure 19.10 *Total internal reflection accounts for the sparkle of diamonds.*

Diamond has a high refractive index and a small critical angle. When light falls on a cut diamond, it undergoes a series of total internal reflections before passing out of the top again. This explains the sparkle (Figure 19.10).

An optical fibre is a thin glass or plastic fibre with diameter between 0.01 and 0.5 mm. The critical angle for glass is 42°. Light entering a fibre at 42° undergoes a series of total internal reflections before emerging (Figure 19.11). Bundles of fibres are used to send images from inside the human body, and in telecommunications.

Figure 19.11 *An optical fibre.*

19.4 Splitting white light into its component colours

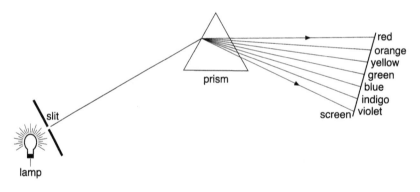

Figure 19.12 *Splitting white light.*

Isaac Newton observed that when he allowed sunlight to fall on a prism in a darkened room, the light spread into a spectrum of different colours (see Figure 19.12). In order to determine the exact cause of a colour, Newton varied the position of the prism relative to the hole which was letting the sunlight in. He also tested the other variables in his experiment, one at a time. He found that a second prism, suitably placed, caused the colours to recombine to give white light.

The process by which a prism splits white light into its component colours is known as **dispersion**. This happens because glass has a different refractive index for each colour.

How to create a spectrum

1 Place a convex lens, of focal length 10 cm say, in front of a lamp so as to focus its light on to a screen 1 metre away.
2 Place a prism in the path of the light.
3 Move the screen (while still keeping it 1 metre from the lamp) so that light falls on it.
4 Turn the prism to obtain the sharpest possible spectrum.

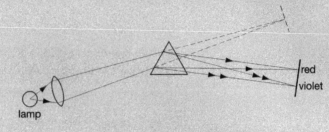

5 Clearly describe the pattern observed.

Despite being continuous, the spectrum can be divided into seven colours.

Note

Many optical devices contain lenses.

Refraction takes place at each surface of a lens.

Note

Beams of light may be:

parallel

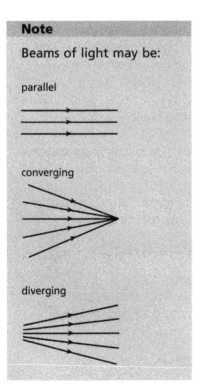

converging

diverging

19.5 Lenses

A **lens** is a curved transparent object which refracts light. Lenses can be thought of as being made up of truncated prisms.

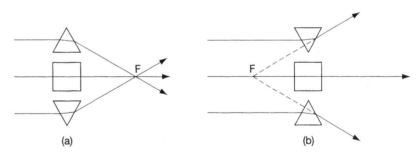

Figure 19.13 *Models of lenses: (a) converging; (b) diverging.*

From Figure 19.13 you can see that lenses cause parallel rays of light either (a) to converge or (b) to diverge. Lenses that cause parallel beams of light to converge to a point or plane are known as **convex** or **converging** lenses (Figure 19.14).

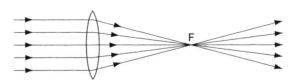

Figure 19.14 *A convex (converging) lens is thicker at its centre than at its ends.*

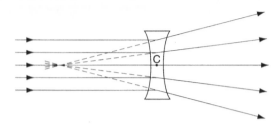

Figure 19.15 *A concave (diverging) lens is thinner at the centre than at its ends.*

A lens that makes a parallel beam diverge or spread out is known as a **concave** or **diverging** lens (Figure 19.15).

Defining the terms used with lenses

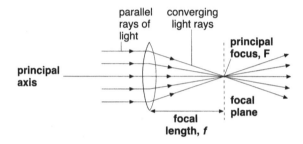

Figure 19.16 *Terms used when describing lenses.*

- The **principal focus** (*F*) is the point on the principal axis to which initially parallel rays converge, or from which they appear to diverge after refraction at the lens (Figure 19.16). There are two principal foci (plural of focus), one on each side of the lens.
- The **focal length** (*f*) is the distance from the centre of the lens to the principal focus.
- The **focal plane** is the plane, of which the principal axis is a part, in which sharp images of the object are formed.

19.6 How lenses form images

Convex lenses form both real and virtual images (Figures 19.17–19.20). Concave lenses form only virtual images (Figure 19.21). The type of image formed by a convex lens depends on:

- the distance of the object from the lens;
- the focal length of the lens.

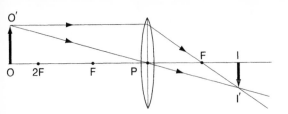 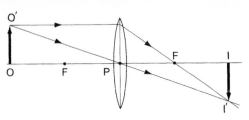

Figure 19.17 *Object (OO') beyond 2F. Image (II') is inverted and diminished (smaller than the object). The image is formed between F and 2F, on the other side of the lens.*

Figure 19.18 *Object at 2F. The image (also at 2F) is inverted and the same size as the object.*

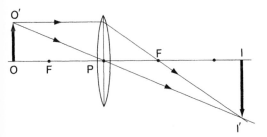 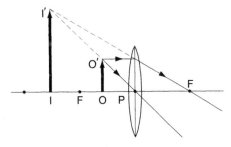

Figure 19.19 *Object between F and 2F. The image is inverted, magnified and formed beyond 2F.*

Figure 19.20 *Object closer to the lens than F. Image is virtual, upright and magnified. Image formed on the same side of the lens as the object.*

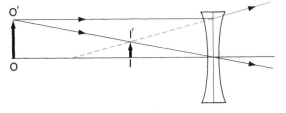

Figure 19.21 *A concave lens causes the rays of light from the object to diverge so that they appear to come from a virtual, upright image smaller than the object.*

Magnification

The linear magnification of an object by a lens is defined as

$$\text{magnification } (m) = \frac{\text{height or size of the image}}{\text{height or size of the object}}$$

In Figure 19.22,

$u =$ distance of object from the lens

$v =$ distance of image from the lens

$f =$ focal length of lens

Note

$\text{magnification} = \dfrac{v}{u}$

If $v < u$, the image is diminished.

If $v > u$, the image is magnified.

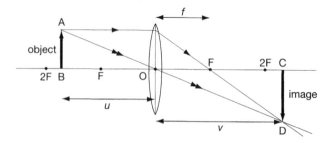

Figure 19.22 *Linear magnification.*

OAB and ODC are similar triangles. It follows that:

$$\frac{CD}{AB} = \frac{OC}{OB}$$

or

$$\frac{\text{height of image}}{\text{height of object}} = \frac{\text{image distance}}{\text{object distance}} = \frac{v}{u}$$

So

$$\boldsymbol{m = \frac{v}{u}}$$

Problem

An object of height 2 cm is placed 15 cm in front of a convex lens of focal length 10 cm. Find, by scale drawing,

a the image distance;
b the height of the image;
c the magnification.
d Use the magnification formula to test the correctness of your answer to **c**.

How to study image formation by lenses

1 Obtain two convex lenses (of different focal lengths) and a concave lens.
2 Point the thinner of the two convex lenses at a well lit, distant object such as a tree or window.
3 Sharply focus the image of the window on a translucent screen (i.e. a screen that lets light through).
4 Measure the distance between the lens and the screen. This distance is the focal length of the lens. Repeat with the other convex lens.
5 a Are the images formed real or virtual? How can you tell?
   b Are the images inverted or upright?
   c Are the images magnified or diminished?
6 Can the concave lens be used to produce an image on the screen? Test your answer by trying it out.

How to determine the focal length of a convex lens by measuring object and image distances

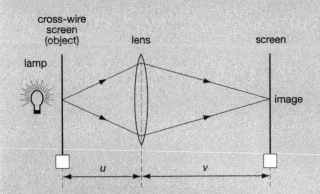

1 Set up the object at a distance u in front of the lens.
2 Obtain a sharp image on the screen. How would you do this?
3 Measure the image distance, v.
4 Vary the distance u between the object and the lens and find the corresponding image distances.
5 Tabulate for each pair of object distance (u) and image distance (v) the values of $\dfrac{1}{u}$ and $\dfrac{1}{v}$.
6 Use the relationship

$$\frac{1}{u} + \frac{1}{v} = \frac{1}{f}$$

to determine a mean value of the focal length (f) of the lens.

19.7 Some applications of lenses

The magnifying glass (hand lens)

A short focal length convex lens may be used as a magnifying glass.

How to estimate the magnification of a lens

1 Take a strip of graph paper and place the lens on it.
2 Count the number of squares in the diameter of the lens.
3 Slowly move the lens away from the paper until it is 10 cm from it.
4 Again count the number of squares you can see across the diameter.
5 Repeat at different distances from the paper.
6 Obtain values for the magnification of the image when the lens is at different distances from the graph paper.
7 Repeat the above steps with a lens of different focal length (Figure 19.23).

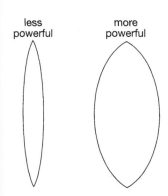

less powerful   more powerful

Figure 19.23 *The fatter the lens, the more powerful it is.*

The projector

A small, high wattage lamp is placed at the focus of a parabolic mirror (why?). This arrangement allows parallel beams of light to fall on the condenser lens which illuminates a transparent slide, which serves as the object for the projector lens. A magnified image of the slide is produced on a screen (Figure 19.24).

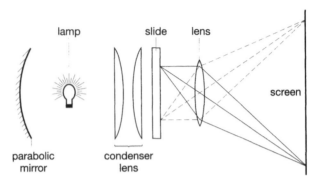

Figure 19.24 *A projector.*

The simple lens camera

The basic parts of the lens camera are shown in Figure 19.25. The camera is a lightproof box with a lens at the front and a light-sensitive film at the back.

The shutter, when opened, allows light to enter. The amount of light which enters is controlled by a stop. Light falls on the film to produce an image. The film is developed to give a negative, from which prints are made.

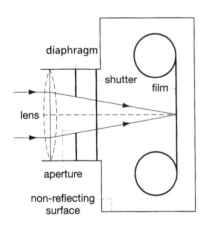

Figure 19.25 *A simple camera.*

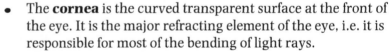 **19.8** **The human eye – the most important optical device**

Figure 19.26 shows the main features of the human eye.

- The **cornea** is the curved transparent surface at the front of the eye. It is the major refracting element of the eye, i.e. it is responsible for most of the bending of light rays.
- The **pupil** is an opening at the front of the eye that allows light to enter. It has a diameter in the range 2–7 mm. Pupil size is greatest in dim light and least in bright light.
- The **ciliary muscles** allow the lens to adjust its focal length so that it focuses equally well on distant and near objects (Figure 19.27).
- The **lens** is a convex lens, connected by suspensory ligaments to the ciliary muscles. The lens is flexible and its shape can be altered by the ciliary muscles so that its focal length changes. This mechanism is called **accommodation**.

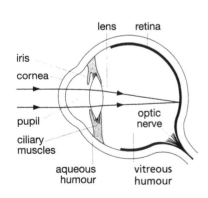

Figure 19.26 *The human eye: light rays are refracted first by the cornea and then by the lens.*

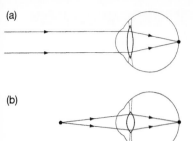

(a)

(b)

Figure 19.27 *Accommodation. (a) Ciliary muscles fully relaxed – lens pulled thin. Long focal length. Distant object being viewed. (b) Ciliary muscles contracted – lens becomes fatter. Short focal length. Close object being viewed.*

- The **retina** is a layer of light-sensitive cells at the back of the eye. Real, inverted images are formed here. The optic nerve connects the retina to the brain.
- The **iris** is the coloured part of the eye. It controls the opening at the front of the eye and, therefore, the amount of light entering.

Defects of vision

A person with 20/20 vision is able, as shown in Figure 19.27, to focus comfortably on both distant and close objects. Some people see distant objects as a blur because rays of light from these distant objects come to a focus in front of the retina. This defect of vision is known as **short-sightedness**.

Correction for short-sightedness is achieved by the use of spectacles with concave or diverging lenses. The diverging lens ensures that images of distant objects are brought to focus on the retina (Figure 19.28).

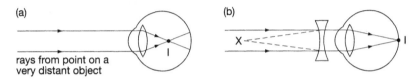

(a)

rays from point on a
very distant object

(b)

Figure 19.28 *(a) Short-sightedness. (b) Correcting short-sightedness. The diverging lens causes rays to diverge as though coming from X.*

A **far-sighted** person sees distant objects clearly but has difficulty focusing on objects nearby. The near point of a far-sighted person is greater than 25 cm (the near point of a person with normal vision). Images are brought to focus behind the retina for those who are far-sighted. Far-sightedness is corrected by the use of a converging lens (Figure 19.29).

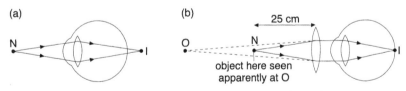

(a)

(b)

25 cm

object here seen
apparently at O

Figure 19.29 *(a) Far-sightedness. (b) Correcting far-sightedness. The converging lens reduces divergence of rays.*

As we get older the ciliary muscles become less able to change the shape of the lens for viewing very close or distant objects. This condition is known as lack of accommodation or **presbyopia**.

Lack of accommodation is 'corrected' by the use of bifocal glasses (two-lenses-in-one). Bifocals contain a converging lens for viewing close-up objects and a diverging lens for viewing distant objects.

1 Make a comparison of the human eye and the lens camera under the following headings:
   • refracting system;
   • nature of the image formed;
   • how the image is detected;
   • light barrier;
   • control of the amount of light entering;
   • how accommodation is achieved.

2 a Find out how the astronomical telescope works. In your answer pay attention to such terms as the objective, the eyepiece, infinity and angular magnification.

   b Describe how you would construct a homemade telescope.

Checklist

After studying Chapter 19 you should be able to:

- define refraction

- discuss the bending of light as it travels through materials of different optical densities

- recall that rays of light which are normal to the interface between two media are not refracted

- recall that in refraction through a parallel-sided glass block, the emergent ray is parallel to the incident ray, but is laterally displaced

- discuss total internal reflection

- define the critical angle

- describe applications of total internal reflection, e.g. optical fibre applications

- discuss dispersion

- recall that a convex lens makes parallel rays converge to a point, known as the focal point (principal focus) of the lens

- recall that parallel rays diverge on being refracted by a concave lens; the refracted rays, when produced backwards, appear to meet at the focal point of the lens

- recall that the principal axis of a lens is that line perpendicular to the lens which passes through its optical centre

- recall that all rays of light passing through the optical centre are not deviated (refracted)

- describe the use of lenses in optical instruments

- recall that the eye consists essentially of a refracting system and a screen (the retina)
- recall the refracting system of the eye
- discuss the eye's focusing mechanism
- discuss near-sightedness, far-sightedness and their correction
- recall that people who have lost the power of accommodation need 'bifocals'
- compare the eye and the camera

Questions

1 Light of wavelength 690 nm enters glass from air. The wavelength of the light in the glass is 460 nm. What is the refractive index of the glass?

2 What is the speed of light in a medium of refractive index 1.8?

3 The distance between an object and the image of it formed by a convex lens is 100 cm. Given that the magnification of the image is 4, what is
  a the object distance;
  b the image distance;
  c the focal length of the lens?

4 Plane wavefronts move towards the transparent block (medium 2) in the direction shown in the diagram. The refractive index of medium 2 relative to medium 1 is 1.2.

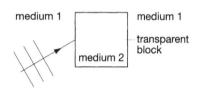

  a Copy the diagram and complete it to show the path of the waves in the block and when they emerge again into medium 1.
  b What is the maximum angle of incidence (for waves travelling from medium 2 into medium 1) for which refraction is possible at the interface between medium 2 and medium 1?
  c What happens if the maximum angle of incidence calculated in **b** above is exceeded? Explain.
  d Describe practical examples of the phenomenon described in **c**.

D   Revision questions for Section D

Multiple-choice questions

Questions 1–3 concern the phenomena associated with waves, labelled A–D.

A dispersion   B diffraction
C refraction   D interference

An example of this phenomenon is:

1 the spreading of sound waves after they pass through a door.

2 the alternate bright and dark fringes observed when light from a distant sodium lamp is viewed through a double slit.

3 the change in direction of a wave which results from a change in speed.

Questions 4–6 refer to the terms A–D which describe parameters of a wave. For each question select the appropriate response.

A frequency   B amplitude
C period   D wavelength

4 The minimum distance within which a wave exactly repeats itself.

5 The number of complete waves passing a given point in 1 second.

6 The maximum displacement of a waveform from its zero or reference position.

7 Which of the following statements concerning sound is **not correct**?
A Sound is transmitted as transverse waves.
B Sound may be produced by vibrating systems.
C Sound does not travel through a vacuum.
D Sound travels more slowly than light.

8 With a transverse wave motion:
A there is no transmission of energy
B particle movement is perpendicular to the direction of propagation
C no reflection or refraction takes place
D no medium is required.

9 In a certain medium the speed of a group of waves has a fixed value. If the frequency of the waves is doubled their wavelengths will be:
A four times their original value
B half their original value
C unchanged
D twice their original value.

10 A certain radio programme is beamed on a wavelength of 1200 m. What is the corresponding frequency of the radio signal?
A 3.6×10^{11} Hz   B 3.6×10^{8} Hz
C 2.5×10^{7} Hz   D 2.5×10^{5} Hz
(use $c = 3 \times 10^{8}$ m s^{-1})

11 Ripples in a ripple tank which travel 0.5 m in 2.5 s have a frequency of 4 Hz. What is the wavelength of the water waves?
A 0.05 m   B 0.25 m
C 0.50 m   D 2.5 m

12 A man standing 300 m from a cliff claps his hands and hears an echo 1.8 s later. What approximate value do these results give for the velocity of sound?
A 165 m s^{-1} B 330 m s^{-1}
C 463 m s^{-1} D 660 m s^{-1}

13 Two particles vibrating with the same amplitude are in phase if they have the same:
A displacement and velocity
B displacement and speed
C displacement only
D speed only.

14 The particles of a medium are in periodic motion perpendicular to the direction of propagation of the energy. Which of the following terms is defined by the above statement?

A longitudinal pulse
B longitudinal wave
C torsional pulse
D transverse wave

15 Which one of the following statements about sound waves is correct? They:
A will not pass through a vacuum
B will not travel through solids
C travel through solids at much reduced speed
D will not travel through liquids.

16 The velocity of sound in air is $330\,\text{m s}^{-1}$. A note of frequency 300 Hz will have a wavelength, in metres, of:
A 0.91   B 1.1   C 11   D 630.

17 Which **one** of the following waves does not belong to the electromagnetic spectrum?
A X-rays   B blue light
C radio waves   D sound

18 Which **one** of the following properties of waves would have to change for the energy carried by the wave to increase?
A frequency   B amplitude
C velocity   D wavelength

19 Two reed instruments are playing notes of the same pitch. Which of the following must be the same for the two notes?
A amplitude
B frequency
C length of string vibrating
D tension of the vibrating string

20 What is the upper frequency detectable by the normal human ear?
A 25 Hz   B 2000 Hz
C 20 000 Hz   D 250 000 Hz

21 Sound travelling through air is an example of a:
A longitudinal progressive wave
B longitudinal stationary wave
C transverse progressive wave
D transverse stationary wave.

22 Which of the following statements **is not** correct?
A Short waves diffract more than long waves.

B Diffraction is the spreading of waves around obstacles.
C Refraction is the change of direction resulting from the slowing down of waves.
D Water waves travel more slowly when the water is shallower.

23 Which of the following radiation causes sunburn?
A infrared radiation
B ultraviolet radiation
C visible light
D cosmic radiation

Questions 24–28 concern types of waves labelled A–D. For each question select the type of wave, by label, which best fits the description. Each label may be used once, more than once or not at all.

A 3 cm microwave
B sound waves
C infrared
D radiowaves

Which type of wave:

24 has the longest wavelength?

25 has a frequency of 1.0×10^{10} Hz?

26 is not electromagnetic in nature?

27 is used in photography under conditions of low light intensity?

28 has a speed of $3.3 \times 10^2\,\text{m s}^{-1}$ in air?

29 Which of the following statements **is not** true of all electromagnetic waves? They:
A travel at the speed of $3 \times 10^8\,\text{m s}^{-1}$ in air
B are longitudinal waves
C can be diffracted
D undergo both reflection and refraction.

30 Which of the following statements concerning light is **true**? Light:
A is one of a whole range of longitudinal waves
B has a speed in a vacuum of $3.3 \times 10^2\,\text{m s}^{-1}$
C has both particle and wave natures
D is easily diffracted because of its long wavelength.

31 A ray of light, on leaving one optical medium and on entering another, changes direction. This process or change is known as:
A diffraction
B refraction
C reflection
D dispersion.

Questions 32–34

A $\dfrac{\text{image distance}}{\text{object distance}}$ B $\dfrac{\sin i}{\sin r} = n$

C angle of incidence = angle of reflection

D $\dfrac{1}{\sin i_c} = n$

Which of the above:

32 defines linear magnification?

33 is an expression of one of the laws of reflection?

34 expresses Snell's law?

Questions 35 and 36

A principal axis
B principal focus
C focal length
D focal plane

Which of A–D is:

35 the point through which light rays pass after refraction at a converging lens?

36 the 'imaginary' line which passes through the optical centre of a lens, at right angles to the lens?

37 Which part of the camera performs a similar function to the pupil of the eye?
A shutter
B diaphragm
C lens
D film

Questions 38–40

An object placed x metres in front of a convex lens produces an image on a screen placed $3x$ metres on the opposite side of the lens.

38 The focal length of the lens, in metres, is:

A $\dfrac{3x}{4}$ B $\dfrac{3}{x}$ C $\dfrac{x}{4x}$ D $\dfrac{4}{3x}$.

39 The magnification of the image is:

A $\dfrac{x}{3}$ B $\dfrac{3}{x}$ C $\dfrac{x}{3x}$ D $\dfrac{3x}{x}$.

40 The image produced on the screen is:
A real and inverted
B real and upright
C virtual and upright
D virtual and inverted.

41 Which of the following is **not** an example of total internal reflection?
A eclipses of the Sun and Moon
B the sparkle of diamonds
C the use of optical fibres in communication systems
D the use of right-angled prisms in binoculars

42 Which part of the eye refracts light to the greatest extent?
A the lens
B vitreous humour
C aqueous humour
D the cornea

43 A ray of white light, entering a prism at a glancing angle, is:
A refracted only
B diffracted only
C refracted and dispersed
D diffracted and dispersed.

Questions 44–46

A visible spectrum
B optical fibre
C shadow
D prism

Which of the labels A–D best fits the description of questions 44–46?

44 It splits white light into colours.

45 It is a thin tube through which laser light is sent.

46 It is formed when an object blocks light.

47 An image is focused on the back of the eye by altering the shape of the:
A cornea
B lens
C ciliary muscles
D blind spot.

48 Which of the following is mainly responsible for the relative ease of transmission of radio waves around the world?

A reflection   B refraction
C dispersion   D diffraction

49 A plane mirror is moved towards a boy at a speed of $10\,cm\,s^{-1}$. The boy is stationary. The boy's image moves:
A towards him at a speed of $10\,cm\,s^{-1}$
B towards him at a speed of $20\,cm\,s^{-1}$
C away from him at a speed of $20\,cm\,s^{-1}$
D away from him at a speed of $10\,cm\,s^{-1}$.

50 The image of a pinhole camera is always inverted because:
A light is bent on entering the camera
B the images in optical devices are always inverted
C light travels in straight lines
D the pinhole camera contains many refracting elements.

51 An eclipse of the Moon occurs when:
A the Moon's shadow forms an area of darkness on the Earth
B the Earth's shadow forms an area of darkness on the Moon
C the Earth comes between the Moon and the Sun
D the Moon comes between the Earth and the Sun.

52 What is the speed of light in a medium of refractive index 2.0?
A $6 \times 10^8\,m\,s^{-1}$ B $3 \times 10^8\,m\,s^{-1}$
C $2 \times 10^8\,m\,s^{-1}$ D $1.5 \times 10^8\,m\,s^{-1}$

53 The apparent bending of a stick which is held below the surface of the water in a pond can be explained by:
A refraction   B reflection
C diffraction   D dispersion.

54 Which of the following is true for light which leaves air and enters glass?

	Speed	Wavelength	Frequency
A	decreases	decreases	stays the same
B	increases	increases	stays the same
C	increases	decreases	decreases
D	decreases	increases	increases

55 A ray of light strikes a plane mirror at an angle of incidence 20°. The angle between the incident and reflected rays is:
A 10°   B 20°   C 40°   D 50°.

56 The refracted index for:

water $= 1.33$ glass $= 1.5$ diamond $= 2.42$

Which of the following is true?

A The critical angle for diamond is larger than that for water.
B Diamonds sparkle because of multiple total internal reflections.
C Water bends a given incident ray through a larger angle than glass.
D A ray of light which falls at right angles on glass is turned through 90°.

57 A girl stands 10 metres in front of a large mirror. If the girl moves 5 metres towards the mirror, what is the new distance between the girl and her image?
A 5 m   B 10 m
C 15 m   D 30 m

58 The image seen in a periscope is:
A upright
B laterally inverted
C real
D larger than the object.

59 The distance between a pinhole and the screen of a pinhole camera is 40 cm. An object 150 cm high is placed 5 m from the pinhole. What is the height of the image on the screen of the camera?
A 0.19 cm   B 12 cm
C 66 cm   D 133 cm

60 I reflection   II refraction
III dispersion   IV diffraction

Prism binoculars depend for their action on:
A I and II   B I and III
C II and III   D II and IV.

61 An object placed 15 cm from a lens produced a real image twice the size of the object. What is the focal length of the lens?
A 20 cm   B 15 cm
C 10 cm   D 5 cm

Structured and free-response questions

1 Distinguish between a transverse progressive wave and a longitudinal progressive wave, giving one example of each.

2 A girl standing 90 m in front of a vertical cliff claps rhythmically at the rate of 25 claps every 15 seconds. Each clap coincides with the echo of the clap before.
  a Use these data to obtain a value for the speed of sound in air.
  b Comment on the accuracy of finding the speed of sound by this method.

3

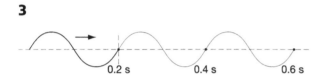

  0.2 s   0.4 s   0.6 s

  a What is meant by the term 'the amplitude of a wave'?
  b What is the amplitude of the wave shown, given that the wave is drawn to a scale of 1 : 10?
  c What is the period of the wave?
  d What is the frequency of the wave?
  e What is the wavelength of the wave, given that wave speed is $10 \, \text{m s}^{-1}$?
  f Will such a wave be easily diffracted? Explain.

4 A graph of displacement (y-axis) against distance from source (x-axis) for a given wave is shown below. The horizontal distance between points B and D is 0.5 m.

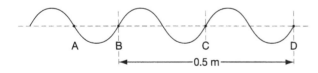

    A   B   C   D
         ◄────0.5 m────►

  a What is the wavelength of the wave?
  b What is the distance between points A and C?
  c If the frequency of the wave is 8 Hz, calculate the wave speed.

d Compare and contrast light and sound as waves.

5 Describe how, using simple materials around the home, you can show for water waves the following phenomena. In each case, give a diagram of your experimental set-up.
  a reflection
  b refraction
  c diffraction
  d interference

6

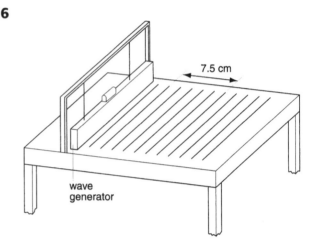

7.5 cm

wave generator

  a What does the distance between two adjacent crests represent?
  b What is the frequency of the generator above if the wave speed is $6.0 \, \text{m s}^{-1}$?
  c Draw a separate diagram to show what would happen if the waves entered shallower water.
  d Describe the modifications necessary to produce, in the ripple tank, (i) a diffraction pattern, (ii) an interference pattern.

7 A signal generator is hooked up to two matching loudspeakers as shown.

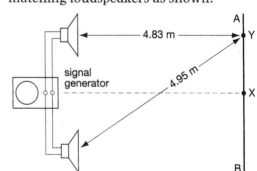

signal generator

◄──── 4.83 m ────►

4.95 m

A
Y

X

B

A detector moved vertically along AB records maximum intensity at X and its first minimum at Y.

a Describe how the intensity of the signal received changes as the detector moves between A and B.

b Use the information provided to obtain a value for the wavelength of the sound emitted.

c Given that the speed of sound is $330\,\mathrm{m\,s^{-1}}$, what is the frequency of the sound emitted?

d What distance apart are the speakers?

8 A fifth-former sets up an experiment to observe interference using the light from the two headlamps of his father's car. To his disappointment, no interference pattern was observed.

a Explain why he did not obtain a pattern. When he brought the headlamps closer, he still did not obtain a pattern.

b By explaining why he did not get a pattern, discuss the conditions necessary for obtaining a good interference pattern.

c Why are diffraction effects more easily observed with sound than with light?

d Would sound be more easily heard on the opposite side of a building if its frequency were increased? Explain.

9 A news item read, in part:
'Disabled people can now operate up to ten appliances without moving from their bed or wheelchair using an ultrasonic system developed by a fifteen-year-old boy and recently accepted for production by a firm of engineers.'

a What is meant by the term 'ultrasonic'?

b Suggest why ultrasonic waves are preferred to light waves for the system above.

c Explain the use of ultrasonics in locating fish.

d Briefly describe four other applications of ultrasonics.

e Comment on the following: (i) 'as blind as a bat', (ii) an ultrasonic generator was turned on and the rats took flight in fright.

10 Sketch a graph to show how the displacement of the particles in a sound wave varies with the distance from the source of sound.

a Label the amplitude and the wavelength on the graph.

b What is the effect on the sound heard if
(i) the wavelength is made shorter,
(ii) the amplitude is increased?

c Why is sound easily diffracted if its frequency is increased ten-fold?

11 a A station broadcasts on a frequency of 100 MHz. What is the wavelength of the waves?

b Microwave ovens use waves of wavelength 0.12 m. Calculate the corresponding frequency of these waves.

12

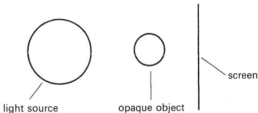

light source   opaque object   screen

a Use a carefully labelled ray diagram to show the area of (i) total shadow.
(ii) partial shadow in the set-up above.

b What important principle of light does your ray diagram in **a** show?

c Use the principle identified in **b** to explain how a pinhole camera works.

13 The diagram shows two $45°$ glass prisms and two rays of light. Copy the diagram and complete the paths of the rays of light through the prisms.

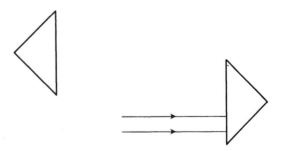

a What important principle of light is illustrated by your drawing?

b Name one optical instrument which uses the above arrangement.

c Why is the use of prisms superior to plane mirrors in such instruments?

14 Consider the diagram below.

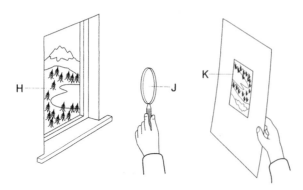

a What is the role of structure J?

b Why is scene K smaller than scene H?

c Why is scene K inverted compared to scene H?

d On what principle of light does structure J work?

e How would you adjust the position of structure J to get scene K larger than scene H? Give details.

15 Copy and label this diagram of the eye.

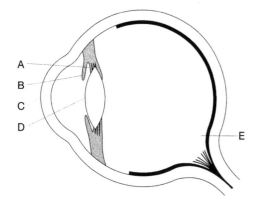

16 Show that, for light travelling from a dense to a less dense optical medium, the critical angle i_c and the refractive index n are related by:

$$\sin i_c = \frac{1}{n}$$

17 Describe an experimental arrangement which shows that white light is made up of a mixture of colours.

18 A pond is 8 m deep. An object viewed from above appears to be 6 m from the surface. What is the refractive index of the water in the pond?

19 The diagram is drawn to a scale of 1 : 10.

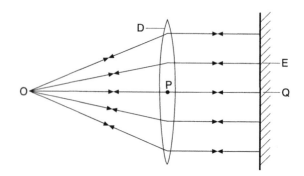

a Measure the distances OP and PQ.

b What is the distance OP a measure of?

c What is the role of structure E in this set-up?

d Describe another experimental method for determining the quantity OP.

20 a What is the name of the object M in the diagram below?

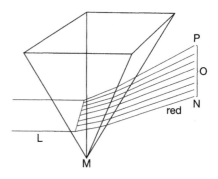

b Give the name of the colour of the light L.

c What is the name given to the region O? Describe the outstanding features of this region.

d This region is part of a larger one. What name is given to the larger region?

e What name is given to the colour P?

f Describe how you would attempt to obtain a clearer pattern of region O.

g What is the name given to the phenomenon illustrated in this experimental set-up?

21 A bright object placed at the bottom of a container appears closer to the surface than it really is.

a Use a ray diagram to explain why this is so.

A fourth-former who investigated this phenomenon obtained the following results of apparent and real depths.

Real depth/cm	9.0	12.0	15.0	18.0	20.0	25.0
Apparent depth/cm	6.8	9.0	11.3	13.4	14.9	?

b Plot a graph of real depth (y-axis) against apparent depth (x-axis).

c Determine the gradient of the graph, stating units if any.

d What is the gradient a measure of?

e Outline possible steps the student could have taken to obtain values of real depth and apparent depth.

22 a You can read these words because this page reflects light quite well. Why then is no image formed as in the case of a plane mirror (also a good reflector)? Explain as fully as you can.

b List two advantages of a fluorescent lighting fixture over an incandescent (hot) bulb of the same power rating.

c Is there any advantage in using a frosted incandescent lamp over a lamp that is made of plain glass? (Assume that both lamps have the same wattage.)

d What condition leads to the use of 'bifocals'? Why are there two lenses in bifocals? What type of lens is at the top of a bifocal? What type of lens is at the bottom? Explain.

e Draw a carefully labelled ray diagram to illustrate the use of a convex lens as a magnifying glass.

23 a Light from a utility pole reaches a convex lens as parallel rays. Why is this so?

b Draw a ray diagram to show what happens to the rays on leaving the lens. Use your diagram to obtain a rough value for the focal length of the lens.

c Describe, in point form, how you would obtain a more accurate value for the focal length of the lens.

24 A lens of focal length 15 cm forms an inverted image, three times as large as the object.

a What kind of lens is it?

b Using a scale drawing, show the object and image positions relative to the lens.

c Determine the object distance.

25 Your younger brother or sister believes that the eye is the source of light.

a List experiment(s) that you could carry out to show that this hypothesis is not valid.

b Can you think of any observation(s) that might have led to this belief?

26 Draw a ray diagram to show how a convex lens could be used to produce:

a a virtual magnified image of an object;

b a real image half the size of the object.

c Name **one** practical device in which the lens is used as in **a**.

27 By referring to at least **four** different instruments, write an essay entitled 'The use of lenses in optical instruments'.

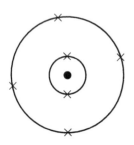

● the nucleus (+)
✕ electrons in their shells (−)

Figure 20.1 *Atoms contain positive and negative charges.*

All matter is made up of atoms. Atoms contain electrons (negatively charged particles) and a nucleus which consists of protons and neutrons (Figure 20.1). Protons are positively charged. Neutrons have no charge. Ordinarily, the atom is neutral. If an electron is pulled off, what is left behind is positively charged. If an electron is added, a negatively charged particle (ion) is formed.

This chapter is concerned with the behaviour of charges at rest (i.e. with static charges), the fields around these charges and the forces between them. It also looks at applications involving static charges, such as lightning, photocopying, crop spraying, electrostatic removal of dust and the working of the van de Graaff machine.

20.1 The fundamentals

Experiments show that:

- there are only two kinds of charge. Only one of these – the negatively charged electron – moves easily;
- an object acquires a positive charge when electrons are removed from it;
- unlike charges attract while like charges repel;
- charge is conserved (i.e. neither destroyed nor created).

Charles Coulomb (1736–1806), using a sensitive balance, showed that the force between two charges is inversely proportional to the square of their distance apart.

$$F \propto \frac{1}{d^2}$$

where F = force and d = distance apart.

Robert Millikan (1868–1953) showed that all charges are whole number multiples of the charge on the electron.

$$q = Ne$$

where q is the charge on the particle, e is the charge on an electron, N is a whole number.

20.2 Producing static charges

Static electricity may be produced by friction (rubbing), by induction and by contact. Rubbing seems the best way of producing static charges.

Note

Static electricity is produced when a few electrons are transferred from the surface of one object to that of another.

Charging by friction

A static charge is obtained by rubbing a hard insulator with a soft insulator. Rubbing causes electrons to move from one rubbed material to the other. The materials so rubbed acquire equal but opposite charges.

Note

Charging by friction leads to a redistribution of charges, but net charge remains constant.

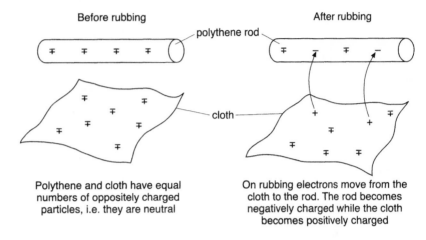

Before rubbing

After rubbing

Polythene and cloth have equal numbers of oppositely charged particles, i.e. they are neutral

On rubbing electrons move from the cloth to the rod. The rod becomes negatively charged while the cloth becomes positively charged

Figure 20.2 *Rubbing a polythene rod.*

Note

The arrows in Figures 20.2 and 20.3 indicate the direction of electron movement when the materials are rubbed.

In Figure 20.2, the polythene and cloth have equal numbers of oppositely charged particles before rubbing, i.e. they are neutral. On rubbing, electrons move from the cloth to the rod. The polythene rod becomes negatively charged while the cloth becomes positively charged.

By contrast, a Perspex or acetate rod becomes positively charged when rubbed with a soft cloth, because electrons move from the rod to the cloth (Figure 20.3).

Note

A rubbed acetate rod acquires a net positive charge.

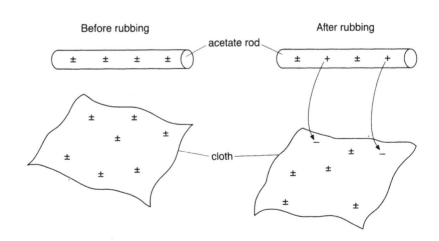

Before rubbing

After rubbing

Figure 20.3 *Rubbing an acetate rod.*

Charging a metal can by induction

A negative charge may be given to an insulated metal can as shown in Figure 20.4.

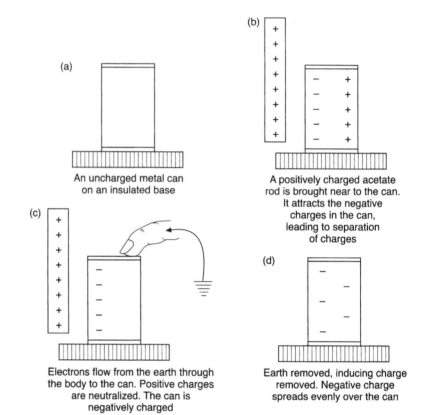

Figure 20.4 *Charging by induction.*

There is no sharing of charge when an uncharged insulator comes into contact with a charged insulator.

Charging by contact

An uncharged conductor which is in contact with a charged conductor acquires the same sign of charge (Figure 20.5).

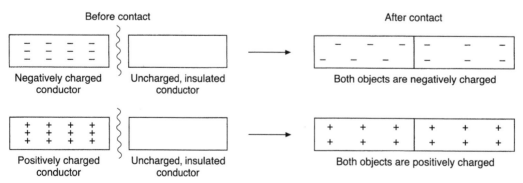

Figure 20.5 *Charging by contact.*

20.3 Electric fields

A charged object has an electric field surrounding it. Electric fields are represented on diagrams by field lines or lines of force. The direction of the electric field at any point is given by the direction of the force on a unit positive charge placed at that point. Electric field lines move outwards from positive charges and inwards towards negative charges (Figure 20.6).

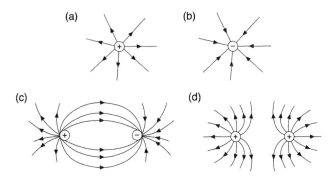

Figure 20.6 *The electric field around point charges.*

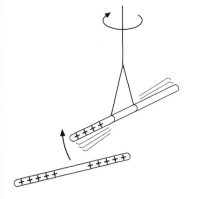

Figure 20.7 *Like charges repel.*

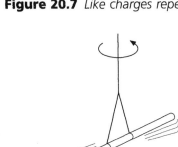

Figure 20.8 *Unlike charges attract.*

20.4 The interaction of charges

Figure 20.7 shows a charged acetate rod held in a stirrup and suspended by a thread. When another charged acetate rod is brought near the suspended one the movement is as shown, indicating that the rods are repelling one another.

When one charged polythene rod is brought near another suspended charged polythene rod, the two rods repel. But when a charged polythene rod is brought near a suspended charged acetate rod as shown in Figure 20.8, the rods attract and the movement of the suspended rod is as shown.

Charged objects may attract an uncharged object by inducing a charge in it (see Figure 20.4).

We can conclude that objects with the same charge repel while objects with opposite charges attract.

It can be shown experimentally that the force of attraction or repulsion between charges depends on:

- the magnitude of the charge – the greater the charge the greater the attraction or repulsion;
- how close the charges are – the closer the charges the bigger the attraction or repulsion.

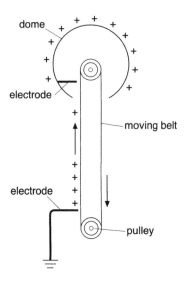

Figure 20.9 *A van de Graaff machine.*

20.5 Getting static charges to work for us

The van de Graaff machine

A van de Graaff machine (Figure 20.9) can generate very high voltages. It contains a hollow conducting sphere – a dome – which rests on an insulating stand (why?). A belt of insulating material – usually rubber – is turned by a motor. The moving belt takes charges from an earthed conductor to the top of the dome (see Figure 20.10).

The main use of van de Graaff machines is to provide very high voltages to accelerate charged particles to high velocities. These high kinetic energy particles are used in nuclear research and in medicine.

Figure 20.10 *A hair-raising experience: the girl stands on an insulated base such as polythene and touches the dome lightly. Charge spreads over the surface of her body. Strands of hair all acquire the same charge and repel each other.*

Electrostatic painting

Figure 20.11 *Electrostatic painting.*

In Figure 20.11, a large voltage is set up between the spray gun and the work to be painted. The paint is charged by friction as it leaves the gun at high speed. The globules acquire the same charge, repel each other and spread out evenly over the entire surface. Coverage is good and there is little wastage of paint.

Crops are sprayed in a similar fashion.

Lightning and lightning conductors

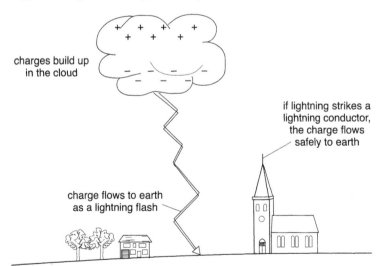

charges build up
in the cloud

if lightning strikes a
lightning conductor,
the charge flows
safely to earth

charge flows to earth
as a lightning flash

Figure 20.12 *A lightning flash between a cloud and the earth.*

Clouds moving through air become charged by friction. These charges can build up to such an extent that an electric discharge in the form of a spark or flash takes place between two neighbouring charged clouds or between a charged cloud and earth. Such a discharge is known as **lightning**.

Unprotected tall buildings could serve as points for the discharge of lightning. The energy released during such a discharge may be enough to set buildings on fire.

Lightning conductors are attached to tall buildings to protect them from damage during lightning discharges. A lightning conductor is a thick copper strip which stretches from the earth up the side of the building, ending in a spike. These conductors allow charge-laden clouds to discharge safely.

The damage done by lightning can be quite severe. Lightning strikes damage houses and electrical installations. Rocks are shattered; sometimes holes are made in them. Lightning strikes burn, and they can kill. The damage that lightning can cause may be gauged from this extract from a newscast on a foreign television station on 12 August 1986.

FIVE KILLED BY LIGHTNING

Lightning struck and killed five persons in separate incidents over the weekend.

Lightning 'hit' four persons huddled under an umbrella at the beach ... killing three and critically injuring the fourth ...

Two teenagers were killed when they sought refuge during a thunderstorm under a big tree in a field near their home. A third was burnt from the waist down.

Removing dust electrostatically

Consider the diagram in Figure 20.13.

- The central fine wire mesh is positively charged relative to the surrounding metal plates.
- The intense electrical field causes gas atoms to be converted to positively charged ions.
- The ions then attach themselves to dust particles in the moving gas stream.
- The now positive particles are repelled from the mesh and attach themselves to the metal plate, from which they are removed.

The electrostatic removal of dust is important to the chemical industry because it is an effective way of removing dust and other fine solids from gas streams. The removal of dust

- reduces the level of environmental pollution;
- prevents expensive catalysts from being poisoned.

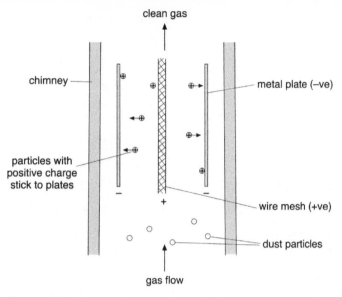

Figure 20.13 *An electrostatic precipitator can remove dust from waste gases in a factory chimney.*

Photocopiers

The dry photocopiers used in offices make use of the electrostatic force that exists between charged particles. Central to the photocopier is an aluminium drum coated with a layer of selenium (Figure 20.14). Aluminium is an excellent electrical conductor. Selenium, on the other hand, is an insulator in the dark but conducts appreciably when exposed to light. The steps in dry photocopying may be summarized as follows.

- The selenium-coated drum is given a positive charge.
- The document to be copied is scanned as the drum turns. Where light reflected from the white parts of the page hits the drum, the positive charge is destroyed.

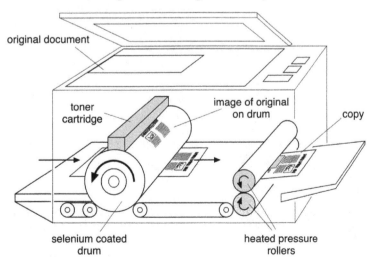

Figure 20.14 *A photocopier.*

- Negatively charged black toner powder from a roller sticks to the positively charged parts of the drum (which correspond to the black parts of the original image).
- The toner is transferred to the paper.
- The finished copy is passed between pressure rollers and the toner melts into the paper.

Checklist

After studying Chapter 20 you should be able to:

- recall that there are only two kinds of charge – positive charge and negative charge
- recall that static charges are charges at rest
- explain what happens when an object becomes charged
- recall that objects with like charges repel, objects with unlike charges attract
- recall that a charged object may also attract an uncharged object
- recall that insulators store charge
- recall that an electric field surrounds each charged object
- define an electric field as a region in which an electric charge experiences a force
- recall that the earth can serve both as an electron source and an electron sink
- recall that rubbing is the best way of producing static charges
- recall that electric field lines show the direction of the electric force on a positively charged particle; they show how a positively charged object would move if it were placed in an electric field
- recall that the further apart two charged objects are, the weaker the force between them
- describe and explain important applications of static electricity.
- explain how lightning develops between a cloud and earth or between two adjacent charged clouds
- explain how tall buildings are protected from lightning damage by conducting (lightning) rods

Questions

1. P and Q are charged conductors which are held in insulating stands. The conductors are connected to earth. Copy the diagram and indicate, by arrows, the direction of electron flow.

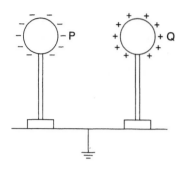

2. Provide a brief explanation for each of the following:

   a Articles of clothing cling together on being removed from the dryer.

   b A light, uncharged metallized sphere which is suspended from a thread is attracted to a charged polythene rod held nearby. After touching the polythene rod, the sphere is repelled.

   c Tiny bits of paper placed between two charged plates bounce back and forth between the plates.

   d Tall buildings made of reinforced concrete (concrete and steel) are often struck by lightning without damage.

21.1 The link between static charges and current electricity

When the van de Graaff generator in Figure 21.1 is switched on, the metallized table-tennis ball moves to and fro. It touches first one plate and then the other. Each time the ball touches a plate the microammeter indicates a current flow.

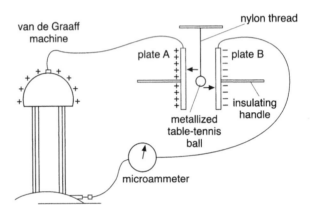

Figure 21.1 *Electric charge is transferred from one plate to the other: an electric current flows.*

Note

Both static and current electricity are concerned with charges and voltage. Whereas static electricity normally involves high voltages and low or intermittent (on-and-off) currents, current electricity mainly involves low to moderate voltages and reasonably small currents.

(1) On touching plate A the ball becomes positively charged by contact. (2) The ball is immediately repelled from plate A, (3) and is subsequently attracted to plate B where (4) it gives up its positive charge, (5) becomes negatively charged and (6) is, in turn, repelled. The cycle is repeated. In this way charge is taken from one plate to the other across the gap – demonstrating a link between static charges and current electricity. The microammeter reading results from the flow of charge from the generator to the plates and then through the connecting wires.

In Figure 21.2 static charges from the dome of the generator flow through the neon lamp. So an electric current is a flow of charge. A charged object has electrical potential energy. When two objects have different charges we say there is a **potential difference** between them.

Alessandro Volta (1745–1827) invented the first battery. He noticed that current flowed continuously from the battery when it was connected in a circuit. This discovery opened up the entire field of current electricity which has many practical modern applications.

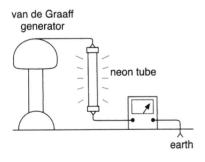

Figure 21.2 *The flow of charge lights the lamp.*

21.2 Conductors, insulators and semiconductors

We can classify solid materials into conductors, insulators and semiconductors, depending on how they conduct electricity.

Conductors

Note

Conductors contain free electrons.

Conductors allow electric currents to flow through them with minimum opposition.

Metals are good conductors of electricity because their atoms have one or more loosely held electrons in their outer shells. These electrons are easily shifted from atom to atom in the metal structure and are considered to be equally shared among all the atoms. However, unless a potential difference is applied, the loosely held electrons drift randomly through the structure and there is no net movement of charge in any one direction.

The free electrons in a metal move in a similar fashion to the individual particles in a gas (see Figure 21.3). Individual electrons have instantaneous velocities (i.e. velocities at any instant) of about 10^5 m s^{-1} but, because of frequent collisions, their average velocity may be as low as 10^{-7} m s^{-1}.

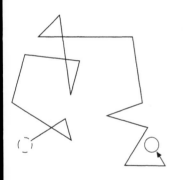

Figure 21.3 *The random movement of free electrons in a conductor.*

Note

Silver, copper and aluminium are among the best conductors.

Good conductors, in the form of wires, are used to carry currents because they have low **resistance** or, more accurately, low **resistivity** (see p. 245).

When a cell is connected across the ends of a conductor (Figure 21.4), there is a net movement of electrons towards the positive end of the conductor, B. At the same time electrons are fed into end A from the negative terminal of the cell. These electrons, in turn, force other outer electrons down the line and so on. This movement of electrons is an electric current.

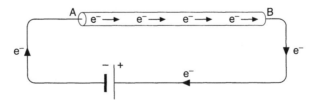

Figure 21.4 *When a cell is connected across the ends of a conductor, all the electrons move in one direction.*

Insulators

In some materials, all the electrons, including the outer ones, are tightly bound and are therefore not free. In these materials, known as insulators, the electrons do not move from atom to atom under normal conditions. Indeed, large amounts of energy must be supplied to free the electrons from the attractive forces of the nucleus. Glass, porcelain, polystyrene and paraffin wax are examples of insulators.

Semiconductors

Silicon, germanium and gallium arsenide are examples of semiconductor materials. Semiconductors have electrical conducting properties which, though better than those of insulators, are not nearly as good as those of conductors.

The electrical conducting properties of semiconductors may be increased many orders by:

- raising the temperature;
- the addition of small quantities of suitable impurities.

Semiconductor materials are used in a wide range of electronic devices (see Chapter 23).

Things to do

Using the apparatus below, group the materials rubber, polythene, copper, iron, zinc, sulphur, graphite, iodine and magnesium into those which conduct an electric current and those which are electrically non-conducting.

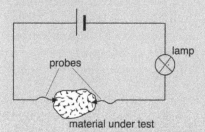

a Which of the materials when touched by the probes cause the bulb to light?
b How should these materials be classified?
c Why do these materials cause the bulb to light?
d How are these materials different from those which do not cause the bulb to light? Explain as fully as you can.

A property known as **resistivity** may be used to classify materials into conductors, insulators and semiconductors. Table 21.1 lists the resistivity values for selected materials. The best conductors have the lowest resistivities, of the order of $10^{-8}\,\Omega\,\text{m}$. Good insulators have resistivities of the order $10^{10}\,\Omega\,\text{m}$ or greater.

Note

Resistivity is a constant which measures the electrical conducting property of a given material. Resistivity (ρ) is related to resistance (R) by the equation:

$$R = \frac{\rho l}{A}$$

where l is the length of the conductor and A is its cross-sectional area. ρ has units $\Omega\,m$ (ohm metre) (R has units Ω – see p. 258).

Table 21.1 Resistivities

Material	Resistivity/$\Omega\,m$	Material	Resistivity/$\Omega\,m$
silver	1.6×10^{-8}	germanium	0.5
copper	1.7×10^{-8}	silicon	640
gold	2.4×10^{-8}	mica	$>10^{11}$
nichrome	150×10^{-8}	rubber	$>10^{13}$
manganin	48×10^{-8}	Teflon	10^{16}
carbon (graphite)	3.5×10^{-5}	wood	3×10^{10}

Types of electric charges and currents

In metals, an electric current is a flow of negative electrons. But in other conducting materials it may be a flow of other charged particles ('charge carriers') or even positive 'holes' (see Chapter 23).

Table 21.2 Charge carriers

Type of charge	Value of charge/C	Polarity	Type of current	Where applicable
electrons	1.6×10^{-19}	negative	electron flow	in metallic conductors, graphite and n-type semiconductors
ions	multiples of the charge on the electron	positive or negative	ion current	in gases and liquids, including electrolytes
holes	1.6×10^{-19}	positive	hole	in p-type semiconductors

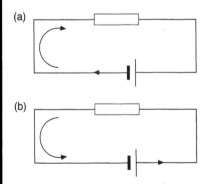

Figure 21.5 (a) Electron flow. Electrons flow round this circuit from the negative terminal to the positive terminal. (b) Conventional current. Positive charges are assumed to move from the positive to the negative terminal.

Electron flow versus conventional current

An electric current in a metallic conductor is a flow of electrons towards a region of greater positive potential. Unfortunately, early researchers assumed that current flowed from a region of positive potential to one of negative potential. This is called **conventional current** to distinguish it from electron flow.

21.3 Charge, current, potential difference, energy and power

Two fundamental principles apply to the study of electric circuits and electric currents. These are the principles of **conservation of charge** and **conservation of energy**. Make every effort to determine how these principles apply to situations encountered in the remainder of Section E.

Note

A junction is a point in an electric circuit where wires carrying currents connect together. The total currents entering a junction equal the total current leaving the junction. This is a statement of the principle of conservation of charge.

Table 21.3	Some important physical quantities and units		
Physical quantity		**Unit**	
Name	Symbol	Name	Symbol
current	I	ampere	A
charge	Q	coulomb	C
potential difference	V	volt	V
power	P	watt	W
energy	E or W	joule	J

Charge and current

- Current is charge on the move: a flow of charge constitutes an electric current.
- The SI unit of current is the **ampere** (A). If 6.25×10^{18} electrons flow past a point in a circuit in 1 second, the current flowing is 1 ampere.
- The quantity of electric charge which passes any point in a circuit depends on the value of the current and the time for which it flows.
- The unit of charge is the **coulomb** (C). A coulomb of charge is transferred in 1 second by a steady current of 1 ampere.

$$\frac{\text{charge } (Q)}{\text{(coulombs)}} = \frac{\text{current } (I)}{\text{(amperes)}} \times \frac{\text{time } (t)}{\text{(seconds)}}$$

$$Q = I \times t$$

therefore

$$I = \frac{Q}{t}$$

and

$$t = \frac{Q}{I}$$

1 ampere = 1 coulomb per second

1 coulomb = 1 ampere flowing for 1 second

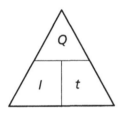

Note

$1\,A = 1\,C\,s^{-1}$

$1\,C = 1\,A\,s$

Things to do

Copy and complete the table:

Q/C	I/A	t/s
20		10
4	2	
	5	6

Example

Given that 1.25×10^{16} electrons pass a given point in an electric circuit in 10^{-3} seconds, calculate: (i) the charge, (ii) the current flowing.

Solution

6.25×10^{18} electrons have a total charge of 1 C

1.25×10^{16} electrons have a charge of 2×10^{-3} C

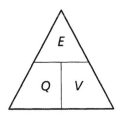

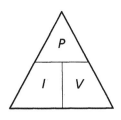

$$\text{current} = \frac{\text{charge (C)}}{\text{time (s)}}$$

$$= \frac{2 \times 10^{-3}}{1 \times 10^{-3}}$$

$$= 2 \, A$$

Potential difference

Electric **potential difference** (**p.d.**) measures the work done by a battery or other source in driving 1 coulomb of charge between two points in an electric circuit. The unit of potential difference is the **volt**, defined as one joule per coulomb.

Potential difference is also commonly called **voltage**. Sources of potential difference include: the cell (or battery) which may be rechargeable, the solar cell, the generator and piezo-electric devices.

Energy and power

$$\text{energy } (E) = \text{charge } (Q) \times \text{potential difference } (V)$$

$$\frac{\text{energy}}{\text{(joules)}} = \frac{\text{current}}{\text{(amperes)}} \times \frac{\text{time}}{\text{(seconds)}} \times \frac{\text{potential difference}}{\text{(volts)}}$$

$$E = Q \times V$$

Therefore

$$Q = \frac{E}{V}$$

and

$$V = \frac{E}{Q} = \frac{E}{It}$$

Remember that **power** is the rate of energy transfer or the rate at which work is done. So:

$$P = \frac{E}{t} = \frac{Q \times V}{t} = \frac{I \times t \times V}{t} = IV$$

$$\frac{\text{power } (P)}{\text{(watts)}} = \frac{\text{current } (I)}{\text{(amperes)}} \times \frac{\text{potential difference } (V)}{\text{(volts)}}$$

Power is related to current and resistance (see p. 260) by:

$$P = I^2 \times R$$

and it is related to potential difference and resistance by:

$$P = \frac{V^2}{R}$$

Example

The starting motor of a car draws a current of 60 amperes from a 12-volt battery for 7 seconds. Calculate **a** the charge flowing, **b** the power dissipated, **c** the energy transferred.

Solution

a charge = current (A) × time (s) = $(60 × 7)$ A s = 420 C

b power = current × potential difference
= $(60 × 12)$ W = 720 W

c energy = power × time = $720 × 7 = 5040$ J

21.4 a.c. versus d.c.

The simple flow of charge when electrons leave the negative terminal of a battery and flow around a circuit to the positive terminal is known as **direct current**, or **d.c.** (Figure 21.6).

- Direct current flows in one direction only.
- Direct current may be steady or fluctuating.

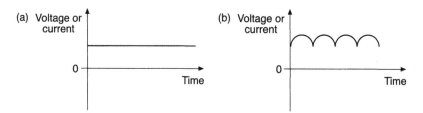

Figure 21.6 *(a) Steady d.c. from a freshly charged battery. (b) Fluctuating d.c., for example from a d.c. generator.*

With **alternating current (a.c.)**, the electrons flow backwards and forwards in the circuit. Alternating current reverses direction periodically. The frequency of the mains supply is 50 Hz (50 cycles per second) in some countries and 60 Hz in others. A 50 Hz supply changes direction 100 times each second whereas a 60 Hz supply changes direction 120 times each second.

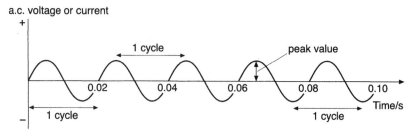

Figure 21.7 *Graph of voltage or current against time for a 50 Hz a.c. supply.*

Note

Low frequency a.c. makes the pointer of a moving-coil meter move to and fro. With high frequency there is no apparent movement.

In Figure 21.7, notice that:

- the a.c. signal varies in value over a cycle;
- the a.c. signal shown reverses direction 10 times every 0.1 s; its frequency is therefore 50 Hz;
- the time for one cycle is known as the **period** T;
- frequency $f = \dfrac{1}{T}$; a 50 Hz a.c. signal has a period of 0.02 s.

Example

An alternating signal has a period of 0.01 s.

a What is its frequency?
b How many times per second does this signal reverse polarity (i.e. positive $\leftrightarrow$ negative)?

Solution

a Frequency $= \dfrac{1}{T} = \dfrac{1}{0.01}$ Hz $= 100$ Hz or 100 cycles/s.
b Since an a.c. signal reverses polarity twice every cycle, a 100 Hz signal reverses $100 \times 2 = 200$ times per second.

Thinking it through

One electrical device operates off a 120 V supply a second identical device operates off a 240 V supply. Your classmate argues that the second device uses twice the power of the first. Is she correct? Explain fully.

Generators can supply both a.c. and d.c.; a.c. is better for the transmission of power.

At present most electricity is generated as a.c.

a.c. and d.c. are equally satisfactory for heating and lighting.

Table 21.4	A comparison of d.c. and a.c.
d.c.	**a.c.**
fixed direction	reverses direction
may be steady in magnitude from a given source	varies between reversals of polarity
its value cannot be stepped up (increased) or stepped down (decreased) by a transformer	can be stepped up or down as required by a transformer
has an associated heating effect	has an associated heating effect
has an associated magnetic effect	has an associated magnetic effect
can be used in electrolysis, electroplating and battery charging	cannot be used in electrolysis, electroplating and battery charging
	can be converted to d.c. using rectifiers.

Things to do

Using suitable resource materials, find out as much as possible about the electrical circuits in a motor car. Explain how all these circuits operate from a single 12 V battery.

21.5 The oscilloscope

Like the computer monitor and the television receiver, the oscilloscope uses electron beams to build up pictures on a screen. The main component of the oscilloscope is the **cathode ray tube** or CRT. A diagram of a CRT is shown in Figure 21.8.

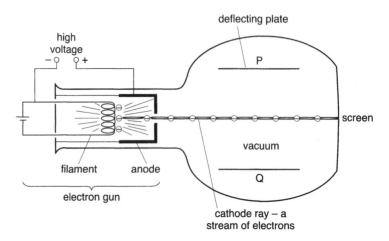

Figure 21.8 *A cathode ray tube.*

- The **electron gun** is a filament heated by passing a current through it, which releases electrons. The electrons pass through a hole in the anode as a narrow beam.
- The **deflecting plates**, P and Q, have an electric field between them. The electron beam is deflected by the electric field and follows a curved path.
- The **fluorescent screen** is coated with a layer of a fluorescent salt. When electrons hit it, it gives off light. If the electron beam changes direction it leaves a glowing trace on the screen.

You can think of the oscilloscope as a CRT with specialized internal circuitry. It measures such physical quantities as potential difference and frequency.

Figure 21.9 shows the overall structure of a cathode ray oscilloscope.

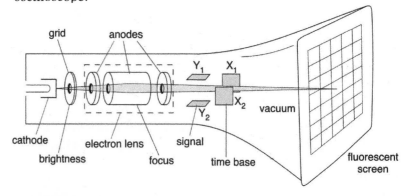

Figure 21.9 *A cathode ray oscilloscope.*

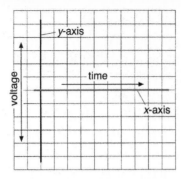

Figure 21.10 *The screen shows potential difference on the y-axis and time on the x-axis.*

Displaying a.c. and d.c. signals on the oscilloscope

(a) 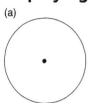 (b) (c) 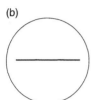 (d) 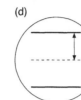 (e) 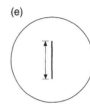 (f)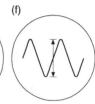

(a)	(b)	(c)	(d)	(e)	(f)
Oscilloscope on. No signal applied. Time base off. A centrally positioned dot is seen on the screen with a.c./d.c. switch off.	Oscilloscope on. No signal applied. Time base on. The spot is swept across the screen giving a line with a.c./d.c. switch off.	Oscilloscope on. A d.c. signal applied. Time base off. The a.c./d.c. switch is set to d.c. The spot moves up or down depending on polarity. The extent of movement of the spot is proportional to the applied voltage.	Oscilloscope on. A d.c. signal applied. Time base on. a.c./d.c. switch set to d.c. Compared with (b) the line moves up or down depending on the polarity of the applied voltage.	Oscilloscope on. An a.c. signal applied. Time base off. a.c./d.c. switch set to a.c.	Oscilloscope on. An a.c. signal applied. Time base on. a.c./d.c. switch set to a.c.

Figure 21.11 *Oscilloscope displays.*

Checklist

After studying Chapter 21 you should be able to:

- recall that electric current is the flow or movement of charge
- recall that charged objects have electric potential energy
- recall that a difference of potential exists between two objects of different charge
- recall that conductors are those materials which, when a potential difference is applied across them, transfer charge
- recall that insulators do not allow the passage of charge through them; however, they can store or hold charge on their surfaces
- distinguish between electron flow and conventional current
- recall that the unit of quantity of charge is the coulomb (C)

- define the coulomb
- recall that $Q(C) = I(A) \times t(s)$
- define potential difference as the energy transferred per unit charge
- define power
- recall that the unit of power is the watt
- recall that power (W) = energy (J)/time (s)
- recall that power $P(W) = I(A) \times V(V)$
- recall that direct current (d.c.), which may be steady or fluctuating, flows in one direction only
- recall that an a.c. signal is characterized by its peak value (of voltage or current), its r.m.s. (root mean square) value and by its frequency
- recall that a.c. is preferable to d.c for the transmission of power but battery charging and electrolysis require d.c.
- recall that the oscilloscope is an important piece of test equipment which is used to display electrical signals.

Questions

1 The charge on an electron is -1.6×10^{-19} C. How many electrons passing a given point in an electric circuit each second will give rise to a current of 4.0 A?

2 Colour-coded resistors are power rated. You are provided with five $1000\,\Omega$ resistors rated at 0.15 W, 0.25 W, 0.5 W and 2.0 W. Determine the maximum voltage that can be applied across $1000\,\Omega$ resistors of **a** 0.25 W and **b** 2 W without exceeding the maximum power rating. $(P = V^2/R)$

3 The average power dissipated in a loudspeaker is 55 W and the speaker is attached to a 120 V supply. Determine the r.m.s. value of the alternating current.

4 The total charge transferred between two points in 10 seconds is 900 coulombs. What is the average current flowing during this time?

22.1 Circuit symbols

It is inconvenient and cumbersome to draw life-like pictures of parts (components) of electric circuits. Instead, we use symbols to represent these components and to show how they are connected.

Table 22.1 Graphic symbols for common electrical components

Component	Symbol	Function/position in circuits/comments
cell		Supplies d.c. Dry cells may be rechargeable or non-rechargeable. Wet cells are rechargeable.
battery of cells		A series combination of n identical cells, each of e.m.f. E. The e.m.f. of the combination is $E_n = nE$.
d.c. supply	+ −	
a.c. supply	~	Usually obtained from the mains. The voltage and frequency are usually specified.
switch		Opens and closes electric and electronic circuits. Switches are placed in the live wire of a.c. circuits.
filament lamp (bulb)	or	Sometimes used as a current detector. Consists of an evacuated glass envelope and a thin filament.
fixed resistor	or	Resistors control current flow in electric circuits. May be connected in series or in parallel.
variable resistor	or	Also called a rheostat. Used to maintain a steady value of current.

Component	Symbol	Function/position in circuits/comments
semiconductor diode		Allows electron flow in one direction only. Used to change a.c. to d.c., i.e. in rectification.
ammeter	Ⓐ	A current-measuring device. It is placed in series with the component, the current through which is being measured. Ideally, ammeters have low resistance so that the potential difference across them is as small as possible.
voltmeter	Ⓥ	A device for measuring potential difference. It is placed in parallel with the component, the potential difference across which is being measured. Ideally, voltmeters have high resistance. The presence of the voltmeter should not affect the value of the current in the circuit.
fuse	or	Protects wires, parts of circuits or whole circuits from damage by excessive currents. Fuses are placed in the live wire of a.c. circuits.
earth		This connection protects users of electrical equipment from electrical shocks caused by stray currents.
junction		Wires are actually connected.
conductors crossing		Wires cross but do not touch.

Table 22.1 Graphic symbols for common electrical components (continued)

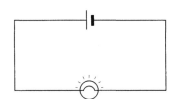

Figure 22.1 *A closed circuit.*

22.2 Types of electric circuit

A **closed circuit** is any continuous path round which an electric current flows. Energy is transferred from the source of potential difference to a device (also called a **transducer**), e.g. a compact disc player or television set, by charges that move through conducting wires or cables.

The lamp in Figure 22.1 lights because because an electric current flows through it. This circuit reveals three features which are common to all electric circuits. All closed electric circuits:

- have a source of potential difference (sometimes called a voltage source or an e.m.f.);
- have a complete conducting path for current flow;
- contain resistance; the lamp provides the resistance in this case.

Note

Open circuits often occur when the conducting path is mechanically broken or is burnt out by excessive currents.

Large amounts of heat are produced by a short circuit. Conducting wires burn out and batteries run down if short circuits last for any appreciable time. Fires may occur.

An **open circuit** is one in which there is a break in the conduction path. The lamp in Figure 22.2 does not light because there is a break in the circuit.

A **short circuit** occurs when a voltage source or potential drop has a closed path of low resistance across its ends. The lamp in Figure 22.3 does not light because almost no current flows through it. Most of the current flows through the short circuit.

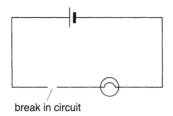

break in circuit

Figure 22.2 *An open circuit.*

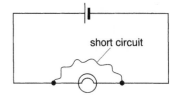

short circuit

Figure 22.3 *A short circuit.*

22.3 Cells and batteries

Electrical energy sources produce potential differences. An electric **cell** is a device which converts the energy of chemical reactions to electrical energy. Each cell consists of two different terminals (usually metals) which are connected to each other, while being in contact with an electrolyte.

Note

A number of cells joined together are called a **battery**.

The dry cell

The common dry cell has a zinc outer case which serves as the negative terminal (Figure 22.4). A central carbon rod is the positive terminal. Ammonium chloride in the form of a jelly is the electrolyte. The dry cell also contains a mixture of powdered carbon and manganese dioxide. The carbon improves conduction, the manganese dioxide reduces polarization (see Note).

Note

Polarization is the process by which hydrogen atoms build up around the positive terminal of a cell or battery. Polarization reduces both the potential difference and the current that the cell can provide.

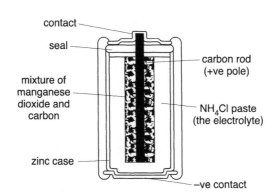

contact

seal

carbon rod (+ve pole)

mixture of manganese dioxide and carbon

NH_4Cl paste (the electrolyte)

zinc case

−ve contact

Figure 22.4 *A carbon–zinc dry cell.*

A size D dry cell supplies a maximum current of 0.15 A (150 mA). Size C, B and A cells supply smaller currents.

The dry cell is an example of a **primary cell**. The reactions in primary cells cannot be reversed. This means that primary cells cannot be recharged.

The lead–acid battery

The lead–acid cell is an example of a **secondary cell**. Secondary cells can be recharged again and again.

The lead–acid cell has a negative plate made of soft (spongy) lead (Figure 22.5). This plate normally looks grey in colour. The positive plate is made of lead dioxide which is red in colour. The electrolyte is sulphuric acid.

Each fully charged lead–acid cell gives about 2.1 volts. Six lead–acid cells are connected in series to produce a 12 volt motor car battery (or accumulator).

Large currents can be drawn from lead–acid cells and batteries. When a battery is in use or is being discharged, both the positive and negative plates are converted to lead sulphate. After a while the battery can no longer give a potential difference because the two plates are covered with the same substance and one of the conditions for a potential difference no longer exists. Also, during discharge the sulphuric acid becomes too dilute to be useful. A lead–acid cell or battery is considered discharged when the relative density of the sulphuric acid drops to about 1.12 (Figure 22.6).

> **Note**
>
> Dry cells do not supply large currents.
>
> Currents of the order of 200 A–400 A are needed to start a car. These currents are needed only for short periods and are provided by a 12 V battery. The battery is recharged by the engine when the car is running.

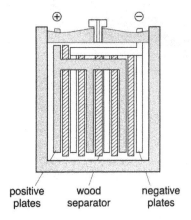

positive plates   wood separator   negative plates

Figure 22.5 *A cross-section of a lead–acid battery or accumulator.*

> **Note**
>
> Electric charges leave batteries with energy, flow through circuits and return to the batteries with less energy.

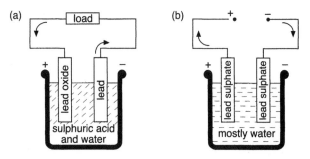

Figure 22.6 *(a) The condition of a battery when fully charged. It is ready to supply a current. (b) This battery is being recharged from an outside d.c. source. The positive of the source is connected to the positive of the battery. Arrows show the direction of electron flow.*

Care of lead–acid cells and batteries

- Avoid heating by excess charging and discharging currents.
- Do not leave the battery in a discharged state for a long time.

- Always use distilled water when topping up cells.
- Do not allow the relative density of the sulphuric acid to fall below 1.15.
- Do not allow the potential difference of any lead–acid cell to fall below 1.8 V.
- Remove corrosion from the terminals, using baking soda.
- Coat terminals with petroleum jelly after they have been cleaned.

Recharging a lead–acid battery

Figure 22.7 shows how a run-down lead–acid battery may be recharged from a d.c. supply.

The following points should be carefully noted:

- the positive end of the d.c. supply is joined to the positive end of the battery;
- low charging currents should be used;
- the variable resistor is used to control the current.

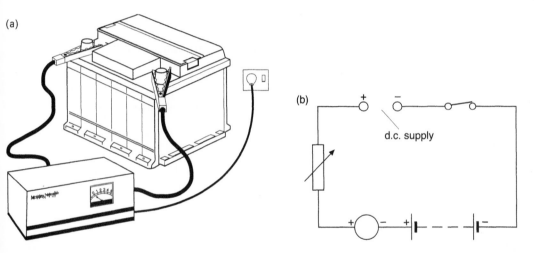

Figure 22.7 *(a) Although connected to the a.c. mains, the output from this battery charger is d.c. (b) A circuit for recharging a battery.*

Primary cells and secondary cells – a comparison

Primary cells are cheaper to produce, easier to install and compare favourably with secondary cells which do the same job, when life and cost are considered.

However, primary cells have higher internal resistance than secondary cells. Because of this, they cannot supply large currents without being damaged. Primary cells cannot supply currents for long periods.

> **Note**
>
> Primary cells are not rechargeable.
>
> Secondary cells are rechargeable.

22.4 Resistance

All the components in an electric circuit provide a certain impediment to current flow. This property of circuit components is known as **resistance**.

The resistance of a conductor increases (see p. 245) if:

- its length increases;
- its diameter decreases.

The SI unit of resistance is the **ohm** (Ω), named after Georg Ohm. The resistance of a device is 1 ohm if the current through it is 1 ampere when the potential difference across its ends is 1 volt.

Special components known as **resistors** conduct electricity but make it difficult for electrons to flow through them. Resistors oppose, control or limit the current in electrical circuits.

(see p. 245)

Note

Resistance either limits the current flowing or produces heat or provides suitable potential drops.

Thinking it through

Resistivity is an inherent property of materials.

- Give an example of another inherent property of materials.
- Silver is the best conductor in the list of Table 21.1 (page 238). Yet copper is widely used to carry currents in electrical circuits. Why?

Nichrome and manganin are widely used in making standard resistors.

- Which elements are present in nichrome and manganin?
- Which of the materials listed in Table 21.1 would best serve as insulation for electrical wires?

Resistors

Resistors are of two types:

- **fixed resistors** have one specific value, for example 5 ohms or 1 million ohms (1 MΩ) (Figure 22.8);
- **variable resistors** either have a range of values or can be adjusted to the specific resistance required in a circuit.

Note

For colour-coded resistors read from left to right. The first two numbers give you the digits, the third number gives the number of zeros.

Black = 0	Brown = 1
Red = 2	Orange = 3
Yellow = 4	Green = 5
Blue = 6	Purple = 7
Grey = 8	White = 9

The fourth band gives the tolerance, which may be 5%, 10% or 20%.

Colour-coded resistors are also power rated.

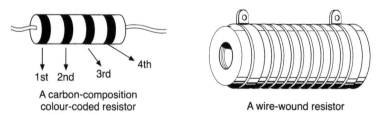

1st 2nd 3rd 4th

A carbon-composition colour-coded resistor

A wire-wound resistor

Figure 22.8 *Two types of fixed resistor.*

The rheostat

A rheostat or variable resistor is a two-terminal device. The resistance between the two terminals may be varied by a sliding contact or other means (Figure 22.9).

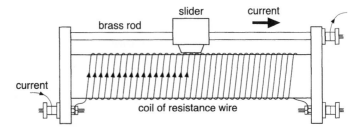

Figure 22.9 *An example of a common laboratory rheostat.*

22.5 Bringing current, potential difference and resistance together

Current, as you already know, is measured by ammeters (Chapter 2, p. 21).

Ammeters should be connected in series with the circuit, or part of the circuit, through which the current is to be measured (Figure 22.10). The positive terminal of the ammeter should be connected, directly or indirectly, to the positive terminal of the voltage source. An ammeter should have low resistance.

Voltmeters are used to measure the potential difference between two points in a circuit (see Chapter 2, p. 22). The voltmeter should be connected in parallel as shown in Figure 22.11. A voltmeter should have high resistance.

All electric circuits have:

- a source of potential difference (V);
- resistance (R);
- current (I).

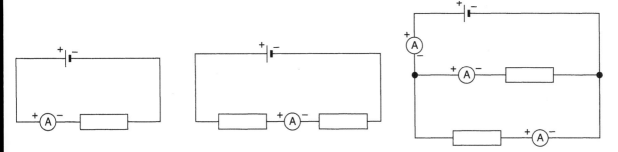

Figure 22.10 *Connecting ammeters in electric circuits.*

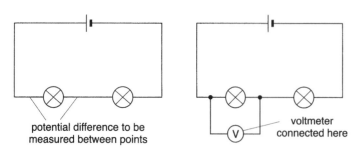

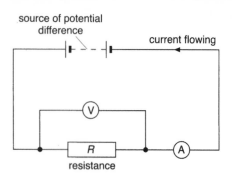

Figure 22.11 *Connecting voltmeters in electric circuits.*

Figure 22.12 *Measuring the resistance of a conductor.*

Note

$\bigotimes$– measures the potential difference across R

$\bigcirc\!\!\!A$– measures the current through R

The resistance of a conductor can be measured by placing it in a circuit, measuring the potential difference across it (using a voltmeter) and the current through it (using an ammeter), Figure 22.12. Then the value of the resistance can be calculated by dividing the voltmeter reading by the ammeter reading.

Potential difference (V) = current $(I) \times$ resistance (R) is the defining equation for resistance. This relationship holds whether the resistance is constant or changing.

Thinking it through

Copy and complete the table:

V/V	I/A	R/Ω
20		5
	7	4
36	3	

22.6 Ohm's law

In the special case where R (resistance) is constant, the relationship:

$$V = I \times R$$

expresses Ohm's law.

Formally, Ohm's law may be stated as follows:

At constant temperature, the potential difference across the ends of a conductor is directly proportional to the current through it.

Devices which obey Ohm's law are called **ohmic devices**. Those devices which do not obey Ohm's law are called **non-ohmic devices**.

Note

Georg Ohm (1787–1854) worked out in 1827 the relationship connecting current, resistance and potential difference. This relationship is called Ohm's law.

1 Set up the circuit below. Device D may be a metal wire, a filament lamp, a diode, a solution of copper sulphate, etc.

2 Use the rheostat to vary the current and potential difference. Record several pairs of *I/V* readings for each device you test.

3 Draw a graph of *I* against *V* for each device.

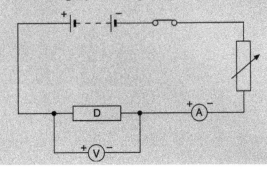

Figures 22.13–22.17 show the current/potential difference graphs for different conductors.

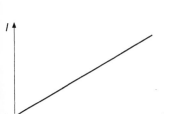

Figure 22.13 *An ohmic device, e.g. a length of resistance wire or a colour-coded or wire-wound resistor. For these types of devices the graph of current versus potential difference is a straight line which passes through the origin.*

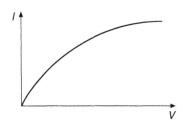

Figure 22.14 *A lamp filament (non-ohmic). This graph is an approximate straight line at low currents. The resistance of the lamp filament increases at higher currents and the graph curves correspondingly.*

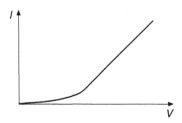

Figure 22.15 *A semiconductor diode (a non-ohmic device).*

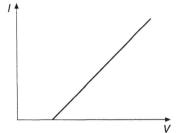

Figure 22.16 *This is the graph obtained if the device D is an electrolytic cell, say of copper sulphate solution with copper electrodes. This system obeys Ohm's law, since the graph is a linear one.*

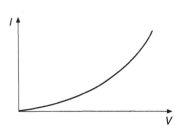

Figure 22.17 *This is the graph for a device whose resistance decreases with increasing temperature.*

Note

A series circuit provides one pathway for circuit current.

22.7 Series circuits

The resistors R_1, R_2 and R_3 in Figure 22.18 are of different values. They are connected in **series**.

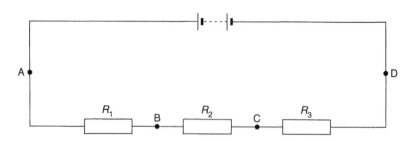

Figure 22.18 *Resistors in series.*

Note

Two or more resistors connected in series offer a greater resistance to current flow than any one of the resistors alone.

For series circuits:

• the same current flows through all components – in this case, through the three resistors, the battery and connecting wires; identical ammeters placed at A, B, C and D give the same reading;
• R_T (the total resistance) $= R_1 + R_2 + R_3$;
• voltmeters placed separately across R_1, R_2 and R_3 give different readings, *but*

$$V_{battery} = V_{AD} = V_{R_1} + V_{R_2} + V_{R_3}$$
$$= IR_1 + IR_2 + IR_3$$

Thinking it through

• Why is it not sensible to connect household appliances in series?
• Two identical lamps were connected to the mains supply, one using a short extension cord, the other using a very long extension cord. Which lamp will light more brightly? Explain how you arrived at your answer.

• if there is a break at any point in the circuit, no current flows;
• the total resistance of a series combination of resistors may be determined experimentally from the potential difference and the current;
• the total power dissipated in a series circuit is equal to the sum of the powers dissipated in the individual resistors.

For the circuit of Figure 22.18,

$$I = \frac{V_{AD}}{R_1 + R_2 + R_3}$$

If the resistors R_1, R_2 and R_3 are replaced by a single resistor of value R_T whose value is $R_1 + R_2 + R_3$, the same current I will be drawn from the battery and will flow through the circuit.

Example

A battery of 20 V is connected to a series arrangement of 5 Ω, 3 Ω and 2 Ω resistors. What is

a the effective resistance in the circuit?

b the current flowing?

c the potential difference across each resistor?

d the total power dissipated in the circuit?

Solution

a $(5 + 3 + 2)\,\Omega = 10\,\Omega$

b $I = \dfrac{V}{R} = \dfrac{20\,\mathrm{V}}{10\,\Omega} = 2\,\mathrm{A}$

c $V = IR$, so $5\,\Omega \times 2\,\mathrm{A} = 10\,\mathrm{V}$, $3\,\Omega \times 2\,\mathrm{A} = 6\,\mathrm{V}$,
$2\,\Omega \times 2\,\mathrm{A} = 4\,\mathrm{V}$

d Power $= V \times I = 20 \times 2 = 40\,\mathrm{W}$

Problem

a What is the e.m.f. of the battery, if the meter in the diagram below gives a full-scale deflection of 10 mA?

b What is the potential difference across each resistor?

c Given that all the resistors are rated at 0.5 W, is any resistor likely to burn out? [*Hint:* find the power dissipated in each resistor.]

Check your answers with your teacher.

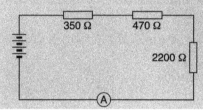

22.8 Parallel circuits

For the parallel arrangement of resistors of Figure 22.19:

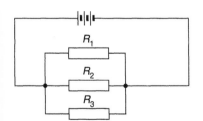

Figure 22.19 *Resistors in parallel.*

- the potential difference across each parallel branch or load resistor is the same and is equal to the potential difference of the battery;
- the reciprocal of the total resistance equals the sum of the reciprocals of the individual resistances:

$$\frac{1}{R_\mathrm{T}} = \frac{1}{R_1} + \frac{1}{R_2} + \frac{1}{R_3}$$

- the total current is the sum of the currents in the individual branches:

$$I_\mathrm{T} = I_1 + I_2 + I_3$$

- the total current divides at each junction leading to parallel branches of a circuit in such a way that the potential difference across each resistor branch is the same;
- the total resistance for a parallel combination of resistors is always less than the value of the smallest resistance;

Note

A parallel circuit is any circuit which provides more than one pathway for the current.

Thinking it through

- Why are household circuits connected in parallel?
- Why is it not wise to connect many appliances which draw large currents in parallel?

- the total power dissipated equals the sum of the powers dissipated in the individual resistors;
- a break in any branch of a parallel circuit does not affect other branches. However, the total current decreases;
- adding an extra branch in parallel with those already present increases the total current taken from the cell or battery.

In Figure 22.20, meter A_3 registers the current taken from the battery. Meter A_4 also registers the current taken from the battery.

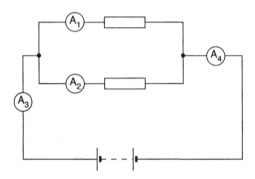

Figure 22.20 *Testing the current law.*

Example

A $12\,\Omega$ resistor and a $6\,\Omega$ resistor are connected in parallel. What is the effective resistance of this combination?

Solution

$$\frac{1}{R_T} = \frac{1}{6} + \frac{1}{12} = \frac{3}{12}$$
$$R_T = \frac{12}{3}\,\Omega = 4\,\Omega$$

reading on A_3 = reading on A_4

= [reading on A_1 + reading on A_2]

This illustrates the current law for parallel arrangements which states that '**the sum of the currents entering a junction is equal to the sum of the currents leaving that junction**'.

In Figure 22.21:

- each appliance is at the voltage of the mains supply;
- each appliance may be switched on or off independently of the others;
- the appliances draw different currents from the mains.

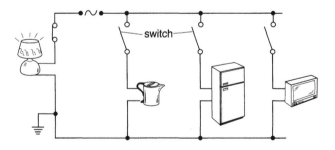

Figure 22.21 *A parallel arrangement in the home.*

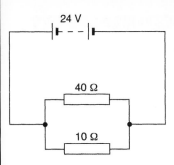

24 V

40 Ω

10 Ω

Example

Determine **a** the total resistance in the circuit on the left, **b** the current drawn from the battery, **c** the potential difference across each of the resistors, **d** the current through the 40 Ω resistor, **e** the current through the 10 Ω resistor.

Solution

a $\quad \dfrac{1}{R_T} = \dfrac{1}{40} + \dfrac{1}{10}$

$\qquad R_T = \dfrac{40 \times 10}{40 + 10} \, \Omega = 8 \, \Omega$

b $\quad V = IR$

$\qquad I = \dfrac{24}{8} = 3 \, \text{A}$

c The potential difference across the resistors = potential difference of the battery (parallel circuit) = 24 V

d Current through the 40 Ω resistor = $\dfrac{24}{40} \, \text{A} = 0.6 \, \text{A}$

e Current through the 10 Ω resistor = $\dfrac{24}{10} = 2.4 \, \text{A}$

How to analyse combined series–parallel circuits

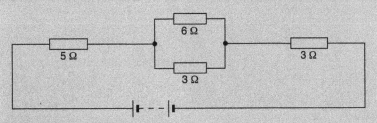

6 Ω

5 Ω

3 Ω

3 Ω

To solve problems on series–parallel circuits, follow these steps.

1 Redraw the circuit so that resistances are easily seen.
2 Reduce the circuit to a simple series (or equivalent) circuit in which parallel segments of the original circuit are replaced by equivalent series resistances.
3 Calculate the total resistance of the circuit.
4 Calculate the total current, etc.

This is the equivalent circuit of the circuit above:

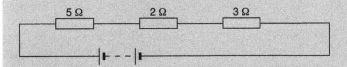

5 Ω   2 Ω   3 Ω

The 2 Ω resistor is the equivalent of the 6 Ω and 3 Ω resistors in parallel.

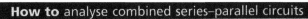

Things to do

Draw equivalent circuits for:

a

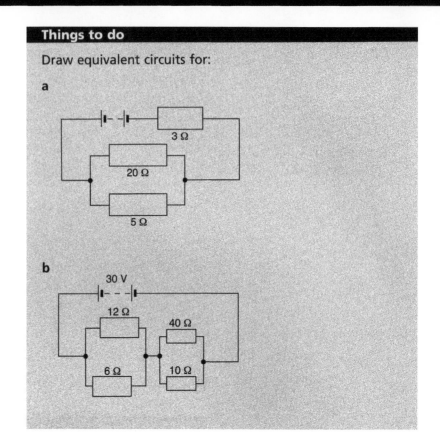

b

Example

You have four ammeters, each of resistance $5\,\Omega$ and with range 0–$1\,\text{mA}$, 0–$10\,\text{mA}$, 0–$50\,\text{mA}$ and 0–$100\,\text{mA}$ respectively. The error in the full-scale reading on each meter is 1%. State, with reason, which meter you would select for use in this circuit.

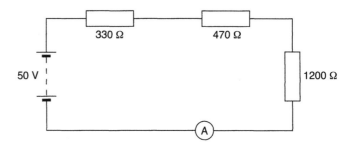

Solution

$$R_\text{T} = 330\,\Omega + 470\,\Omega + 1200\,\Omega = 2000\,\Omega$$

$$I = \frac{V}{R_\text{T}} = \frac{50}{2000} = 0.025\,\text{A} = 25\,\text{mA}$$

The 0–$50\,\text{mA}$ ammeter would be most suitable. (Why?)

Checklist

After studying Chapter 22 you should be able to:

- recall the graphic symbols for circuit elements
- recall that all electrical circuits consist of a source of potential difference, a conducting path for the current and resistance
- recall that large currents flow during short circuits, which causes much heat and may lead to fires
- recall that the main parts of a dry cell are the zinc case (negative terminal), a central carbon rod (positive terminal) and an electrolyte (ammonium chloride)
- recall that secondary cells are rechargeable; primary cells are not rechargeable and are discarded once used
- recall features of primary and secondary cells
- recall that the defining relationship for resistance is

$$\text{resistance} = \frac{\text{potential difference}}{\text{current}}$$

- recall that ammeters have low resistance, ideally; they are connected in series with the device, the current through which is being determined
- recall that voltmeters have high resistance, ideally; they are connected in parallel with the component, the potential difference across which is required
- recall that for metallic conductors and electrolytes at constant temperature ('ohmic' devices) current is directly proportional to potential difference
- recall that for non-ohmic devices, such as a filament lamp or a diode, potential difference is not proportional to current
- recall that circuit components may be connected in series, in parallel and in series–parallel arrangements
- recall that each branch of a given parallel arrangement is at the voltage of the supply
- recall the characteristics of series circuits
- recall the characteristics of parallel circuits
- set up electrical circuits, giving due consideration to the suitability of components and to correct polarity

Questions

1 Comment on the significance of the statement 'resistivity is a property of a given material while resistance is a property of the material and its dimensions'.

2 A number of light bulbs are to be connected to the same mains supply. Will the bulbs be brighter when connected in series or in parallel?

3 Four identical resistors, each of value $R\,\Omega$ are available. How can these all be connected to obtain an effective total resistance of $R\,\Omega$?

4 Predict the effect on the current in a circuit if a voltmeter is mistakenly connected as for an ammeter.

5 In the circuit below the resistance of lamp L_2 is twice that of lamp L_1. Which lamp draws the larger current? Which lamp dissipates greater power?

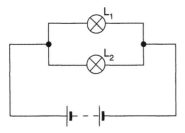

6 A resistor (resistance R) is connected first in parallel and then in series with a $5.0\,\Omega$ resistor. A battery delivers four times more current to the parallel combination than it does to the series combination. Determine possible values for R.

7 For an ohmic device at constant temperature, the readings of an ammeter (current) and of a voltmeter (potential difference) suitably connected to it were:

V/V	I/A
1.65	0.60
2.40	I_x
3.00	1.25
3.75	1.45
4.50	1.80
V_y	2.10
5.25	2.20
8.00	3.30

a What is meant by the term 'an ohmic device'?

b Plot a graph of voltmeter reading, V (y-axis), against ammeter reading, I (x-axis).

c Use the graph to find the values of (i) V_y and (ii) I_x.

d Given that there was no zero error in the ammeter show that there must be a zero error in the voltmeter and find its value.

e Why is it more sensible to set meters to zero rather than correct readings from graphical data?

f Use your data to determine the resistance of the ohmic device.

Electronic circuits

In this chapter we move from electricity to basic electronics. We discuss, however briefly, solid state devices such as diodes and logic gates. There are many of these devices and new ones are being developed all the time. In the last 30 years smaller and smaller electronic components capable of performing increasingly complex tasks at higher speeds have been manufactured. These microelectronic devices have been used widely. They have made certain tasks more reliable, more reproducible and much less expensive.

23.1 Semiconductors revisited

As pointed out in Chapter 21, the conducting properties of pure silicon (Figure 23.1) and pure germanium can be improved appreciably by adding a small amount of either a Group III or a Group V element as impurity. Silicon or germanium to which the impurity has been added is said to be 'doped'.

Silicon doped with boron or aluminium or gallium (Group III elements) is one electron short for each silicon atom replaced (Figure 23.2(a)). Materials doped with Group III elements have **p-type** properties, meaning that they have an electron deficiency. The missing electrons can be thought of as **holes**. An electron moves from a neighbouring atom to fill the hole, thereby creating another hole in the position just vacated by the electron. The process continues down the line – as electrons advance, holes recede, as shown in Figure 23.2(b).

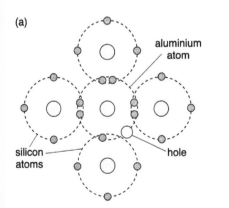

Figure 23.1 *One silicon atom linked to four others. Observe that the central atom uses all its valence electrons in bonding.*

(a)

(b)

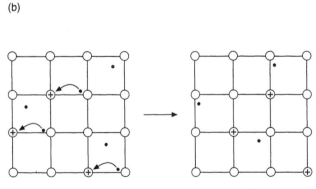

Figure 23.2 *(a) A p-type semiconductor. (b) Note that the holes are advancing, as the electrons move.*

On the other hand, silicon doped with a Group V element, such as phosphorus or arsenic, acquires an excess of electrons. Such materials are know as **n-type** semiconductors. The bonding

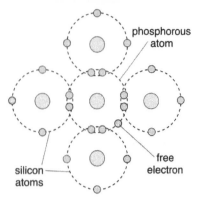

The symbol for a diode.

Figure 23.3 *An n-type semiconductor.*

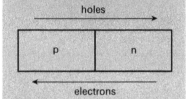
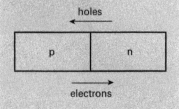
situation in these materials is shown in Figure 23.3. Electrons are the majority carriers in n-type semiconductors (i.e. electrons carry most of the current; holes are the minority carriers).

23.2 Semiconductor diodes

Semiconductor diodes are made by growing n and p-type crystals together. The point at which the two crystals join is called a p–n junction.

Semiconductor diodes function as one-way electrical valves – they allow current to flow in one direction only. If the diode is connected as in Figure 23.4, the lamp lights. The diode is said to be **forward biased** in this mode. Diodes conduct readily when forward biased because electrons and holes move easily across the p–n junction.

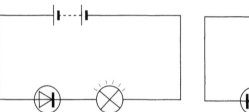

Figure 23.4 *The diode is forward biased. The lamp lights.*

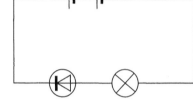

Figure 23.5 *The diode is reverse biased. The lamp does not light.*

If the diode is connected as in Figure 23.5, the lamp does not light. In this mode, the diode is described as being **reverse biased**.

Diodes, when forward biased, have reasonably low resistance and conduct an electric current. The reverse-biased diode has very high resistance and negligible current passes. From this you should appreciate why diodes are said to act as 'one-way electrical valves'.

Overheating and other unsuitable treatment will 'destroy' diode properties.

- An 'open diode' has infinite resistance in both directions. It does not conduct in either direction.
- A diode which has been shorted has low resistance in both directions. It conducts equally well in both directions.

A diode does not conduct unless the forward voltage exceeds a certain minimum value. For silicon, this voltage is 0.6 V.

The diode as a rectifier

Diodes convert a.c. to d.c., a process known as **rectification**. A single diode will give you half-wave rectification (Figure 23.6).

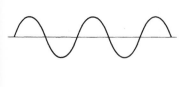

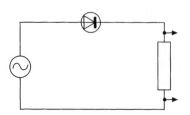

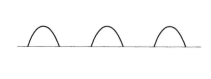

The input is an
a.c. signal

Half-wave rectification of the
a.c. signal by the diode

The output from the diode displayed on
the screen of an oscilloscope.
Note that the negative half cycles of the input
have been clipped off

Figure 23.6 *Half-wave rectification.*

The negative half cycles of the alternating current have been clipped off in half-wave rectification because the diode does not conduct on these half cycles. During these half cycles the diode is effectively reverse biased.

You can achieve full-wave rectification by using a network of four diodes as shown in Figure 23.7. This network is referred to as a 'bridge rectifier'.

A **capacitor** is used to smooth the fluctuating d.c. so that the value of the current is approximately constant.

(a)

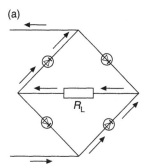

(b)

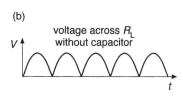

voltage across R_L
without capacitor

(c)

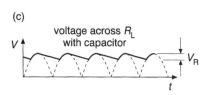

voltage across R_L
with capacitor

Figure 23.7 *(a) The bridge rectifier. (b) Full-wave rectified output, as displayed on the screen of an oscilloscope. (c) A rectified signal that has been 'smoothed'.*

23.3 Logic gates

Much of modern technology uses **digital** electronics. In this area of electronics, the electrical signal is either high or low, ON or OFF. There is no middle ground. The ON or high state is generally represented as 1 or a potential difference of $+5$ V and the OFF or low state represent by 0 or 0 V. The numbers 0 and 1 are the only numbers used. This allows the system to record the signals in simple binary notation, e.g. 1101, thereby making computations and operations of circuits easy.

The fundamental component in digital electronics is the **logic gate**, which is the simplest form of any digital circuit. Using logic gates, many complex digital circuits can be designed to perform a wide variety of functions. There are five basic logic gates: NOT, AND, OR, NAND and NOR.

When you have several combinations of gates it is very easy to confuse the inputs and their respective outputs. The **truth table** is a useful tool which analyses the inputs and outputs of one or more gates by logging them in an orderly manner.

NOT gate or inverter

This gate has one input and one output. The output of a NOT gate is the inverse of the input, hence the name inverter. If the input is high or 1 the output is low or 0. In Figure 23.8(b) input 1 corresponds to the switch being closed, and output 1 means the lamp is on.

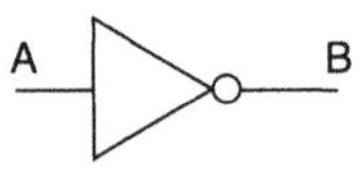

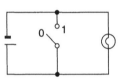

Input	Output
0	1
1	0

(a) NOT gate symbol *(b) Electrical circuit representation* *(c) Truth table*

Figure 23.8

AND gate

This gate has two or more inputs and one output. The output is high or on (lamp on in Figure 23.9(b)) *if and only if* all the inputs are high or on (both switches closed in Figure 23.9(b)). Under any other conditions the output is low or off.

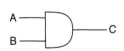

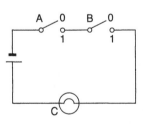

Input A	Input B	Output C
0	0	0
0	1	0
1	0	0
1	1	1

(a) AND gate symbol *(b) Electrical circuit representation* *(c) Truth table*

Figure 23.9

OR gate

This gate has two or more inputs and one output (Figure 23.10). The output is high or on as long as *one or more* of the inputs are high or on. Under any other conditions the output is low or off.

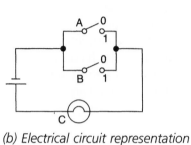

Input A	Input B	Output C
0	0	0
0	1	1
1	0	1
1	1	1

A ⎯⎯⎯▷⎯ C
B ⎯⎯⎯

(a) OR gate symbol *(b) Electrical circuit representation* *(c) Truth table*

Figure 23.10

NAND gate

This gate has two or more inputs and one output (Figure 23.11). The output is low or off *if and only if* all the inputs are high or on. Under any other conditions the output is high or on.

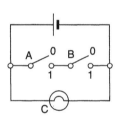

Input A	Input B	Output C
0	0	1
0	1	1
1	0	1
1	1	0

A ⎯⎯⎯
B ⎯⎯⎯ C

(a) NAND gate symbol *(b) Electrical circuit representation* *(c) Truth table*

Figure 23.11

NOR gate

This gate has two or more inputs and one output (Figure 23.12). The output is high or on *if and only if* all the inputs are low or off. Under any other conditions the output is low or off.

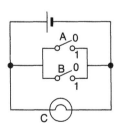

Input A	Input B	Output C
0	0	1
0	1	0
1	0	0
1	1	0

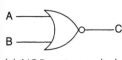

A ⎯⎯⎯
B ⎯⎯⎯ C

(a) NOR gate symbol *(b) Electrical circuit representation* *(c) Truth table*

Figure 23.12

Combining logic gates

Although uses for single logic gates can be found, much more can be accomplished when gates are used in combinations. The use of several gates of one or more types together to perform given functions is known as **combinational logic**.

It is more cost-effective for a manufacturer to produce one or two types of logic gate, usually the NAND and NOR gates, and assemble these to perform the functions of the other types. This is one area in which combinational logic is important. Figures 23.13 and 23.14 show how NOT, AND and OR gates are created from NAND and NOR gates respectively.

Combinational logic at work

Some examples of the uses of combinational logic are in calculators, burglar alarm systems, automatic lighting and heating systems.

Gate	Diagram	Truth table
NOT		
OR		
AND		

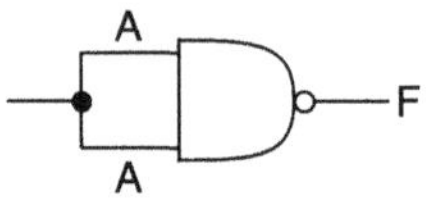

NOT truth table:

A	F
0	1
1	0

OR truth table:

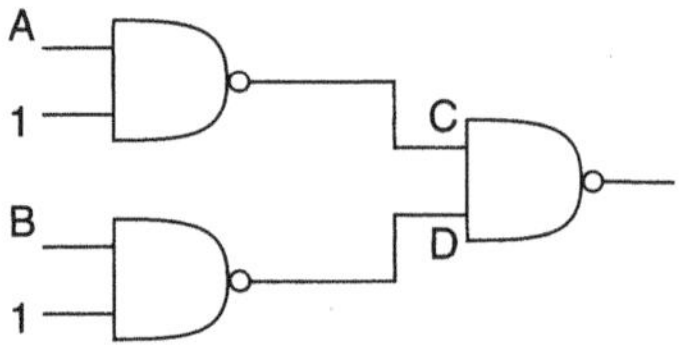

A	B	C	D	F
0	0	1	1	0
0	1	1	0	1
1	0	0	1	1
1	1	0	0	1

AND truth table:

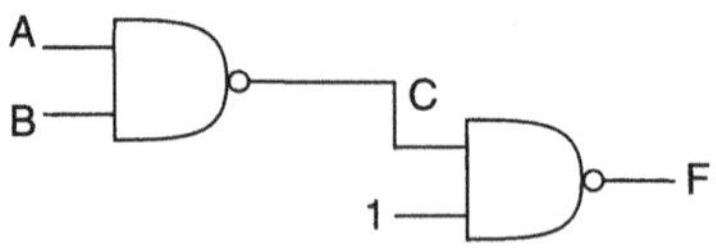

A	B	C	F
0	0	1	0
0	1	1	0
1	0	1	0
1	1	0	1

Figure 23.13 *Using NAND gates to construct NOT, OR and AND gates.*

Gate	Diagram	Truth table
NOT		

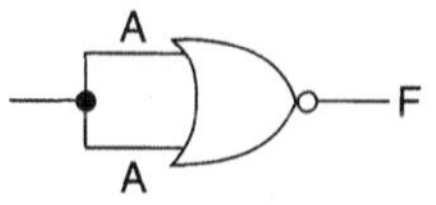

A	F
0	1
1	0

Gate	Diagram	Truth table
OR		

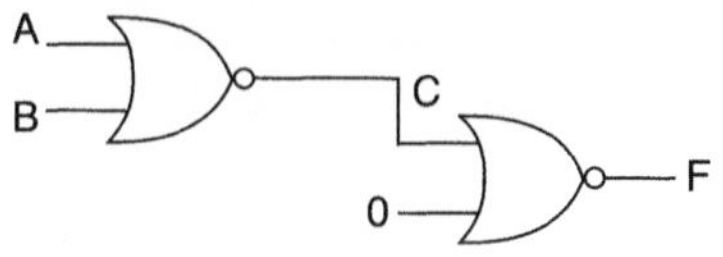

A	B	C	F
0	0	1	0
0	1	0	1
1	0	0	1
1	1	0	1

Gate	Diagram	Truth table
AND		

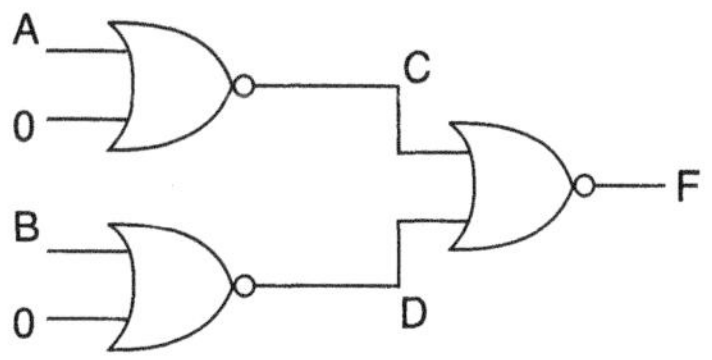

A	B	C	D	F
0	0	1	1	0
0	1	1	0	0
1	0	0	1	0
1	1	0	0	1

Figure 23.14 *Using NOR gates to construct NOT, OR and AND gates.*

Note

An LED is a light-emitting diode.

In the combinational circuit of Figure 23.15, two NAND gates, a capacitor, two resistors and an LED are used to produce a 'flasher'. The LED flashes about twice a second if the input at A is HIGH(1) and glows continually if the input is LOW(0).

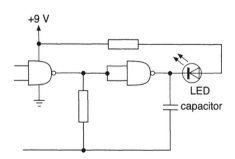

Figure 23.15 *A flasher.*

Some other examples of the uses of combinational logic are given below (Figures 23.16 and 23.17). It is very useful to note that when using logic gates, there are 2^n possible input combinations, where n is the number of initial inputs.

A	B	C	D	E	F	Output G
0	0	0	0	0	0	1
0	0	0	1	0	1	1
0	0	1	0	0	1	1
0	0	1	1	0	1	1
0	1	0	0	0	0	1
0	1	0	1	0	1	1
0	1	1	0	0	1	1
0	1	1	1	0	1	1
1	0	0	0	0	0	1
1	0	0	1	0	1	1
1	0	1	0	0	1	1
1	0	1	1	0	1	1
1	1	0	0	1	0	1
1	1	0	1	1	1	0
1	1	1	0	1	1	0
1	1	1	1	1	1	0

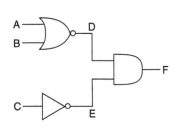

Figure 23.16 $n = 4$ so that the number of possible combinations $= 2^4 = 16$.

A	B	C	D	E	Output F
0	0	0	1	1	1
0	0	1	1	0	0
0	1	0	0	1	0
0	1	1	0	0	0
1	0	0	0	1	0
1	0	1	0	0	0
1	1	0	0	1	0
1	1	1	0	0	

Figure 23.17 $n = 3$ so that the number of possible combinations $= 2^3 = 8$.

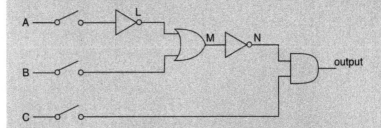

In this diagram an open switch corresponds to a LOW(0) input and a closed switch corresponds to a HIGH(1) input.

- What combinations of A, B and C lead to a logic 1 state for the final output?
- Suggest a possible use for this combination of logic gates.

Things to do

1 Using two AND gates, design a system which turns on the motor (state 1) of a washing machine when the following are in logic state 1: the on/off switch, the door switch and the water level.
2 How can the same task be achieved using NAND gates?
3 Using combinational logic, design a burglar alarm system which is set automatically when both the front and back doors of your home are closed.
4 Using combinational logic design, using NOR gates only, a system to enable a housewife to safely operate a microwave oven.

Checklist

After studying Chapter 23 you should be able to:

- recall that semiconductors have resistances which are intermediate between those of conductors and insulators
- recall that silicon and germanium are typical semiconductor materials
- recall that the conducting properties of silicon and germanium are improved considerably by doping with Group III or Group V elements
- recall that doping with a Group III element produces a p-type material and doping with a Group V element produces an n-type material

- recall that semiconductor diodes are made by joining p-type and n-type materials; the point of contact of the p and n materials is called a p–n junction
- recall that diodes act as one-way electrical valves: they conduct well when forward biased, but poorly when reverse biased
- recall that defective diodes either conduct current in both directions or do not conduct at all
- discuss how semiconductor diodes are used to rectify a.c. to d.c.
- recall the symbols for NOT, AND, OR, NAND and NOR logic gates
- recall the output states and truth tables of NOT, AND, OR, NAND and NOR gates for different input states
- use truth tables to work out the output from combinations of logic gates

Questions

1 Distinguish between n-type and p-type semiconductors, citing one example of each.

2 Discuss, with appropriate diagrams, rectification.

3 Define the following:
  a truth table
  b logic gate
  c combinational logic.

4 Using truth tables, determine the final output of each of the combinations of gates below.

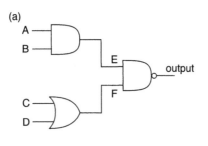

(a)

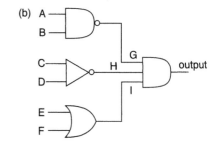

(b)

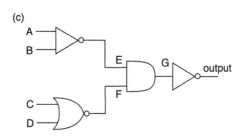

(c)

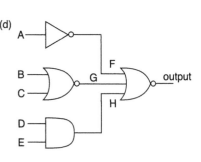

(d)

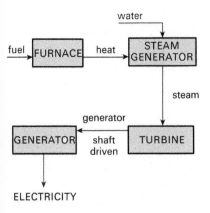

Figure 24.1 *Process flow at the power plant.*

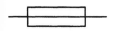

The symbol for a fuse.

Most domestic and industrial electricity comes from the a.c. mains. From the power plant (Figure 24.1), electricity goes to substations in cities, towns and villages. High tension (high voltage) cables then take the electricity to streets, factories, schools, homes and wherever else it is needed (see Figure 26.15, p. 309). Smaller cables inside buildings take the electricity to the outlets.

24.1 Live, neutral and earth

The cables which connect electrical appliances to the mains supply have three insulated wires. Why are they insulated? These insulated wires, which are colour-coded, are the live, the neutral and the earth.

- The live wire has brown insulation.
- The neutral wire has blue insulation.
- The earth wire has yellow-green insulation.

The live and neutral wires carry the current, with the neutral one providing the return path. This ensures that the circuit is complete. The live wire is alternately positive and negative with respect to the neutral wire. These changes of polarity (i.e. from positive to negative and back again) occur 50 or 60 times each second, depending on the frequency of the supply.

Switches

Switches are an integral part of electric circuits. They open electric circuits by interrupting the path of current flow and close circuits by completing the path of current flow. Switches are connected in the live part of the circuit (Figure 24.2).

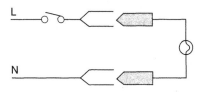

Figure 24.2 *Connecting a switch.*

Fuses and MCBs

Fuses use the heating effect of an electric current to break or open a circuit when an excessive current flows. Fuses reduce the danger of overheating by limiting the size of the currents which can flow through wires and whole circuits.

(a) Fuse of the type used in older buildings

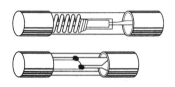

(b) Standard fuses

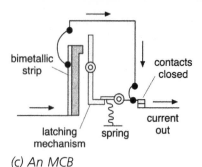

bimetallic strip

contacts closed

latching mechanism

spring

current out

(c) An MCB

Figure 24.3

Fuses may be made of 'tinned' copper wire. They are current rated, i.e. rated according to the maximum safe current they can carry. The fuse should 'blow' when a fault develops so that this safe current is not exceeded.

MCBs (miniature circuit breakers) are resettable switches that protect equipment from excessive currents by opening or tripping when a specified safe current is exceeded. MCBs perform the same function as fuses.

Example

Three appliances are connected in parallel to a 240 V supply. The toaster has a power rating of 1500 W, the iron a power rating of 900 W, and the microwave oven 1800 W. Calculate the current used by each device and the fuse required.

Solution

The current used by each device may be calculated using $I = \dfrac{P}{V}$:

$$I_{\text{toaster}} = \frac{1500}{240} \qquad I_{\text{iron}} = \frac{900}{240} \qquad I_{\text{microwave}} = \frac{1800}{240}$$
$$= 6.25\,\text{A} \qquad\qquad = 3.75\,\text{A} \qquad\qquad = 7.5\,\text{A}$$

If the circuit were fitted with a 15 A fuse or circuit breaker, the fuse or breaker would blow or open when all appliances were used at the same time. A 20 A fuse or breaker is suitable for such a circuit to prevent damage to the appliances.

The earth

One of the connections on a power socket is called earth. This connection is often joined to a metal pipe or the the earth connection on the supply cable. The earth terminal on an electrical appliance is connected to its outer casing.

The earth wire provides protection for the user. It offers a low-resistance pathway to earth (the ground) for stray currents. All appliances should be either earthed or 'double insulated'.

Double insulated appliances are so made that current cannot leak to the outside to cause shock.

If an appliance is neither earthed nor double insulated and there is 'trouble' in the circuit, e.g. a short circuit develops between the live wire and the casing, then your body becomes part of the circuit to earth and a current may flow through your heart. This is what we call an electric shock.

24.2 Oh, how shocking!

You can be killed (electrocuted) by electricity from the mains or by contact with heavy equipment at as low as 40 volts. The extent of shock depends not so much on the voltage but more on the current which passes through the body. Currents as low as 50 mA have caused death.

From Ohm's law, you know that the current I is related to the potential difference V and resistance R by the equation:

$$I = \frac{V}{R}$$

This means that the current flowing depends on both the voltage and the body resistance. The body resistance, however, is not constant; it varies considerably with:

- the area of contact with conductors carrying the current;
- the condition of the skin. For example, the resistance of dry skin may be as high as $0.5\ \text{M}\Omega$ whereas that of wet skin may be of the order of $500\ \Omega$.

The extent of shock depends also on the path the current takes through the body. You can easily be electrocuted if the heart is a part of the circuit.

24.3 Using the correct voltage

Electrical equipment should be operated at the voltage specified by the manufacturers. A 120 V lamp which is plugged into a 240 V mains supply burns four times as brightly but will burn out (blow) in a short time. On the other hand, a 240 V lamp which is plugged into a 120 V supply burns only one quarter as brightly. Clearly, it will not provide the required brightness.

For lamps, small differences from the required voltage lead to variations in brightness and in the life of the lamp. The same argument may be extended to other items of electrical equipment. Additionally, some equipment, such as record turntables, will not function properly if the frequency of the supply is not matched to the frequency at which the device was designed to work.

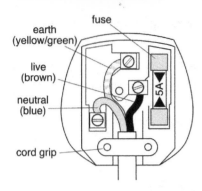

earth
(yellow/green)

fuse

live
(brown)

neutral
(blue)

cord grip

5A

Figure 24.4 *A correctly wired 3-pin plug.*

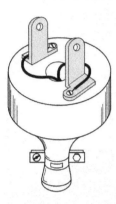

Figure 24.5 *A wired 2-pin plug.*

24.4 Wiring that plug

Some do's and don'ts

- Don't use three-wire flex on two-pin plugs.
- Don't use the earth terminal if connecting two-wire flex to a three-pin plug.
- If the appliance has a metal case and is not double insulated, use three-wire flex and a three-pin plug.

24.5 The cost of electricity

The higher the power rating of an appliance, the greater the cost of running it per unit time. Provided that both appliances are on for the same time, the cost of running the radiant heater is 10 times that of running the fan.

Energy is usually measured in joules (J). However, the joule is too small to be a convenient unit for measuring commercial electricity. The electricity used for both domestic and industrial purposes is measured in **kilowatt hours** (**kW h**).

$$\frac{energy\ (E)}{(kilowatt\ hours)} = \frac{power\ (P)}{(kilowatts)} \times \frac{time\ (t)}{(hours)}$$

Example

A 750 W electrical appliance is used continuously for 12 hours. How many units of electrical energy does it consume during that time?

Solution

Step 1. Convert 750 W to kilowatts = 0.75 kW

Step 2. Multiply kilowatts × hours = 0.75 × 12 = 9.0 kW h

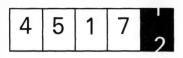

Figure 24.6 *The readout on a digital meter.*

Reading your electricity meter

Recently built homes have digital-type electricity meters (Figure 24.6), but many homes still have the analogue meters (Figure 24.7).

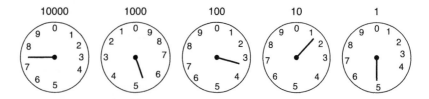

Figure 24.7 *Clock dial (analogue) type meters look like this.*

To read the meter shown in Figure 24.7:

- Start with the dial on the extreme left. Here, the pointer lies between 7 and 8.   Record   70 000
- The second dial from left reads between 5 and 6.   Record   5000
- The third dial reads between 3 and 4.   Record   300
- The fourth dial reads between 1 and 2.   Record   10
- The fifth dial is on 5.   Record   5

The overall reading = 75 315

This meter reads 75 315 units, i.e. 75 315 kW h.

> **Things to do**
>
> Look at the electricity meter at your home or school.
>
> **1** Record the present reading.
> **2** Read the meter every two weeks for three months.
> **3** Determine the average number of units of electricity per week used over the three month period.

24.6 Household electricity

Electricity Boards (light and power companies) distribute power to separate households using a pair of thick, insulated power cables. Each building is connected in paralled with these cables, one of which is live, the other the neutral. Additionally, each user needs to operate lighting and power circuits independently. Parallel circuits, therefore, are used within a household.

For a 120 V supply, electricity from the live wire passes through a series connection of the mains fuse or circuit breaker, the service meter and main switch to the distribution box. The neutral wire is grounded, i.e. it is at zero potential. The thickness of the supply cable and the current rating of the fuse or circuit breaker are chosen to 'handle' the maximum safe current expected to flow through the circuit(s).

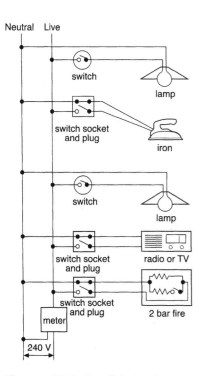

Figure 24.8 *Parallel circuits.*

Thinking it through

Justify the need for parallel connections in household circuits. Why are series connections not suitable?

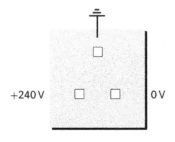

Figure 24.9 *A 3-pin socket.*

For heavy duty appliances such as water heaters, electric ranges, washing machines, spin dryers and power tools, which operate off 240 V, a third wire (line) at earth potential is used. With this arrangement, the live wire is at +240 V, the neutral wire at 0 V, the third wire is connected to the ground and the metal casing of the appliance (Figure 24.9). The available potential difference is 240 V. The function of the earth is to allow the current to escape if the casing becomes live.

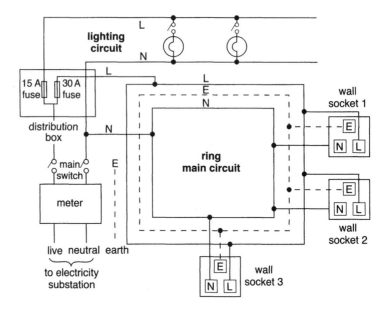

Figure 24.10 *Domestic ring main and lighting circuit.*

The power circuits in modern households are usually on **ring mains**. A ring main provides two separate pathways for the live and neutral currents (Figure 24.10).

With ring circuits it is possible to use twice the current a single line can carry. It also enables you to use one socket without all the others being switched on.

Thinking it through

A device operates off 240 V. A similar device of the same power rating working off a 120 V supply requires twice the current and thicker cables.

- Explain, fully, why this is so.
- Comment on the current rating of the fuse or circuit breaker needed for a 240 V as compared with a 120 V supply, assuming that devices of similar power rating are run off the two supplies.

24.7 Some energy conversions around the home

Electric cookers

- The elements are made from nichrome – an alloy of nickel and chromium. The elements glow red-hot when the cooker is at its maximum setting.
- Power ratings of these cookers range from 3 kW to 15 kW.
- They are protected by a 60 A fuse and run off 230 (or 220) volts.

Electric iron

- Power rating 750 W

Problems

1 Your electric cooker is rated at 3.5 kW. What current does it draw from the 230 V domestic mains, when it is at maximum setting? How many units of electrical energy will it use up in four hours of continuous use?
2 If an iron is used continuously for six hours, calculate the number of units of electrical energy it uses up. In fact, the energy the iron uses in this time is less than that which you calculated above. Suggest a reason for the difference.

Filament lamp

- A metal filament lamp contains a fine coil of tungsten (melting point 3650 K) which is coiled a second time to make it more efficient.
- The glass envelope contains argon (or nitrogen). These gases are unreactive and prevent oxidation of the metal.
- Filament lamps are not very good at converting electrical energy into light. Most of the energy is wasted as heat.

Refrigerator (see Chapter 13)

- Electrical energy drives the compressor pump which moves the refrigerant through the refrigerator.
- The energy conversions which take place inside the device can be summarized as follows:

EVAPORATION	COMPRESSION
(latent heat taken in from surroundings inside the fridge)	(latent heat given out into the air outside the fridge as vapour condenses)

liquid ⟶ vapour ⟶ liquid

24.8 Wasting electrical energy

What follows is a list of the ways in which we waste energy.

1. Leaving security lights on all day

If two 250 W security lights are left on for 10 hours, the energy wasted can be calculated as follows:

Power rating (combined) $= 2 \times 250\,W = 0.5\,kW$

Energy wasted $=$ power (kW) $\times$ time (hours) $= 0.5 \times 10 = 5\,kWh$

Problems

1 Identify the energy transformations that take place in each of a to e.
  a Electric kettle, 1.5 kW, 120 V
  b Washing machine, 1 kW, 230 V
  c Toaster, 1.2 kW, 12 V
  d Food mixer, 300 W, 120 V
  e Vacuum cleaner, 250 W, 120 V
2 How many units of electricity will each use in 5 hours?

Note

Plug-in timers may be used to turn lights and other appliances on and off automatically.

Note

$1\,\text{kW} = 1000\,\text{J}\,\text{s}^{-1}$

Crude oil has an energy equivalent of $4.3 \times 10^7\,\text{J}\,\text{kg}^{-1}$.

$$1\,\text{kW}\,\text{h} = 1000 \times 3600\,\text{J} = 3.6 \times 10^6\,\text{J}$$

$$5\,\text{kW}\,\text{h} = 5 \times 3.6 \times 10^6\,\text{J} = 1.8 \times 10^7\,\text{J}$$

$$\therefore \quad \text{kg equivalent of oil wasted} = \frac{1.8 \times 10^7}{4.3 \times 10^7}$$

$$= 0.42\,\text{kg}$$

This may not seem to be a lot of oil, but think of the many households throughout your country that waste energy in this way.

2. Overlighting homes and business places

- High wattage bulbs are used where lower wattage bulbs would serve.
- Incandescent bulbs are used where lower wattage fluorescent fixtures would serve.

3. Using inefficient lighting systems

- Lampshades and bulbs covered with dust and grit reduce the amount of light received.
- Many lampshades are not sufficiently translucent. (The more translucent the shade the more light is let through.)
- Many interior walls and ceilings are dark coloured. Light colours reflect more light than dark colours.

4. Keeping hot water temperatures unnecessarily high

Many water heaters use more energy than an air-conditioner, a refrigerator, a stereo and a TV combined.

5. Keeping the hot water running unnecessarily

6. Overworking the refrigerator

- Opening the refrigerator more frequently than necessary allows warm air to get in. The appliance has to work harder under these conditions. Energy is wasted.
- Putting hot things in a refrigerator makes it work harder.
- Dirt on condenser coils at the back of the refrigerator make it difficult for the refrigerator to exchange heat with its surroundings.

7. Running the washing machine/spin dryer at less than maximum efficiency by:

- using less than full loads;
- using hot water when cold water would do the job;
- keeping the dryer on longer than necessary.

Note

Wherever possible use one large bulb rather than several small ones.

Never use bulbs of higher wattage than the fixtures were designed to take.

Note

Keep the refrigerator well-stocked, but allow air to circulate freely. The freezer, on the other hand, should be packed full.

8. Keeping rooms/buildings too cold

- Research has shown that most people prefer to work at temperatures in the range 24–26°C. Yet many rooms are cooled to 20°C and lower. It takes 60% more energy to keep the room at 22°C than to keep it at 26°C.
- Running an air-conditioning unit when it is not needed wastes energy. On the other hand it is also energy wasting to turn it on and off for short runs.

24.9 Summary of safety rules for electricity

It is very important to observe safety rules whenever electricity is involved.

Outside

- Do not fly kites or model airplanes near transmission lines.
- If your kite gets tangled with power lines, do not hold on to the string.
- Do not climb utility poles to retrieve kites.
- Do not use electric tools in the rain, or in damp areas, unless these tools are grounded (earthed) through the supply cord.
- Do not place radio or TV antennae close to power lines.

Inside

- Turn off the main switch when replacing a fuse.
- Do not use extension cords to connect high wattage appliances to power sources. Connect these directly to the wall socket.
- Do not place a cord under a rug or where people can trip over it.
- Keep extension cords away from water and damp areas.
- Do not wrap cords around metal objects.
- Do not place cords near stoves or heaters.
- When disconnecting an appliance, pull it out holding the plug, not by the cord.
- Do not use wet hands to unplug an appliance.
- Do not stand in water when unplugging an appliance.
- Never touch anything electrical when taking a bath.
- Do not use water to put out an electrical fire.
- Should the electricity service be disrupted, disconnect or shut off any appliance which would go on automatically when the power is restored.
- Replace damaged flex and plugs immediately.
- Electrical energy is always looking for low-resistance paths. Be sure *you* do not become the path. View all voltages above 50 V with suspicion. Always proceed with caution.

Note

In California (1977) a kite string wrapped two high voltages wires, bringing them together. The result – loss of 700 acres of timber and 200 homes.

Note

When you are wet, e.g. when you perspire, your body resistance gets smaller and you are more easily shocked. Explain why this is so. (Hint: use Ohm's law)

Allow heat from electrical equipment to escape. This is especially applicable to TVs, amplifiers, refrigerators.

If a number of appliances are used at the same time, they may overload the circuits leading to a serious fire.

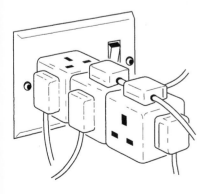

Figure 24.11 *An overloaded power point is a recipe for fires!*

Checklist

After studying Chapter 24 you should be able to:

- describe the general steps in the commercial production of electricity
- recall that the international code for wires and cables is live – brown, neutral – blue, earth – yellow-green
- recall that the fuse and circuit breaker protect circuit components from damage by excessive current
- recall that the ordinary fuse is made of a metal alloy of low melting point which is designed to melt if a specified current is exceeded
- recall that MCBs, miniature circuit breakers, perform the same function as fuses; MCBs use electromagnets or bimetallic strips
- recall that switches and fuses are connected in the live parts of electrical circuits
- recall that fuses are current rated; the correct fuse should be connected to each electrical device
- wire plugs correctly
- recall that, in house-wiring systems, every circuit is connected in parallel with the supply
- calculate the cost of domestic electricity measured in kilowatt hours (kW h)
- recall that many of the electrical appliances in the home and industry are energy guzzlers
- discuss steps which should be taken to save energy
- recall that electricity, though useful, can be dangerous. We should be safety-conscious around electricity
- explain the bad effects of connecting electrical appliances to incorrect or fluctuating voltages

Questions

1 An electric device has a power rating of 2.2 kW. It is connected to a 120 V supply. What is a suitable fuse for the device?

2

17/07/99

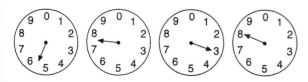

17/05/99

a Write down the meter reading of (i) 17/07/99 (ii) 17/05/99.
b How many units of electricity were used over the two month period?
c Convert the value obtained in **b** to joules.
d Calculate the cost of the electricity used in this period at the flat rate of $0.15 per unit.

3 How are fuses useful in circuits?

4 An electric heater (power rating 1250 W), an electric iron (power rating 1000 W and a microwave oven (power rating 900 W) are connected to a common 120 V supply.

a How much current does each appliance draw from the supply?

b Is a 20 A fuse adequate for this situation?

$(P = P_1 + P_2 + P_3 \text{ and } P = V^2/R)$

5 Mr Forde left the following lights on for 12 hours:

2 security lights rated at 60 W each
1 kitchen light rated at 100 W
1 bathroom light rated at 75 W

a What is the combined power rating?
b Calculate the wasted energy in kW h.
c Determine the kg equivalent of oil, given that crude oil has an energy equivalent of 4.3×10^7 J kg^{-1} and that $1 \text{ kW h} = 3.6 \times 10^6$ J.

6 List five ways in which energy is wasted in the home, and the methods by which this wastage could be reduced.

Magnetism and the magnetic effects of electric currents

We have been fascinated by the 'mysterious' behaviour of magnets for centuries. Magnetism is a force that acts at a distance and makes a magnet:

- attract and repel other magnets;
- attract iron, nickel, cobalt and a few other substances.

25.1 Permanent magnets

Magnetic poles

Iron filings cling mostly to the ends of a bar magnet (Figure 25.1). This indicates that the magnetic effect is greatest at the ends or **poles** of the magnet. Each magnet has two poles – a north pole and a south pole. A freely suspended magnet aligns itself in a north–south direction. The end of the magnet which points in the northerly direction is its north pole (or its north-seeking pole); the other end is the south pole.

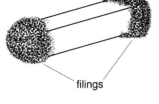

filings

Figure 25.1 *Iron filings stick to the ends of a bar magnet.*

Field lines

A **magnetic field** is the region or space around a magnet where the effects of the magnet can be experienced.

How to map a magnetic field

A Using iron filings
1 Lay a sheet of paper over a bar magnet.
2 Sprinkle iron filings on the paper.

The filings will lie along the lines of force in the magnetic field.

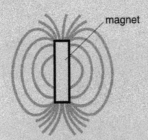

magnet

B Using a plotting compass
1 Place a strong bar magnet on a sheet of paper.
2 Place a plotting compass near one end of the magnet.
3 Use a pencilled dot to mark the position of the end of the compass needle.

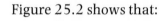
Figure 25.2 shows that:

- magnetic field lines describe a curved loop from the north pole to the south pole of the bar magnet;
- the concentration of field lines is greatest near the poles.

In reality, field lines indicate the path a 'unit north pole' would take if it moved in a magnetic field. Figure 25.3 shows the field between two strong magnets.

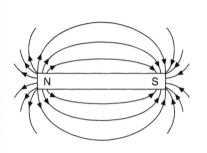

Figure 25.2 *Field lines move out of north poles into south poles.*

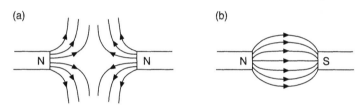

Figure 25.3 *The field between two bar magnets.*

Magnetic flux

Field lines represent the **magnetic flux**, which measures the 'flow' of magnetism around a magnet. A strong magnet has a greater magnetic flux than a weak one. However, the best measure of the strength of a magnet is the **magnetic flux density**. Magnetic flux density is the magnetic flux per unit area.

Uniform magnetic fields

In a uniform magnetic field, the field lines are parallel and evenly spaced (Figure 25.4).

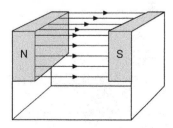

Figure 25.4 *A uniform magnetic field.*

Note

Like magnetic poles repel.

Note

Unlike magnetic poles attract.

25.2 Magnetic forces

A force exists between magnets. This force acts at a distance (it is a non-contact force).

How to investigate the forces between magnets

1 Place a fairly strong magnet, A, on a watch glass.
2 Bring the north pole of a hand-held magnet, B, near the north pole of A.
3 Observe what happens.
4 Repeat, bringing the north pole of magnet B near the south pole of magnet A.
5 Repeat, bringing the south pole of B near the north and south poles of magnet A.

In the experiment above you should have seen that:

- the north pole of magnet B repels the north pole of magnet A;
- the south pole of magnet B repels the south pole of magnet A;
- the north pole of magnet B is attracted to the south pole of magnet A;
- the south pole of magnet B is attracted to the north pole of magnet A.

Note

The force of attraction between magnets depends:

- directly on the strengths of the magnets;
- inversely on the square of their distance apart.

The closer two magnets are the greater the force between them.

How to study the effect of distance on the force between magnets

1 Place a 'strong' magnet, A, north pole facing up, on a top pan balance.
2 Record the balance reading.
3 Hold a second magnet, B, north pole down, 10 cm directly above magnet A.
4 Observe that the balance reading increases, indicating a repulsive force between the magnets.
5 Decrease the distance between the magnets. You should see that, as the magnets get closer, the repulsive force between them increases.

25.3 A theory of magnetism

Magnetic materials can be thought of as consisting of tiny magnets. These tiny magnets are called **domains**. Each domain has its own north and south pole.

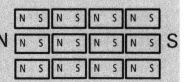
In the unmagnetized state, the domains of a given material are randomly oriented, as shown in Figure 25.5. In this random orientation, the domains cancel each other's effects.

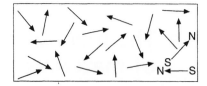

Figure 25.5 *Domains in an unmagnetized bar.*

When the material is magnetized the domains align themselves in an orderly fashion. This produces a north pole at one end and a south pole at the other as shown in Figure 25.6.

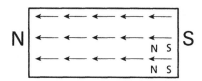

Figure 25.6 *Domains in a magnetized bar.*

Although each domain is in itself a weak magnet, there are millions of them. When all the domains line up in an orderly fashion, they give the bar or material a reasonable magnetic strength.

25.4 Magnetic induction

The process by which a magnet can induce a magnetic field in an unmagnetized iron bar or other material, without the two touching, is know as **magnetization by induction**. The magnetic field of the magnet causes the internal molecular magnets or domains of the iron bar to align themselves in the same direction (Figure 25.7).

Recall that magnetic field lines move out of north poles and into south poles. In Figure 25.7, the north pole of the magnet *induces* a south pole in the end of the bar which is

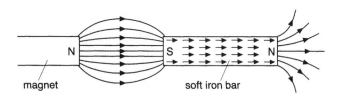

Figure 25.7 *Magnetic induction.*

closer to it by attracting the south poles of the domains. The north pole of the magnet then attracts the induced south pole of the bar.

25.5 The magnetic properties of materials

Permeability is a measure of the ease with which magnetic flux passes through a substance. Magnetic flux passes more easily through iron than through steel, say. Iron is sometimes described as being magnetically *soft*. Iron has a high permeability. Steel, on the other hand, is described as being magnetically *hard*. Steel has a low permeability. What this means is that it is easier to magnetize iron than it is to magnetize steel.

However, steel retains magnetism better than iron. Steel is said to have a higher **magnetic retentivity** than iron.

Table 25.1	Some magnetic materials and their uses	
Material	**Composition**	**Uses**
Alnico	An alloy of Al, Ni, Co, Cu and Fe. This material has a high retentivity	In making permanent magnets, e.g. bar magnets for general laboratory use
Mumetal	An alloy of Ni, Fe and Cu It has a high permeability	Excellent for shielding electrical equipment
Magnadur	Ceramic type magnets made by subjecting ferrite to intense heat and pressure	Makes the strongest of permanent magnets. Used, for example, in electric motors

25.6 The magnetic effect of an electric current

When an electric current passes through a wire, a magnetic field is produced around the wire. If the wire is wound into a coil the resulting magnetic field is similar to that of a bar magnet with north and south poles.

1 Pass a straight wire through a hole in a sheet of card and connect it in a circuit as shown.

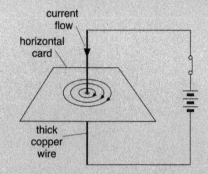

2 Pass a current through the wire.
3 Sprinkle iron filings on the card round the wire.
4 Observe that the filings on the card assume a definite pattern.

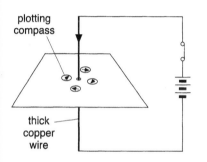

Figure 25.8 *Mapping the field lines round a conductor.*

Note

Hans Oersted (1777–1851) was the first to show that an electric current has a magnetic effect.

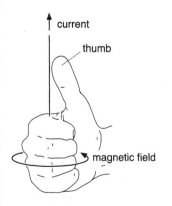

Figure 25.9 *The right-hand grip rule.*

You should see that:

- the field lines make a pattern of concentric circles, symmetrical around the conductor at the centre;
- the pattern is densest nearest the conductor (wire);
- the magnetic field is strongest near the conductor, but gets progressively weaker with distance;
- the magnetic field is in a plane perpendicular to the direction of current through the conductor.

The direction (clockwise or anticlockwise) of the field lines can be mapped using a plotting compass (Figure 25.8). This was first done by Oersted in 1819.

Predicting the direction of the field around a current-carrying conductor

Maxwell's screw rule:

Imagine that a right-handed screw is turned so that it moves forward in the direction of conventional current. Then, its direction of rotation indicates the direction of the magnetic field due to the current.

Alternatively, you can use the '**right-hand grip rule**' (Figure 25.9) to predict the direction of the magnetic field:

If a conductor carrying a current is gripped with the right hand, with the thumb pointing along the conductor in the direction of conventional current, the curl of the fingers around the conductor indicates the direction of the magnetic lines of force.

(a)

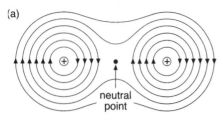

(b)

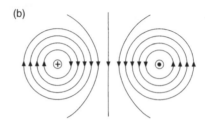

⊕ conventional current down   ⊙ conventional current up

Figure 25.10 *(a) Field lines associated with two adjacent straight conductors carrying currents in the same direction. No net magnetic effect is experienced at the neutral point. (b) Field lines associated with two adjacent straight conductors carrying currents in opposite directions.*

The magnetic field due to current in a loop

If a straight conductor is bent into a loop and a current passed through it, the magnetic field would be as shown in Figure 25.11.

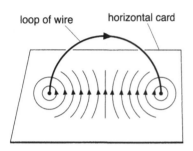

Figure 25.11 *Magnetic field round a current-carrying loop of wire.*

> **Note**
>
> Identical flat coils, called Helmholtz coils, placed one radius apart are often used to provide uniform magnetic fields.

Points of interest:

- The magnetic field inside the loop is dense.
- All the lines of force inside the loop reinforce each other, since they are in the same direction.
- The magnetic field associated with the loop is effectively the same as for a bar magnet with opposite poles at opposite faces.

The field inside a flat coil is not uniform. This disadvantage can be overcome by using a solenoid.

Solenoids

A solenoid is an arrangement in which a wire is looped into a helix or spiral (Figure 25.12). The ideal solenoid is longer than it is wide (at least twenty times).

Like a single loop, the solenoid concentrates magnetic field lines inside the coil. Opposite magnetic poles exist at the ends of the solenoid. The magnetic effect of the solenoid is greater than that of the single coil since many turns of wire are involved and many more magnetic field lines are aiding each other inside the solenoid.

Whereas the field inside the solenoid is uniform, the field outside the solenoid is like that of a bar magnet. The north and south poles of a solenoid can be determined using the following modification of the right-hand grip rule.

Imagine a solenoid gripped in the right hand in such a way that the curled fingers are indicating the direction of conventional current, then the thumb points to the north pole.

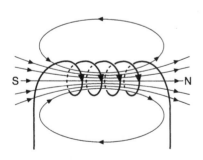

Figure 25.12 *The field of a solenoid.*

25.7 Electromagnets

An electromagnet is a temporary magnet which has magnetism only when current is passing through a coil of wire. The strength of the magnetic field associated with an electromagnet can be increased by:

- increasing the current;
- increasing the number of turns of coil per unit length;
- including a soft iron core.

Electromagnets are used:

- to lift iron rods and steel bars and sheets in steelworks;
- as cranes in scrap yards;
- in hospitals, to remove iron or steel splinters from eyes;
- in the electric bell;
- in the cores of transformers;
- in control systems.

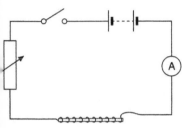

Figure 25.13 *A simple electromagnet.*

How to make a simple electromagnet

1. Wind about 1.5 m of SWG 26 covered copper wire on an iron core, e.g. a 12 cm iron nail. Keep the turns close and wind in the same direction.
2. Connect the ends of the wire as shown in Figure 25.13.
3. Turn on the switch and count the number of paper clips or iron staples that the electromagnet will hold.
4. Adjust the current and note the new number of clips or staples which are now supported.

From the above experiment you should see that the bigger the value of the current flowing the stronger the electromagnet.

25.8 The magnetic relay

The electromagnetic relay is a current-operated device that opens or closes contacts and, in so doing, controls the flow of current in one or more separate circuits.

Figure 25.14 shows a simple electromagnetic relay. This relay uses a small current to switch on a much larger one.

If switch S_1 is opened, no current flows and the contacts S_2 are also opened. The lamp does not light.

The lever system has three interconnected parts L, M and N. If switch S_1 is closed, a current flows through the coil, a magnetic field is set up and L, which is made of soft iron, is attracted to the electromagnet E. N then makes contact with S_2, closing it. A current flows in circuit C_2 and the lamp lights.

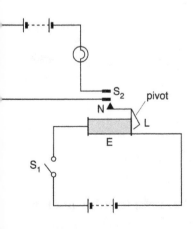

Figure 25.14 *A simple electromagnetic relay.*

Electromagnetic relays are used, for example, in the switching system of a car, to switch on X-ray machines, in telephone exchanges and for switching on cameras on rockets. These cameras are connected to relay contacts which are closed by radio signals which are sent from Earth.

25.9 The force on a current-carrying conductor in a magnetic field

How to show there is a force on a current-carrying conductor in magnetic field

1. Set up the arrangement shown in Figure 25.15.
2. Observe that there is no movement of the aluminium strip with the switch open.
3. Close the switch and observe what happens.

The strip moves because a force acts on it. This force arises from the interaction between the current and the magnetic field.

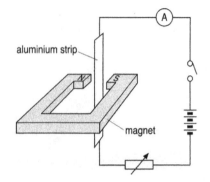

Figure 25.15 *Demonstrating the force on a conductor.*

Remember that a moving charge (current) has an associated magnetic field. Effectively, what we have is the interaction of two magnetic fields.

The direction of the force which results from the interaction of a current and a magnetic field can be predicted from '**Fleming's left-hand rule**' (Figure 25.16), which states that:

If the forefinger, second finger and thumb of the left hand are held at right angles to one another, and if the forefinger points in the direction of the field and the second finger in the direction of the current, then the thumb points in the direction of the force which is acting on the conductor carrying the current.

The size of the force depends on:

- the strength of the magnetic field;
- the value of the current.

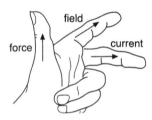

Figure 25.16 *Fleming's left-hand rule.*

A beam of charged particles can be deflected by a magnetic field

Just as a current-carrying conductor in a magnetic field experiences a force, when a beam of charged particles enters a magnetic field it experiences a force. The beam is accelerated and describes a curved path (Figure 25.17). A beam of electrons entering a magnetic field in the direction shown is equivalent to conventional current in the opposite direction (Figure 25.17(b)).

Things to do

Experiment with the apparatus of Figure 25.15 to show that the force is greatest when the current and the magnetic fields are at right angles and zero when they are in the same or opposite direction (i.e. when the angle between them is zero).

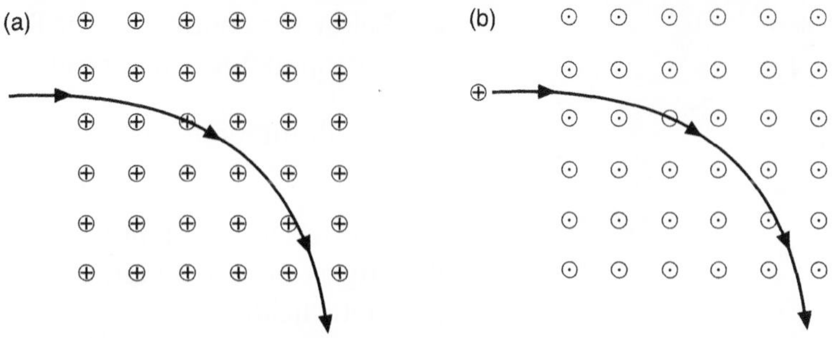

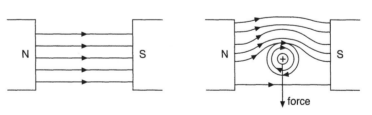

Figure 25.17 *(a) The magnetic field is directed into the plane of the paper. The beam of electrons describes a circular path. (b) The magnetic field is directed out of the plane of the paper. A beam of positive particles is deflected downwards.*

Things to do

Use Fleming's left-hand rule to show that the electron beam in Figure 25.17(a) curves downwards.

Fleming's left-hand rule can be used to predict the direction of motion of the beam. For example, if the directions of the beam of electrons and of the magnetic field are as shown in Figure 25.17(a), this can be taken to be equivalent to conventional current as shown in Figure 25.17(b).

What happens to the magnetic flux when a current-carrying conductor is placed in a magnetic field?

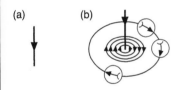

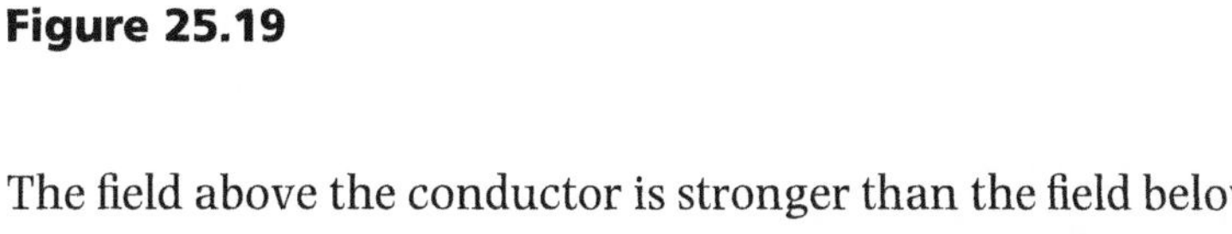

Figure 25.18 *The magnetic field round a current-carrying conductor.*

If conventional current flows in a straight conductor in the direction shown in Figure 25.18(a), the direction of the associated magnetic field, as predicted by the right-hand grip rule, is as shown in Figure 25.18(b).

If the conductor is now placed in the uniform magnetic field shown in Figure 25.19(a), the resulting magnetic field is as in Figure 25.19(b).

(a) Uniform field *(b) The interaction of the two fields*

Figure 25.19

The field above the conductor is stronger than the field below it.

- Above the conductor, the field of the conductor and that of the magnet are acting in the same direction. The field is strengthened.

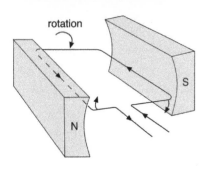

Figure 25.20 *A couple acts on a rectangular coil in a magnetic field.*

> **Note**
>
> A **couple** is a pair of forces equal in size but opposite in direction which cause an object to rotate.
>
> F
>
> F

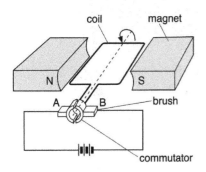

Figure 25.21 *An electric motor.*

> **Note**
>
> A **commutator** is a metal ring which is split into halves. It reverses the current direction in the coil on each half cycle – ensuring that the motion of the coil is continuous.

- Below the conductor, the field of the conductor and that of the magnet are in opposition. The field is weakened.

The resultant force pushes the conductor in the direction shown.

Now consider a rectangular coil with conventional current flowing as shown in Figure 25.20. The coil is in a uniform magnetic field.

If you apply the left-hand rule, you will see that the left-hand side of the coil moves down, whereas the right-hand side of the coil moves up. The coil rotates clockwise. We say that:

A couple acts on a rectangular current-carrying coil which is placed in a uniform magnetic field.

Devices which use coils and magnetic fields include the electric motor, the electric generator, loudspeakers, microphones and measuring instruments.

25.10 The electric motor

A motor is a rotating device which converts electrical energy into mechanical energy.

The d.c. motor and the d.c. generator (see Chapter 26) have the same basic parts: coils, uniform magnetic fields and split rings (commutators). Given the right conditions, most generators can operate as motors.

A d.c. motor operates from d.c. only. All d.c. motors have commutators.

In Figure 25.21, current flows from the d.c. supply through brush A, to the coil and back to the supply through brush B. It creates a magnetic field round the coil.

Current flow is such that the north pole of the coil is next to the north pole of the permanent magnetic field. Likewise, the south pole of the coil is next to the south pole of the magnetic field.

The magnetic field of the coil opposes that of the permanent magnet. Repulsion takes place. Since the permanent magnet is fixed and the coil moveable, the coil rotates. As the coil rotates its north pole is attracted to the south pole of the permanent magnet.

At the instant when two unlike poles are about to line up, the commutator changes the polarity of the coil. Repulsion takes place and the coil continues to rotate.

The simple d.c. motor in Figure 25.21 can be improved by:

- using an electromagnet in place of the permanent magnet;
- using more than one loop in the coil;
- winding the coils around a soft iron core.

The soft iron core on which the coils are wound, plus the coils, is called an **armature**.

25.11 The loudspeaker

The loudspeaker is a device for converting electrical signals into audible sound. In the commonest type of loudspeaker a changing current flows through a coil which is suspended between the poles of a specially shaped magnet (Figure 25.22).

The varying current produces a varying magnetic field in the coil. A varying force is exerted on the coil, which moves in and out. The coil is attached to a cone made of stiff paper. The motion of the coil is translated to the cone which moves in unison with the electrical signal. The vibrating cone converts the electrical signal to audible sound.

Note

The action of a loudspeaker is the reverse of that of the microphone.

Speaker systems may contain one or more speakers. At least one of the speakers has a large cone: this speaker responds better to low (bass) notes. Speakers with smaller cones respond better to higher frequency notes.

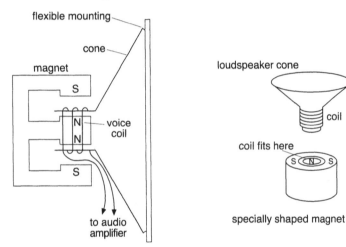

Figure 25.22 *The loudspeaker.*

Checklist

After studying Chapter 25 you should be able to:

- recall the concepts of magnetic poles and magnetic field lines
- recall the directions of magnetic field lines
- describe an experiment to identify the magnetic poles, i.e. north and south poles, and recall the deductions made from the experiment
- define magnetic flux and magnetic flux density
- recall that forces which exist between magnets are either attractive or repulsive

- discuss the variables which these forces depend on
- recall the concept of 'domains'
- discuss magnetic induction
- define permeability as it relates to magnetism
- differentiate between 'hard' and 'soft' magnetic materials, their properties and uses
- discuss the magnetic effects of electric currents
- recall and use the Maxwell screw rule
- recall and use the right-hand grip rule
- discuss magnetic fields due to currents in a single loop and a solenoid
- recall that a current-carrying conductor placed in a magnetic field experiences a force
- recall and use Fleming's left-hand rule
- define an electromagnet and discuss its applications
- recall variables which affect an electromagnet's strength
- construct a simple electromagnet
- discuss the advantages and disadvantages of electromagnets
- map flux patterns around current-carrying conductors
- discuss the d.c. motor
- discuss the loudspeaker

Questions

1 When bar X was placed in a solenoid and the current switched on, it supported six paperclips. The paperclips fell off once the current was switched off. On the other hand, bar Y, when placed in a solenoid and the current switched on, was able to support only three paperclips. One of these paperclips fell off when the current was switched off. What do these observations tell you about X and Y?

2

The diagram shows the direction of conventional current through a wire.
a On a copy of the diagram, sketch the magnetic field around the wire (within the dotted lines) and use the field pattern to explain why the two sections of the wire repel each other.
b What can be done to increase the force of repulsion between the two sections, A and B, of the wire?

3

The diagram shows a top view of the turns of a solenoid as they pass through a cardboard support. The current direction (conventional) is shown.
a Use your knowledge of the magnetic field around current-carrying conductors to show how a compass needle would point at the points A, B, C and D.
b Suggest a practical way of exploring the magnetic field inside a solenoid.

In Chapter 25 you learned that the movement of electric charges generates associated magnetic fields. In this chapter you will discover that a moving or changing magnetic field which cuts across a conductor generates a movement of charge and induces an e.m.f., or voltage, between the ends of the conductor.

Note

Both the generator and the electric motor operate on the principle of electromagnetic induction.

26.1 Induced e.m.f.

An e.m.f. is induced when a conductor moves relative to a magnetic field.

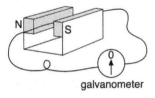

Figure 26.1 *The magnetic field and the conductor are stationary. A voltage (e.m.f.) is not induced. The galvanometer reading is zero.*

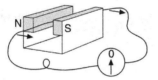

Figure 26.2 *The conductor is moved parallel to the magnetic field. No voltage is induced.*

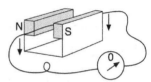

Figure 26.3 *The conductor is moved vertically down. The galvanometer needle deflects to the right while the conductor is moving but returns to zero once motion ceases.*

Consider the diagrams in Figures 26.1–26.3.

The direction of the induced current can be predicted from **Fleming's right-hand rule** (Figure 26.4):

If the thumb and first two fingers of the right hand are held mutually at right angles, with the first finger in the direction of the magnetic field and the thumb in the direction of motion, then the second finger shows the direction of the induced current.

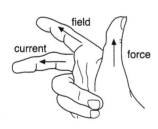

Figure 26.4 *Fleming's right-hand rule.*

An e.m.f. is also induced if a magnet moves relative to a coil or solenoid (Figure 26.5).

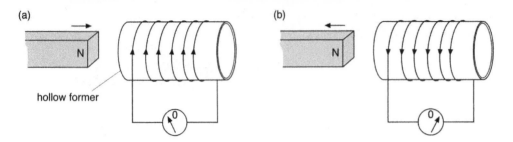

Figure 26.5 *The direction of the induced current depends on the direction of motion of the magnet.*

Using two solenoids to induce an e.m.f.

Consider Figure 26.6. If the primary solenoid (the one attached to the battery) is moved towards the secondary solenoid, the galvanometer needle deflects as shown, i.e. a voltage is induced.

Once relative motion ceases the galvanometer needle returns to zero.

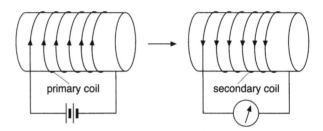

Figure 26.6 *Moving the primary coil towards the secondary coil induces a current.*

The same effect may be achieved by closing the switch in Figure 26.7. The solenoids in Figures 26.6 and 26.7 are said to be **inductively coupled**.

When the switch in Figure 26.7 is just closed, the magnetic field associated with the primary solenoid expands and 'couples' the secondary solenoid, inducing an e.m.f. there. The galvanometer needle deflects to the right. The moment the expansion of the magnetic field ceases, the meter reading returns to zero, i.e. an e.m.f. is no longer induced in the secondary coil.

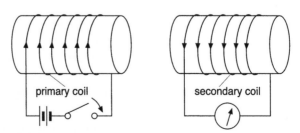

Figure 26.7 *Closing the switch induces a current in the secondary coil.*

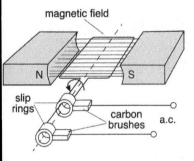

Figure 26.8 *A simple a.c. generator.*

If the switch was closed and is then suddenly opened, the magnetic field around the primary solenoid contracts. A contracting field around the primary induces an e.m.f. across the secondary solenoid. The galvanometer deflects to the left. Once the contraction is complete the induced e.m.f. disappears.

What factors determine the size of the induced e.m.f.?

The value of the e.m.f. (voltage) induced across a conductor which is cut by magnetic field lines depends on:

- the strength of the magnetic field;
  The stronger the magnetic field, the bigger the induced e.m.f.
- the number N, of turns of coil per unit length (for solenoids);
  The greater the value of N, the bigger the induced e.m.f.
- the speed with which the magnetic field lines cut the conductor (or solenoid);
  The greater the speed, the bigger the induced e.m.f.
- the type of core, if any.
  A soft iron core increases the value of the induced e.m.f. Soft iron concentrates field lines and makes an electromagnet stronger.

26.2 The simple a.c. generator

An a.c. generator is a mechanical device which converts rotational energy into electrical energy.

The simple a.c. generator (Figure 26.8) consists of a loop or coil of wire which is rotated in a magnetic field. As the coil rotates, it cuts the magnetic field and a voltage is induced across it. This voltage causes a current to flow. Circular **slip rings**, attached to the ends of the coil, lead the current away. Carbon brushes ensure that good electrical contact is made with the slip rings.

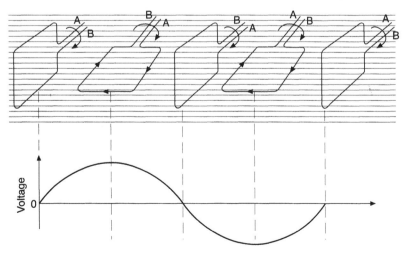

Figure 26.9 *The output from a simple a.c. generator.*

As the coil rotates in the magnetic field an alternating current is produced, as shown in Figure 26.9. On one half cycle the coil cuts the magnetic field in one direction. On the other half cycle the coil cuts the magnetic field in the opposite direction.

26.3 The simple d.c. generator

The d.c. generator, or dynamo, is similar to the a.c. generator except that the slip ring is segmented, i.e. split into two or more parts. A segmented slip ring is called a commutator (see page 300).

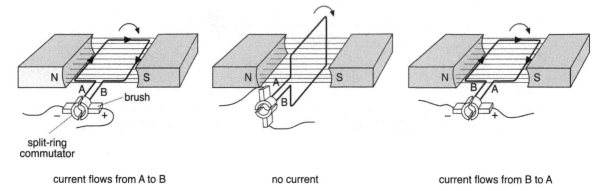

| current flows from A to B | no current | current flows from B to A |

Figure 26.10 *A simple d.c. generator. The commutator effectively converts a.c. to d.c.*

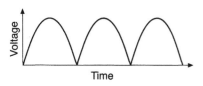

Figure 26.11 *The output from a simple d.c. generator.*

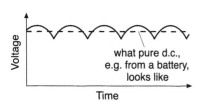

Figure 26.12 *The output from a real d.c. generator.*

The current in the coil reverses with each half turn, but the brush marked '+' in Figure 26.10 is always in contact with whichever half of the commutator is positive at any moment. This means that the output current flows in one direction at all times.

The output from a simple d.c. generator is shown in Figure 26.11. It is not pure d.c. However, all waveforms are above the reference line (the zero line). Such an output is described as *pulsating*. The pulsating d.c. shown is not suitable for most practical purposes. For example, a lamp would keep dimming and brightening alternately.

The armature of a practical d.c. generator is made up of many coils connected in series. The ends of each coil are connected to the segments of the commutator. If many coils (and a more segmented commutator) are used, the voltage approximates to d.c. voltage (see Figure 26.12).

26.4 The transformer

The transformer consists of two coils wound on a common soft iron core (Figure 26.13). The two coils are inductively coupled – mounted so that a varying magnetic field around one will

Note

A transformer consists of two multi-turn coils of wire wound on a common iron core. It works on the principle of mutual electromagnetic induction.

Transformers transfer blocks of power from one circuit to another using a.c.

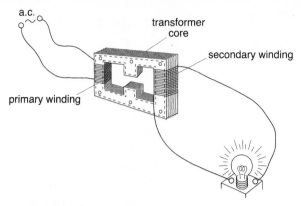

Figure 26.13 *A simple transformer.*

induce a voltage in the other. There is no contact between the coils other than through their magnetic fields.

Alternating current is fed to one coil, called the **primary**. The output from the second coil (the **secondary**) is also a.c. When a.c. is fed to a solenoid the magnetic field expands and contracts on alternate half cycles. The expanding and contracting magnetic field induces an a.c. voltage in the secondary.

The output voltage may be bigger or smaller than the input voltage. If the output voltage (V_s) is greater than the input voltage (V_p) the transformer is functioning in its step-up mode. If output voltage is smaller than input voltage the transformer is in step-down mode.

For step-up transformers:

The symbol for a transformer.

N_s, the number of turns of wire in the secondary coil, is greater than the number of turns, N_p, in the primary coil.

$$N_s > N_p$$

For step-down transformers:

$$N_p > N_s$$

For an ideal transformer:
    power in the primary = power in the secondary

so:

$$V_p \times I_p = V_s \times I_s$$

Also:

$$\frac{V_p}{N_p} = \frac{V_s}{N_s}$$

and so:

$$\frac{V_s}{V_p} = \frac{N_s}{N_p} = \frac{I_p}{I_s}$$

This relationship is consistent with the principle of conservation of energy.

Note

The main use of the transformer is to lower or raise voltage as required.

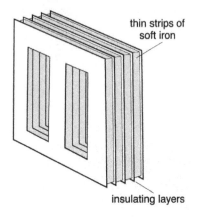

Figure 26.14 *Transformer cores are laminated.*

The design of efficient transformers

Power losses should be kept to a minimum in transformers. The main losses are:

- **Copper losses:** The coils of a transformer have finite resistance. When current flows through them energy is lost as heat. This loss is partially reduced by using moderately thick copper windings.
- **Eddy currents:** These are losses due to variations in the flux linkages. Eddy currents circulate in the iron core, reducing the efficiency of energy transfer. This loss is reduced by **laminating** the soft iron core: thin strips of iron are used instead of a solid core (Figure 26.14). The individual strips are insulated from one another by varnish or an iron oxide coating.

26.5 The transmission and distribution of electrical power

The output from generators is a.c. A network of overhead cables (in some countries underground cables) takes the electricity to towns and villages. The network of cables is known as a 'grid system'. The grid system uses a.c. (and not d.c.) because a.c. voltages can be increased or decreased as required by transformers.

Mains electricity is delivered to our homes at 110 volts or 220 volts, but is transmitted over the national grid at 66 000 volts or 132 000 volts. The voltage is stepped up at the transmission substations and taken by high voltage cables to the distribution substations, where it is stepped down before being sent to the householder or other customer (Figure 26.15, opposite). Often there are several step-down transformers at the distribution substation, each having a different voltage output.

Example

A transformer has a primary coil of 30 turns and a secondary coil of 120 turns. A 15 V a.c. source is applied across the primary. What is:

a the voltage across the secondary load,

b the ratio of the primary current to the secondary current? (You may assume that the transformer is ideal).

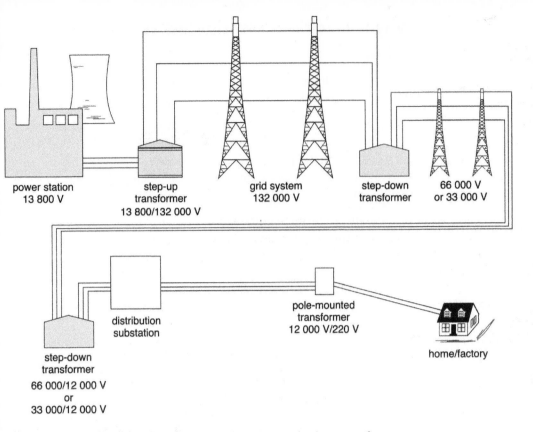

Figure 26.15 *Electricity from the power station to the home or factory.*

Solution

a $\dfrac{N_s}{N_p} = \dfrac{120}{30} = 4$

$\dfrac{V_s}{V_p} = \dfrac{N_s}{N_p} = 4$

$V_s = 4V_p = 4 \times 15$

$= 60\,\text{V}$

b $\dfrac{I_p}{I_s} = \dfrac{N_s}{N_p} = 4$

Problem

An ideal transformer has a primary coil of 250 turns and secondary of 50 turns. The input to the primary is 250 V a.c. Calculate:

a the secondary voltage
b the ratio of the secondary current to the primary current
c the secondary current, given that the power in the primary is 1000 W.

Note

Electricity is transmitted at high voltages to reduce

- voltage losses across cables;
- power losses.

Why is electricity transmitted at high voltages?

Transmission losses are reduced when electricity is carried on the grid system at high voltages.

Consider transmitting a block of power of 10 000 W through cables of resistance 1 Ω using a voltage of **a** 100 V, **b** 10 000 V. Let us calculate the power and voltage losses if the power is transmitted at these two different voltages. Before doing so, recall that:

- power = current × voltage;
- power (heat) losses are given by the product: (current)^2 × resistance (I^2R);
- voltage losses in the cables = current through the cables × the resistance of the cable

a At 100 volts

The current drawn is given by $\dfrac{10\,000\,\text{W}}{100\,\text{V}} = 100\,\text{A}$

The voltage drop across cables = 100 A × 1 Ω = 100 V

The power loss in cables = $100^2 \times 1 = 10\,000\,\text{W}$

This means *all* the power will be dissipated as heat!

b At 10 000 volts

The current drawn is given by $\dfrac{10\,000\,\text{W}}{10\,000\,\text{V}} = 1\,\text{A}$

The voltage losses across the cables = $1^2 \times 1 = 1\,\text{V}$

The power dissipated as heat in this case is now only $1^2 \times 1 = 1\,\text{W}$

At 10 000 V, the voltage losses equal 1 V and the power losses equal 1 W.

Clearly, it is better to transmit electrical power at higher voltages!

Problem

Compare the voltage and power losses when a 100 000 W block of electrical power is transmitted at **a** 1000 V and **b** 5000 V through cables of resistance 1 Ω.

Checklist

After studying Chapter 26 you should be able to:

- recall that the movement of a conductor across a magnetic field induces an electromotive force (e.m.f.)
- use Fleming's right-hand rule
- recall that the movement of a magnet relative to a coil or solenoid induces an electromotive force (e.m.f.)
- discuss the concept of inductive coupling
- discuss factors which affect the size of the induced e.m.f.

- discuss the simple a.c. generator
- discuss the modification of the a.c. generator to produce d.c.
- discuss the transformer and its uses
- differentiate between the 'step-up' and 'step-down' transformer
- recall and use the equations $V_p \times I_p = V_s \times I_s$ and $\dfrac{V_p}{N_p} = \dfrac{V_s}{N_s}$
- discuss the ways in which power losses occur in a transformer
- discuss ways in which transformer efficiency can be increased
- recall that electrical transmission at high voltages minimizes power and voltage losses, i.e. increases efficiency

Questions

1 The input of an ideal transformer is 240 V. The output is 48 V. The current in the primary coil is 0.5 A.

a If there are 250 turns in the primary coil, how many turns are there in the secondary coil?

b What is the power in the primary coil?

c What is the current in the secondary coil?

2 An unmagnetized steel rod is placed in a solenoid as shown.

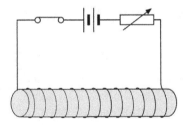

a What effect would the rod now have on a plotting compass?

The entire procedure was repeated with a rod of soft iron, instead.

b What effect, if any, would the soft iron rod now have on a plotting compass?

c Account for the differences, if any, in **a** and **b** above.

3

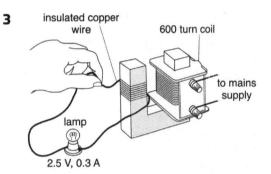

insulated copper wire   600 turn coil
to mains supply
lamp
2.5 V, 0.3 A

a Explain how a 120 V mains supply can be used to light the 2.5 V lamp. In your answer, focus on all underlying physical principles.

b Calculate the power dissipated in the lamp when it is lighting normally.

c What is the power input to the 600 turn coil?

d What is the current flowing through the primary coil?

e How many turns of insulated copper wire are there in the secondary? ·

E Revision questions for Section E

Multiple-choice questions

1 Through which of the following substances would electrons move freely?
A glass   B Perspex
C acetate   D aluminium

2 A rod was rubbed and became negatively charged. The rod:
A gained electrons   B gained protons
C lost electrons   D lost protons.

3 The unit of charge is the:
A coulomb   B volt
C joule   D ampere.

4 Which of the follow operations produces a negatively charged rod? Rubbing:
A a glass rod with fur
B a Perspex rod with a dry cloth
C a polythene rod with a soft cloth
D an acetate rod with silk.

5 The potential difference across the ends of a device is 72 V. The current through the device is 8 A. The resistance of the device, in ohms, is:
A 9   B 64   C 80   D 576.

6 The peak voltage of a mains supply is 340 V. What would a voltmeter, placed across the outlet, read?
A 85 V   B 120 V
C 240 V   D 340 V

7 A battery sends a current of 3.5 A through a resistance of 12 Ω. This battery also sends a current of 10.5 A through another resistor. What is the value of this resistor?
A 42 Ω   B 15.5 Ω
C 4 Ω   D 2 Ω

8 Which of the following is the usual unit for electricity consumed by householders?
A joule   B kilojoule
C watt minute   D kilowatt hour

9 The resistance of a cylindrical metallic conductor is directly proportional to its:
A density   B diameter
C length   D mass.

10 A motor rated at 1800 W operates from a 120 V mains supply. What is the current through the motor?
A 10 A   B 12 A   C 15 A   D 60 A

Questions 11 and 12

A battery of negligible internal resistance supplies 100 J of energy while moving 10 C of charge in 2 seconds through a resistor.

11 The e.m.f. of the battery is:
A 5 V   B 10 V   C 20 V   D 50 V.

12 The current flowing through the resistor is:
A 50 A   B 10 A   C 5 A   D 2.5 A.

13 If the potential difference across a device (of constant resistance) is halved, the current through it is:
A unchanged   B halved
C doubled   D quadrupled.

14 A household consumes 3000 units of electricity over a 2 month period. The Light and Power rates are as follows: $0.15 per unit for the first 400 units; $0.10 per unit for the next 300 units; $0.06 per unit for the next 300 units; $0.03 per unit for units in excess of the first 1000.
What is the electricity bill for this household for this period?
A $230.00   B $206.00
C $168.00   D $128.00

15 Three identical resistors in parallel have an effective resistance of 15 ohms. What would their effective resistance be if they are connected in series?
A 15 ohms   B 45 ohms
C 135 ohms   D 210 ohms

16 The body of an electric appliance is connected to earth in order to:
A complete the circuit
B prevent the breaker from tripping
C protect the user from shock
D limit the current flowing.

17 Each of the following statements concerning alternating voltage is true, except:
A it reverses polarity
B it has a steady value
C it can be stepped up using a transformer
D it can be rectified using a diode.

18 Which of the following statements concerning circuit breakers is correct? Circuit breakers:
A limit the current flowing in electric circuits
B should be connected to the neutral of the a.c. supply
C allow current to flow readily to earth in the event of a fault developing in a circuit
D should be made of alloys of high melting temperature.

19 What applied voltage produces a current of 5 mA through a 2 MΩ resistor?
A 10 MV   B 10 kV
C 10 mV   D 10 μV

20 Two 50 Ω, $\frac{1}{2}$ W resistors are connected in parallel. Their combined resistance and wattage (power) is:
A 100 Ω, 1 W   B 100 Ω, $\frac{1}{2}$ W
C 25 Ω, 1 W   D 25 Ω, $\frac{1}{2}$ W.

21 Four resistors are connected in series to a 100 V source. The values of the resistors are 28 Ω, 46 Ω, 52 Ω, and x Ω. The current drawn from the battery is 0.5 A. The value of x is:
A 52 Ω   B 74 Ω C 100 Ω D 200 Ω.

Questions 22–25

A motor car head lamp is rated 12 V, 36 W.

22 What is the working resistance of the device?
A 24 Ω   B 4 Ω   C 3 Ω   D 0.33 Ω

23 What current flows through the device when it is connected to a 6 V supply?
A 1.5 A   B 4 A   C 6 A   D 72 A

24 How many joules of energy would the device consume in 2 min when operating normally?
A 72   B 144   C 1080   D 4320

25 What safety resistance must be included in the circuit for the device to operate safely from a 36 V supply?
A 0.75 Ω   B 3.0 Ω   C 6.0 Ω   D 9.0 Ω

26 Which of the following is an ohmic device?
A a diode
B a thermistor
C a colour-coded resistor
D a transistor

27 A 2 V lead–acid cell has an internal resistance of 0.05 Ω. What current would flow if the cell is short-circuited?
A 1600 A   B 0.5 A
C 25 A   D 40 A

Questions 28 and 29

An ammeter in series to a cell of e.m.f. E and negligible internal resistance and a 30 Ω resistor, reads 0.3 A.

28 The e.m.f. of the cell is:
A 1.5 V   B 3.0 V
C 9 V   D 12 V.

29 If the 30 Ω resistor is replaced by a 45 Ω resistor, the ammeter would read:
A 0.2 A   B 2.0 A
C 5 A   D 10 A.

Questions 30 and 31 refer to the diagram below.

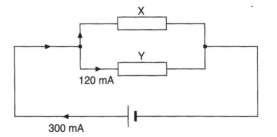

30 The current flowing through resistor X is:
A 120 mA   B 210 mA
C 180 mA   D 300 mA.

31 The ratio of the resistance of X to the resistance of Y is:
A 120/300   B 180/300
C 180/120   C 120/180.

Questions 32–34

An electric fire consists of two identical 2.5 kW elements, X and Y, connected to a 'fused' 250 V supply shown in the diagram below:

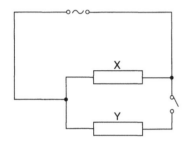

32 The current initially drawn from the supply with the switch closed is:
A 5 A   B 10 A   C 20 A   D 50 A.

33 The current through element X when the switch is open is:
A 5 A   B 10 A   C 20 A   D 50 A.

34 A suitable fuse for the mains supply has current rating:
A 10 A   B 15 A   C 30 A   D 60 A.

35 A defective semiconductor diode:
A conducts only when forward-biased
B conducts only when reverse-biased
C conducts when forward-biased *and* reverse-biased
D does not conduct in either forward-biased or reverse-biased modes.

36 Which of statements I–IV are true? An ammeter:
I is connected in series with the current to be measured
II has a low value resistor in parallel with its operating coil
III significantly affects the p.d. across other circuit components

IV has a fixed range or ranges, which cannot be extended.

A I and II   B II and III
C III and IV   D I and IV

Questions **37** and **38** are related to the diagram below.

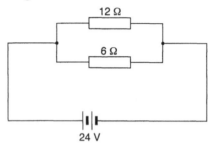

37 What is the effective resistance, in ohms, in this circuit?

A $\dfrac{12 \times 3}{12 + 6}$ B $\dfrac{12 \times 6}{12 + 6}$

C 12×6 D $12 + 3$

38 The current drawn from the battery, in amperes, is:
A 6   B 4   C 2   D 1.33.

Questions **39** and **40**

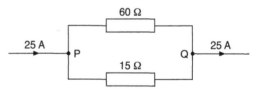

39 What is the value of the current flowing throught the 60 Ω resistor?
A 0.5 A   B 5 A   C 20 A   D 30 A

40 What is the potential difference across the points PQ?
A 5 V   B 25 V   C 50 V   D 300 V

41 Which of the following statements is **not** true?
A Semiconductor diodes obey Ohm's law.
B Conductors have free electrons.
C Germanium is an example of semiconductor material.
D Semiconductors conduct better at higher temperatures.

42 Which of A–D gives the correct location of both the fuse and the switch?

	Switch	Fuse
A	live	neutral
B	live	live
C	neutral	live
D	neutral	neutral

43

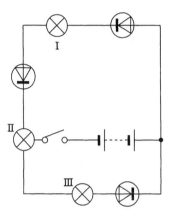

Which of the lamps labelled I to III would light if the switch was closed?
A I only   B I and II
C III only   D I and III

44 Which of the following, added to silicon, gives it n-type properties?
A beryllium   B aluminium
C tin   D phosphorus

45 Which of gates A–D has the following truth table?

L	M	O
0	0	1
0	1	1
1	0	1
1	1	0

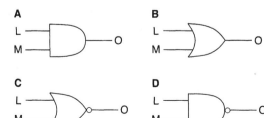

Questions 46 and 47

A sample of a newly fabricated alloy has a resistance of $10\,\Omega\,\mathrm{m}^{-1}$ and is capable of safely carrying a current of $16.67\,\mathrm{A}$ at $750°\mathrm{C}$.

46 What length of this wire would you use to construct a 2 kW heating coil which operates normally at $750°\mathrm{C}$? (Hint: power, $P = I^2 R$.)
A 0.36 m   B 0.72 m
C 2.77 m   D 10 m

47 At what voltage will this coil operate normally?
A 120 V   B 169 V
C 240 V   D 340 V

48 Which logic gate 'corresponds' to the electrical circuit below?

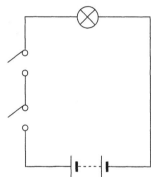

A NAND   B OR   C NOR   D AND

49 A battery of e.m.f. 40 V is connected across the series combination of fixed resistors as shown. What is the potential difference between points I and III?

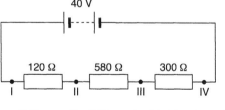

A 3 V   B 7 V   C 12 V   D 28 V

50 The alarm system in a 2-door car is controlled by a logic gate. The effective truth table for this gate is shown overleaf. What type of gate is present in the control system?
A AND   B OR   C NAND D NOR

Input		Output
Door L	Door R	Buzzer
0	0	0
1	0	1
0	1	1
1	1	1
0 = door closed 1 = door opened		1 = buzzer on 0 = buzzer off

51 Which of the following is an equivalent NOR gate?

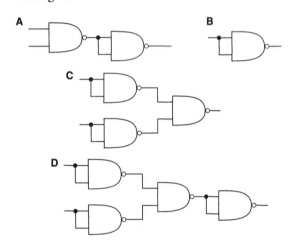

A

B

C

D

52 Analogue and digital refer to two ways in which electricity can:
A carry information
B be switched
C be made safe
D change direction.

53 Which of the following materials is attracted to a magnet?
A aluminium   B copper
C graphite   D iron

54 Which of the following is most suitable for making magnets?
A aluminium   B copper
C iron   D steel

55 For which of the following would an electromagnet be unsuitable?
A the construction of electric bells
B refrigerator doors
C lifting iron sheets
D the recording head of a tape recorder

56 Which of the following beams of particles is not deflected by a magnetic or an electric field?
A alpha particles   B beta particles
C neutrons   D protons

57 The magnetic field associated with a solenoid is:
A different from that of a bar magnet
B stronger outside than inside
C uniform inside the solenoid
D weakened if a soft iron core is included.

58 Magnetic shielding may be achieved by using:
A polythene   B hardened steel
C soft iron   D graphite.

59 Which of the following statements is **true**?
A Magnetic field lines run from the south pole to the north pole of a magnet.
B Magnetic field strength is weakest where the field lines are most closely spaced.
C Bar magnets eventually settle in an east–west direction, if free to rotate.
D Magnets lose some or all of their magnetism if heated or hammered.

Questions 60–62 concern the devices A–D.

A a diode   B a relay
C a transformer   D a motor

Which one of the devices:

60 contains an electromagnet and uses a small current to switch on larger currents?

61 converts a.c. to d.c.?

62 converts electrical energy to mechanical energy?

63 Which of the following materials is **not** magnetic?
A cobalt   B copper
C iron   D nickel

64 The south pole of a magnet attracts:
A soft iron
B the south pole of other magnets
C copper
D a sheet of magnesium.

65 The region around a magnet is called:
A a magnetic domain
B lines of force
C a magnetic field
D magnetic flux.

66 Which of the following materials is most suitable for the core of an electromagnet?
A lead   B iron
C beryllium   D silver

67 The strength of an electromagnetic field can be increased by:
A changing the insulation on the wire of the solenoid
B increasing the number of turns of wire
C replacing the iron core by a copper core
D increasing the size of the iron core.

68 The polarity of an electromagnet depends on:
A the type of core used
B the direction of the current
C the number of turns of coil
D the strength of the current.

69 Which of A–D is the correct alignment of a plotting compass placed at X?

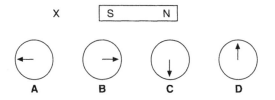

Questions 70–72

Four magnetic fields are represented below. The fields are in a horizontal plane.

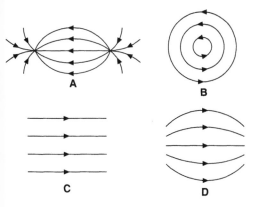

Which of these fields is produced or formed:

70 in the centre of a long straight solenoid which is carrying a current?

71 by a horse-shoe shaped magnet?

72 by a straight conductor carrying a current upwards into the plane of the paper?

Questions 73–75 concern the devices A–D.

A a loudspeaker   B a microphone
C a diode   D a transformer

Which of the devices:

73 is used by power companies in the transmission of electrical power?

74 is used to convert electrical energy to sound?

75 is used in the detection of radio waves?

Questions 76–78

Four magnetic fields A–D are shown below.

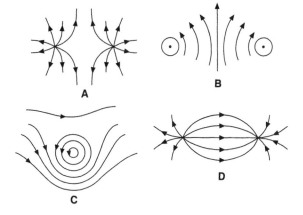

Which of these fields is formed:

76 by two magnets with adjacent like poles facing, horizontally?

77 by two magnets with adjacent unlike poles facing, horizontally?

78 by a vertical coil through which a current is passing?

79 Which of the following statements best describes the role of a commutator in an electric motor? To:
A reverse the current in the coil as it rotates
B ensure that the magnetic field remains uniform
C ensure that a current of steady value flows in the coil
D convert alternating current to direct current.

80 Which of the following scientists was most closely associated with discoveries in electromagnetism?
A Ampère   B Volta
C Thomson   D Faraday

81 A split ring commutator is part of:
A a loudspeaker
B an electric bell
C an a.c. generator
D a d.c. motor.

82 Which of the following statements is **true**? A generator is a device which converts:
A chemical energy into electrical energy
B electrical energy into chemical energy
C mechanical energy into electrical energy
D electrical energy into mechanical energy.

83 Electricity ('Light and Power') companies supply a.c. This is so because:
A a.c. may be used in electrolysis and battery charging
B a.c. is more easily converted to other forms of energy than d.c.
C a.c. produces a heating effect whereas d.c. does not
D a.c. is more easily generated and transmitted than d.c.

84 Transformers:
I   have cores which are made of steel
II have cores which are laminated
III may be used to step up or step down voltages
IV function on both a.c. and d.c.

Which of the above statements are true for transformers?

A I and II   B II and III
C I and III   D II and IV

85 When the forefinger, the thumb and second finger are held mutually at right angles, which of A–D correctly states Fleming's **right-hand** rule?

	The magnetic field is represented by	The induced current is represented by	The motion is represented by
A	thumb	forefinger	second finger
B	forefinger	second finger	thumb
C	second finger	forefinger	thumb
D	thumb	second finger	forefinger

86 Conventional current flows through a thick insulated wire as shown.

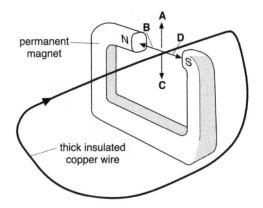

In which of the directions A–D does the wire move when the current is flowing?

87 A transformer can operate on d.c. only if:
A the input to the primary is pulsating d.c.
B the d.c. voltage exceeds a certain minimum value
C the transformer is in the step-down mode
D all power is dissipated as heat in the core.

88 I   Generates electrical power by changing voltages and currents.
II Has a primary coil which is inductively coupled to a secondary coil.
III Transfers electrical energy from one a.c. circuit to another.

IV Converts low current, high voltage d.c.
to high current, low voltage a.c.
Which statements I to IV are true of the
transformer?

A I and II
B I and III
C I and IV
D II and IV

Questions 89 and 90

100 kW of power is transmitted by cables
whose total resistance is 0.5 Ω. The power is
transmitted at 20 000 V.

89 The current through the cables is:

A 0.2 A
B 5 A
C 20 A
D 25 A.

90 What is the power loss in the cables? Power
loss is given by I^2R:

A 2.5 W
B 5 A
C 12.5 W
D 25 W.

91 A transformer is used to step down the
voltage from 20 000 V to 1000 V. What is
the ratio of the number of turns of coil in
the primary to the number of turns of coil
in the secondary?

A 1 : 20
B 20 : 1
C 1 : 400
D 400 : 1

92 Which of the following is true for the thick
conductor when the switch in the circuit
below is closed? It moves:

A vertically down
B vertically up
C to the right
D to the left.

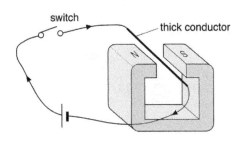

Questions 93 and 94 concern the situations I
to III in the following diagram.

 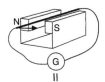

I
conductor stationary

II
conductor moving in
direction of arrow

III
conductor moving in direction of arrow
G = galvanometer

93 For which of the situations I to III does the
galvanometer show a zero reading?

A I only
B I and II
C III only
D II and III

94 For which of situations I to III does the
galvanometer needle move clockwise?

A I only
B I and II
C III only
D II and III

95 Fleming's left-hand rule may be used to
predict:

first finger

second finger

A the direction of the voltage induced in a
coil which is being moved in a magnetic
field
B the location of the north pole of a
solenoid through which a direct current
is flowing
C the motion of a coil which is carrying a
current and is placed in a magnetic field
D the direction of the magnetic field
through a straight conductor which is
carrying a current.

96 Which of the following accurately describes
what happens when the coil in the system
below is quickly pulled out of the magnet?

A The magnetic field weakens.
B The magnet moves horizontally.
C A current is induced.
D The meter is unaffected.

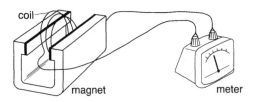

coil

magnet

meter

Structured and free-response questions

1 The force (F) between two charges (q_1 and q_2) is given by the relationship:

$$F = \frac{k_1 q_1 q_2}{d^2}$$

where k is a constant and d the distance of separation of the charges.

a What is the unit of charge?

b What is the unit of force?

c Predict the change in the value of F, if:
(i) the value of q_1 is doubled, while the values of q_2 and d remain fixed,
(ii) the value of d is halved, with the values of q_1 and q_2 fixed.

d Draw diagrams to show the direction of the force between q_1 and q_2 if:
(i) $q_1 = +2$ units, $q_2 = +3$ units, $d = 4$ units, (ii) $q_1 = +3$ units, $q_2 = -1$ unit, $d = 3$ units.

In which of the situations, **d**(i) or **d**(ii) is the force between the charges the greater? Explain how you arrived at your answer.

2 Consider the diagram below:

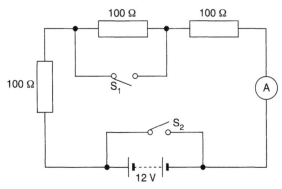

a With S_1 and S_2 opened, calculate (i) the total resistance in the circuit, (ii) the ammeter reading, (iii) the potential difference across each resistor.

b With S_2 opened and S_1 closed, what is (i) the total resistance in the circuit, (ii) the ammeter reading?

c Predict all that is likely to happen if S_2 is closed.

3 A household circuit is wired for 30 A at 120 V.
a Can this circuit carry a 2.5 kW device?
b Can this circuit safely carry, at the same time, the 2.5 kW device and a 1.5 kW device?
c What are the possible consequences of trying to run the devices from a 240 V supply?

4 With the aid of a suitable diagram, explain how a car is painted electrostatically. Comment on the efficiency of the method.

5 Briefly discuss safety precautions that should be taken when using electricity.

6 Describe an experiment that you could set up to demonstrate the link between static and current electricity. Explain the principle(s) underlying the experiment.

7 In a demonstration at a 'science fair' a boy stands on an insulating base. He places his hands on the dome of a van de Graaff generator which is turned on. His hair stands on end.

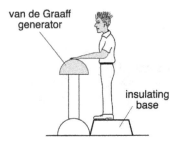

a Why does the boy stand on the insulating base?

b Explain why his hair stands on end.

c Explain why he experiences a shock when he lets go of the generator and steps onto the floor.

8 Explain the principles underlying each of the following situations.
a A balloon was filled with air, then rubbed and placed against a ceiling. The balloon clung to the ceiling as if in defiance of gravity.

b The spray from a can of spray paint spreads out (diverges) from the central hole when the button is pressed. The droplets of paint spread evenly and adhere well to the surface to be painted.

9 The combined resistance R_T for different series combinations of resistors is given is the table below.

Combination of resistors	R_1/Ω	R_2/Ω	R_3/Ω	R_4/Ω	R_5/Ω	R_T/Ω
1	9.8	–	3.2	–	–	13.0
2	–	47	–	33	–	80.0
3	5.6	–	330	–	8.6	344.2
4	220	330	470	–	–	1020
5	220	330	470	840	1500	3360

a Based on the results in the table above, write a general formula for series connected resistors. Show clearly the link between the formula and the data.

b What is the effect, if any, on R_T for combination 3 if the positions of the resistors are interchanged?

c Design an experiment which involves the use of an ammeter and a voltmeter to test the validity of the formula you arrive at in **a** above.

d Calculate the series resistance R_T of $1.5\,M\Omega$, $470\,k\Omega$ and $0.5\,M\Omega$.

e Two resistors connected in series have a combined resistance of $200\,\Omega$. One of the resistors has a value of $132\,\Omega$. What is the value of the other?

f A certain lamp has a resistance of $36\,\Omega$. What is the combined resistance of 10 such lamps? How many such lamps, connected in series, have an effective resistance of $252\,\Omega$?

g The external resistance of a circuit consists of a $450\,\Omega$ and a $560\,\Omega$ resistor connected in series. What is the value of the single resistor which must be added to this combination to make the effective resistance $1500\,\Omega$?

10 a With switch S closed, calculate: (i) the effective resistance in the circuit, (ii) the current supplied by the battery, (iii) the current through the 12 ohm resistor, (iv) the potential difference across the 2 ohm resistor, (v) the power dissipated in the 24 ohm resistor.

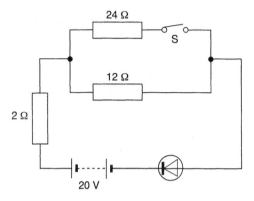

b What is the effect of opening the switch S on (i) the effective resistance in the circuit, (ii) the current supplied by the battery?

c How is the circuit affected if the diode is reversed, with switch S closed? Account for any effect described.

11 The following data concerning constantan resistance wire is taken from a catalogue: 'Diameter/m $= 5.6 \times 10^{-4}$. Resistance per metre $= 1.947\,\Omega$. Resistance per kg $= 913\,\Omega$'.

a Use the data to determine a value for the density of constantan wire in $kg\,m^{-3}$.

Part **b** of this question concerns the resistance wire nichrome. 'Nichrome is an alloy of nickel (80%) and chromium (20%). The density of nichrome is $8.56\,g\,cm^{-3}$.'

Diameter/mm	1.219	0.914	0.711	0.560	0.457	0.315
Current-carrying capacity/A	15	10	7.8	4.8	3.7	2.0

b (i) Determine a likely value for the density of pure chromium in $g\,cm^{-3}$. Show all steps in your calculation

(density of nickel $= 8.9\,\mathrm{g\,cm}^{-3}$).
(ii) What do you think is meant by the term 'current-carrying capacity'?
(iii) What is the relationship between the thickness of nichrome wires and their current-carrying capacity? Clearly show how you arrived at your answer.
(iv) Based on your answer to **b** (iii), justify the use of thin connecting leads for stereo sets and thicker cables for power drills and electric ranges.

c The resistances per unit length for three different resistance wires of the same diameter are given below.

Resistance per unit length	0.025	0.727	2.70
Wire type	copper	constantan	nichrome

(i) State the unit(s) of resistance per unit length. (ii) Which of the wires would you select as connecting leads in electrical circuits? Justify your choice.

12 Using an ohmmeter, a student obtained the following results for constantan wire of diameter 0.315 mm.

Length of constantan wire/m	Resistance/Ω
0	0
0.25	2.01
0.50	4.05
0.75	6.06
1.00	8.08
1.25	10.10
1.50	12.12
2.00	16.17

a Plot a graph of resistance against length for the wire used. State, in precise terms, the relationship between resistance and length. Indicate what factors should be controlled if the stated relationship is to be valid.

b Use the graph to determine (i) the resistance of 1.75 metres of this wire,

(ii) what length of this wire has a resistance of 9.0 Ω.

c Determine the gradient of the graph, stating the units. What is the significance of the gradient?

13

Resistance per metre/$\Omega\,\mathrm{m}^{-1}$	Diameter of wire/mm
2.30	1.626
4.09	1.219
7.27	0.914
12.73	0.711
19.47	0.559
43.06	0.376
61.32	0.315

Use the data above to make a precise statement linking resistance per unit length and diameter for a given material (type of resistance wire).

14 A device operates normally from a 9 volt source. It is rated at 15 W.
a When operated normally, what
(i) current flows through the device,
(ii) is the resistance of the device?
b What current flows through the device if it is now connected to a 6 V supply?
c How can this device be operated normally from a 15 V supply? Give precise value(s) of any additional component(s) needed. Draw a circuit diagram of the arrangement that can be used.

15 The diagram gives a possible circuit for recharging a run-down battery.
a Why is a.c. not used, directly, to recharge the battery?

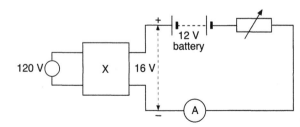

b State two components that device X is likely to contain.

c Explain the function of each component identified in **b**.

The battery should be charged at a low steady current.

d How is the current controlled in this circuit?

e Why does the current tend to decrease as the battery is being charged?

f Why are the charging supply and the battery to be recharged connected with opposing polarities?

g Assuming that the battery is a lead–acid battery, outline five steps for its care and maintenance.

h State two ways in which a lead–acid battery differs from a dry cell.

16

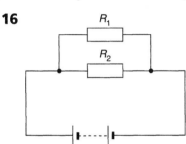

Consider the circuit above. Suppose that the current through R_1 is 0.25 A and that the potential difference across R_2 is 2.5 V.

a Find the value of R_1.

b If the power dissipated in R_2 is 0.375 W, what is the current through R_2? Hence, or otherwise, determine the value of the resistance R_2.

c What is the total current drawn from the cell?

d What is the effective resistance of the combination of R_1 and R_2?

17 Device A: 3000 W, 240 V
Device B: 1200 W, 120 V
Device C: 180 W, 120 V
Device D: 100 W, 120 V

a Which device draws the smallest current from its power supply? Show all steps in your calculations.

b Which device uses the most energy in 2 hours?

c A colour TV and a black and white TV were on for the same time. It costs three times as much to run the colour TV. Comment fully on the significance of this information. State all assumptions.

18 A student connects three 120 V, 60 W devices as shown.

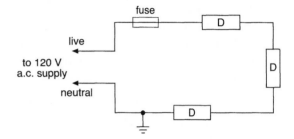

a Why is the fuse connected to the live wire?

b What is the function of the earth connection?

c What is the total resistance in the circuit, as connected?

d What current does each device need to function normally?

e What current is flowing in the circuit as connected?

f In the light of your answers to **d** and **e** criticize the circuit arrangement.

g Draw a suitable diagram of the circuit arrangement that would work. Include in your diagram components which would enable the devices to operate independently.

19 Briefly explain the difference(s) between a permanent magnet and an electromagnet.

20 How, using a magnetic compass, would you determine the north and south poles of an unmarked magnet?

21 What is meant by the term 'magnetic domains'?

22 Use the idea of domains to account for the difference(s) between an unmagnetized material and magnetized material.

23 Iron is described as 'magnetically soft' while steel is described as 'magnetically hard'.

Account for the difference in the behaviour of iron and steel as magnetic materials.

24 What is meant by the term 'magnetic shielding'? Which material is best for magnetic shielding?

25 Which kind of material – soft iron or steel – is better suited for the following purposes?
   a magnetic keepers
   b permanent magnets
   c electromagnets
   d the diaphragm of a telephone earpiece
   e the recording head of a tape recorder

26 Justify the statement 'magnetic induction always leads to attraction'. Use diagrams to support your answer.

27 You are provided with a low voltage, high current d.c. supply, a rheostat, insulated copper wire and a switch. Describe how you would magnetize a small steel bar.

28 Describe how you would compare the strength of the magnet made in question 27 with another (laboratory) magnet.

29 Describe how you would use a plotting compass to map the field lines around a bar magnet.

30 Draw a horseshoe magnet, with its magnetic field. Label the magnetic poles, indicate the air gap, and show the direction of flux (magnetic field lines).

31 Why is an iron nail attracted to both ends of a bar magnet? Give a full explanation.

32 Draw diagrams to show the pattern and direction of the magnetic field produced in a solenoid carrying a current.

33 Describe an experiment that you would carry out to show that a force exists between two magnets which are placed near one another. Indicate the readings you would take and how you would use these readings.

34 Draw a large, clearly labelled diagram of a magnetic relay and explain how it works.

35 a Two pieces of 'different metals' were separately magnetized in a solenoid.

Describe how this can be done in practice.
   b After switching off the current it was found that one piece retained its magnetism, while the other did not. Which of the pieces of metals would be suitable for use as the core of an electromagnet and which piece for use as the permanent magnet in an electric motor? Give a reason for your choice.

36 Describe how you would use the apparatus provided to make a simple current measuring meter. Other pieces of apparatus may be used.

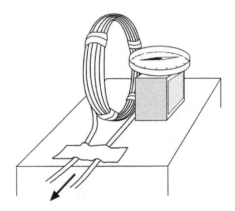

Is your meter expected to give reliable readings? Explain.

How would you set out to calibrate your meter (i.e. establish a scale on it)?

37 A soft iron bar is placed in the centre of the coil as shown and the switch is closed.

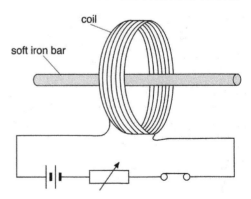

   a What is the effect, if any, on the bar?
   b State **two** ways in which the effect described in **a** can be increased.

c Will the effect described in **a** be greater or smaller if the bar is placed at a position other than along the centre? Explain.

d Predict what will happen if two iron bars are placed side by side inside the coil, with current on.

e Suggest one application for the effect described in **d**.

f If alternating current were used instead of direct current in **d** would the effect observed be the same? Justify your answer.

38 The diagram shows an electromagnetic device.

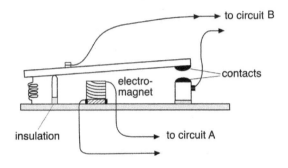

a Name the device.

b Explain how the device works.

c Discuss three widely different applications of the device.

d Relate the use of 'soft iron', in this device, to its magnetic properties.

39 A boy wanted to operate a 12 V, 24 W device Z at his work bench. The only available source of electricity was a 120 V mains a.c. supply 100 metres away. He obtained what he thought was a suitable transformer and appropriate lengths of cable (total resistance 5 Ω). He arranged the components as shown below:

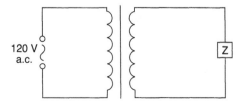

To his dismay, the boy found that the device Z did not function normally when hooked up as shown.

a Suggest a possible reason for the failure of Z to function normally.

b Give an alternative circuit diagram, using the same components, in which Z is more likely to function normally.

c Suggest a suitable value for the ratio of the number of turns of coil in the primary to the number of turns in the secondary for the transformer used.

40 A small electrical (power) plant has a total generating capacity of 200 MW. The voltage at the generator terminals is 13.8 kV. A part of the 200 MW is transmitted at 33 kV to the distribution substations. From the distribution substations, 12 kV feeders move out in all directions. The consumer receives a 230 V supply from the utility poles.

a Express in standard form:
(i) 200 MW, (ii) 33 kV.

b Describe, in outline, how the 13.8 kV at the generator terminals is raised to 33 kV for transmission.

c Why should large blocks of power be transmitted at high voltages?

d Why is low voltage occasionally experienced at certain times of the day?

e The primary fuel used at this power station is natural gas. (i) Describe all the energy transformations which take place from natural gas to electricity.
(ii) Natural gas – which is mainly methane – <u>has an energy value of 37.2 MJ m^{-3}</u>. Explain what is meant by the underlined words.

f On a given day the station operated at 60% total capacity for two hours. Calculate the total energy supplied to all customers in this time, assuming no transmission losses.

g Calculate the theoretical volume of natural gas that was burnt to provide this energy.

h In practice the volume of natural gas required is greater than that calculated in **g**. Why is this so?

The early Greeks believed that matter was made up of indivisible particles. The word atom, derived from the Greek *atomos*, means indivisible.

The idea of the indivisibility of the atom persisted into the twentieth century. However, the discovery of radioactivity and other phenomena indicate that the atom is in fact divisible, given the right circumstances.

Note

Joseph John Thomson (1856–1940) discovered the electron and laid the foundations of mass spectrometry. He was awarded the Nobel Prize in Physics in 1906.

27.1 Enter J. J. Thomson

Starting with Faraday and continuing throughout the nineteenth century, many workers studied the passage of electricity through gases at low pressure in discharge tubes (Figure 27.1). The results of these studies showed that the atom was made up, at least in part, of negatively charged particles.

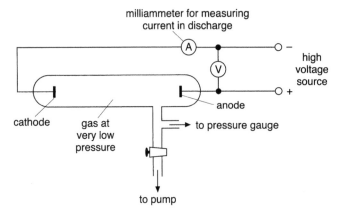

Figure 27.1 *Apparatus like this was used to study the passage of electricity through gases.*

Thomson, after carefully studying these negative particles, concluded that they were:

- present in all matter;
- two thousand or so times lighter than hydrogen – the lightest atom;
- identical in nature with the electrical charge which flowed in the external circuit during electrolysis.

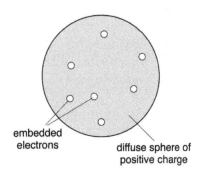

Figure 27.2 *Thomson's plum pudding model.*

As atoms are neutral, the negative charge on the electrons must be balanced by a positive charge. So Thomson suggested a model of the atom in which negative electrons 'were embedded in a diffuse sphere of positive charges' (Figure 27.2). This model,

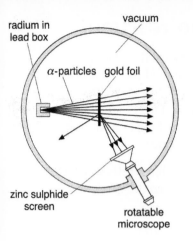

Figure 27.3 *The apparatus used by Rutherford's team.*

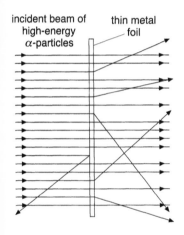

Figure 27.4 *The results of the Geiger–Marsden experiment.*

commonly called the 'plum pudding' model, was not satisfactory because it did not account for the ease with which electrons could be knocked off from atoms.

27.2 Rutherford's planetary model emerges

Rutherford, having studied the phenomenon of radioactivity in depth, realized that alpha (α) particles were highly energetic and penetrated matter easily, knocking electrons from atoms in their path. Armed with such good projectiles, Rutherford had two of his students, Geiger and Marsden, fire away at bits of materials (Figure 27.3). They fired alpha particles at thin metal foils with the following results (Figure 27.4).

- Most of the alpha particles passed through without any change of direction. Clearly, these particles were passing through empty or charge-free space.
- A few particles were deflected through small angles.
- Still fewer particles (about 1 in 10 000) were deflected through a right angle or more. These particles behaved as if they had approached an area of dense positive charge and been repelled.

Based on these observations, Rutherford proposed a nuclear model of the atom in which:

- the bulk of the atom is empty or charge-free space;
- 99.9% of the mass is concentrated in a small region of positive charge at the centre of the atom. This region is the **nucleus** of the atom;
- electrons are found in the space outside the nucleus.

Rutherford pictured the atom as a tiny solar system in which the nucleus plays the role of the Sun, and the electrons the role of orbiting planets (Figure 27.5).

Rutherford's model, although superior to Thomson's plum pudding model, did not address one important question: how were the electrons arranged in the outer atom?

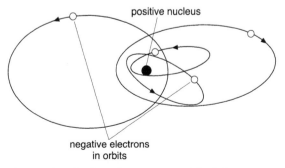

Figure 27.5 *The planetary model of the atom.*

Note

Niels Bohr (1886–1962) was a great theoretical physicist. He was awarded a Nobel Prize in 1922. He explained the spectra of hydrogen and helium in terms of Rutherford's nuclear model of the atom.

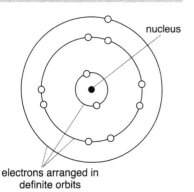

electrons arranged in definite orbits

Figure 27.6 *The Bohr atom.*

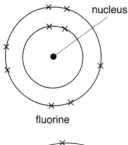

fluorine

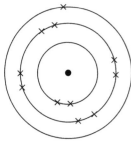

sodium

Figure 27.7 *The electronic configurations of fluorine and sodium.*

Thinking it through

The aluminium atom has 13 electrons. What is its electronic configuration?

27.3 Niels Bohr to the rescue

Rutherford was aware when he put forward the planetary model of the atom that according to 'classical theory' the electrons ought to fall into the nucleus. Bohr was able to show that they did not and explain why they did not. Using the 'quantum theory' first formulated by Max Planck and used successfully by others, Bohr worked out that:

- electrons occupied definite or permissible **shells** or **orbits**;
- an electron, in its permitted orbit, circled the nucleus without loss of energy.

In this way, Bohr was able to firm up and extend Rutherford's picture of the atom (Figure 27.6). The electrons in an atom occupy permissible shells, each of which can hold a maximum number of electrons. The first shell (that nearest the nucleus) holds a maximum of 2 electrons, the second shell a maximum of 8, the third shell a maximum of 18, and so on.

The outer occupied shell is known as the **valence shell**. Electrons in the valence shells are the ones involved when atoms combine.

27.4 Electronic configuration

The arrangement of electrons in an atom is commonly called its **electronic configuration**. The fluorine atom, which contains 9 electrons, has electronic configuration 2, 7 (Figure 27.7). The sodium atom has 11 electrons. Its electronic configuration is 2, 8, 1 – there are 2 electrons in the first shell, 8 in the second and 1 in the third shell.

27.5 Chadwick tracks down the neutron

Rutherford knew that protons alone could not account for the calculated mass of the nucleus and he predicted that the nucleus contained another particle – a neutral particle with mass about that of the proton. Without success, Chadwick and others searched for the neutron during the 1920s. In 1930, two German scientists found that a highly penetrating 'radiation' was given off when beryllium was bombarded with alpha particles. They thought that the new radiation was a type of high-energy gamma radiation. In 1932, French scientists repeating the above experiment obtained results which led Chadwick to conclude that the so called 'new radiation' consisted of particles of mass 'nearly equal to that of the proton and with no charge'. At last, the neutron had revealed itself.

Note

James Chadwick (1891–1974) is credited with the discovery of the neutron. He was awarded the Nobel Prize in Physics in 1935.

To summarize:

- The atom has a tiny nucleus which contains protons and neutrons.
- Electrons move around the nucleus, while occupying definite shells.
- Ordinarily, the atom is neutral became it contains as many protons (positive charges) as electrons (negative charges).
- An atom can become an ion by either gaining or losing electrons.

Table 27.1	The relative masses and charges of subatomic particles		
Particle	**Symbol**	**Relative mass**	**Relative charge**
electron	e	1/1840	−1
proton	p	1	+1
neutron	n	1	0

27.6 Atomic number, mass number, isotopes

Note

Atomic number (Z) = number of protons.

A **nucleon** is a proton or a neutron.

The **atomic number** (Z) of an element is the number of protons in the nucleus of the atoms of that element. For a neutral atom, the atomic number is equal to the number of orbiting electrons.

Mass number (A) is the number of protons plus the number of neutrons in the nucleus. Mass number is also called the nucleon number. It follows that:

$$\text{neutron number} = \text{mass number} - \text{atomic number}$$
$$= A - Z$$

Each atomic species is uniquely defined by its symbol, its atomic number and its mass number. A species so defined is called a **nuclide**. The species carbon-12 is written for short as:

$$^{12}_{6}\text{C}$$

6 is the atomic number, 12 is the mass number and C is the symbol of the element. Chlorine-37 is written as:

$$^{37}_{17}\text{Cl}$$

There are species which have the same atomic number but which differ in their mass numbers. That is, they have the same number of protons but different numbers of neutrons. Such species are called **isotopes**. Carbon-12, carbon-13 and carbon-14 are isotopes of carbon (Figure 27.8). They have the same chemical properties.

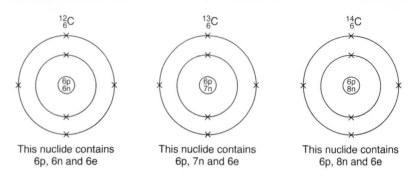

Figure 27.8 *Three isotopes of carbon.*

27.7 The Periodic Table

Early attempts to arrange the elements so as to highlight similarities between them were only partially successful. In 1869, Mendeleev proposed that a relationship existed between the atomic masses of elements and their chemical properties. When Mendeleev arranged the elements in order of increasing atomic mass and in rows, he found that chemically similar elements appeared beneath each other in columns. Mendeleev left gaps where he felt elements were missing (not yet discovered) and correctly predicted the physical and chemical properties of these missing elements. Unfortunately, the atomic masses of a few elements did not fit the pattern.

In 1913 Moseley, using data obtained from his studies with X-rays, proposed that the elements be arranged in order of increasing atomic *number* and not atomic mass. With this arrangement all the elements which seemed out of place in Mendeleev's table fell into their correct vertical sets (or groups). In this way the modern Periodic Table was born.

The Periodic Table, then, is an arrangement of the elements in order of increasing atomic number. Each element in the Periodic Table has one proton more than the element before it. When arranged in the Periodic Table, the elements fall into vertical sets called **groups** and horizontal sets called **periods**.

The atoms of elements in the same group have the same number of electrons in their outer shells. However, on going down the group the number of inner shells increases. Fluorine, chlorine and iodine all have seven electrons in their outer shells. They are members of the same group. But note that, whereas fluorine has only one inner shell, chlorine has two and iodine four inner shells. Their electronic configurations are:

fluorine 2, 7 chlorine 2, 8, 7 and iodine 2,8,18,18,7

Table 27.2

Elements in period 3	Na	Mg	Al	Si	P	S	Cl	Ar
Electronic configuration	2, 8, 1	2, 8, 2	2, 8, 3	2, 8, 4	2, 8, 5	2, 8, 6	2, 8, 7	2, 8, 8

From the electronic configuration of the elements Na to Ar (Table 27.2) you can see that:

- elements in a given period have the same number and type of inner shells;
- elements in a given period differ in the number of electrons in their outer shells.

Each time a new outer shell is started, a new period begins.

The Periodic Table:

- summarizes information about the elements;
- links seemingly unrelated facts;
- provides a basis for explaining the behaviour of elements;
- enables us to make predictions about the physical and chemical properties of elements.

The Periodic Table divides the elements into metals, non-metals and semi-metals. Non-metals are found to the right of the dark 'staircase'. Metals are found to the left. Semi-metals are shaded.

Note

The noble gases (group 0 of the table) were discovered by Lord Rayleigh and Sir William Ramsay. In 1895 Mendeleev adjusted his original Periodic Table to accommodate these.

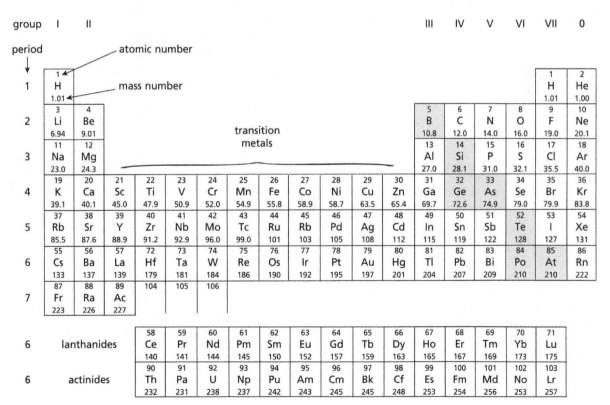

Figure 27.9 *A modern version of the Periodic Table.*

Things to do

An element has mass number 131. The nucleus of an atom of this element contains 77 neutrons.

1 What is the atomic number of this element?
2 Use the Periodic Table to identify this element.
3 To which group and which period does this element belong?

Checklist

After studying Chapter 27 you should be able to:

- discuss the contributions of Thomson, Rutherford, Bohr and Chadwick to our knowledge of the atom
- describe the Geiger–Marsden experiment, and the importance of its results
- recall the contents of the atomic nucleus
- recall the arrangement of the atom's particles
- recall the relative masses, charge and location of the atom's particles
- define atomic number, mass number and nucleon number
- define a nucleon
- recall and use the relationship $A = Z + N$
- represent the structure of simple atoms
- define an isotope
- represent nuclides using standard notation
- relate an atom's shell model to its position in the Periodic Table
- discuss the significance of the Periodic Table

Questions

1 An atomic species X contains 15 protons, 15 electrons and 17 neutrons.
   a What is its mass number?
   b What is its atomic number?
   c Give the symbol for this species.

2 How many neutrons are there in the nucleus of $^{208}_{82}$Pb?

3 a What evidence led Thomson to propose the 'plum pudding' model of the atom? List the evidence.
   b Why is the 'plum pudding' model for the structure of the atom not satisfactory?

4 Briefly suggest how the results of the Geiger–Marsden experiment would have been different if electrons were in the nucleus and protons were located in shells around the nucleus.

5 What was Niels Bohr's contribution to the discovery of atomic structure?

6 An atomic species Y has 14 nucleons, 6 protons and 4 electrons.
   a Draw a model of species Y.
   b Is species Y positively charged, negatively charged or uncharged? Why?

7 Write short notes on the structure of the atom.

8 Draw diagrams to show the atomic structure of each of the following. For each indicate the number of protons, neutrons and electrons.
   a $^{32}_{16}$S b $^{11}_{5}$B c $^{28}_{13}$Al d $^{40}_{20}$Ca

28 Radioactivity

Note

Radioactivity is the process by which the nuclei of unstable atoms disintegrate or decay.

Whenever the words radioactivity and radiation are mentioned some people think immediately of the 1945 bombing of Hiroshima and Nagasaki, or the 1986 Chernobyl nuclear power plant disaster . . . or of the possibility of war in which nuclear missiles might be used. However, radiation and radioactivity have yielded important information about the structure of matter and have served us well in many other ways: for example, in medical diagnosis, in the treatment of diseases and in improving the quality of manufactured goods.

28.1 Becquerel discovers radioactivity

Note

Henri Becquerel (1852–1908) was jointly awarded (with the Curies) the Nobel Prize in Physics in 1903.

Becquerel accidentally discovered radioactivity in 1896, when he observed that photographic plates, secured in black paper, became 'fogged' when placed near a uranium-containing compound. He concluded that: 'a source of energy exists deep inside atoms', since coming out of uranium compounds was 'something' which was more penetrating than light and which ionized the surrounding air. Becquerel's discovery aroused the curiosity of other scientists who eagerly sought answers to such questions as:

- what caused radioactivity?
- why did radiations ionize the surrounding air?
- was there more than one type of radiation?
- why was it difficult to detect this radiation directly?
- what were the best methods for detecting radiation?

28.2 Zeroing in on the nature of radioactivity

Marie Curie was one of the early workers in the field of radioactivity. Among other things, she found that:

Note

Marie Curie (1867–1934) and her husband Pierre did most of their pioneering work in an ill-equipped shed. Marie Curie was awarded two Nobel prizes – in Physics in 1903 and in Chemistry in 1911.

- the element thorium was more active than uranium;
- natural radioactivity was not affected by either changes of conditions or by chemical combination (in other words, the 'activity' of a radioactive element remained unchanged when it formed bonds);
- the mineral pitchblende caused a greater blackening of photographic plates than expected – a fact which caused her to wonder whether the mineral contained radioactive elements other than uranium and thorium.

Other workers found that other elements were also radioactive. Radioactivity was then recognized as a widespread phenomenon.

There were found to be three types of radiation: α-particles, β-particles and γ-radiation (photons).

α-particles

This type of emission was deflected to a small extent in magnetic fields. The direction of the deflection indicated that α-particles were positively charged. In 1909, Rutherford 'proved' that the α-particle was identical with the helium nucleus.

β-particles

These were deflected in a magnetic field but in the direction opposite to that of α-particles. β-particles were later shown to be fast-moving electrons.

γ-radiation

This radiation, being uncharged, was not affected by electric or magnetic fields. It was thought to be wave-like in nature. This was confirmed by Rutherford and Andrade in 1914.

Figures 28.1 and 28.2 show the effect of magnetic and electric fields on radiation. Note that:

- These are composite diagrams. It is not possible to investigate all three radiations at the same time.
- Because of their smaller mass, β-particles are deflected more than α-particles by both magnetic and electric fields.
- The behaviour of radioactive emissions in electric fields can be explained on the basis that like charges repel and unlike charges attract.
- The direction of deflection of α and β-particles in magnetic fields can be predicted from Fleming's left-hand rule (p. 298).

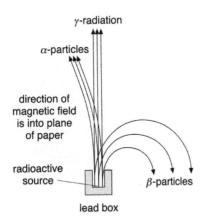

Figure 28.1 *The effect of a magnetic field on radioactive emissions.*

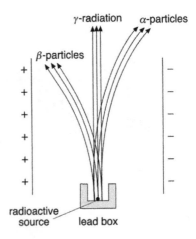

Figure 28.2 *The effect of an electric field on radioactive emissions.*

28.3 Comparing the properties of α, β and γ-radiation

Table 28.1 Properties of radiation

	α-particles	β-particles	γ-radiation
Identity	helium nuclei $^{4}_{2}He^{++}$	fast-moving electrons $^{0}_{-1}e$	electromagnetic radiation of high frequency $\sim 3 \times 10^{19}$ Hz
Charge	++	-1	none
Mass compared with the proton	4	$\sim\dfrac{1}{1840}$	none
Can it affect a photographic plate?	yes	yes	yes
Does it cause fluorescence?	yes	yes	yes
It is stopped by	5 cm of air or a sheet of paper	a few mm of aluminium	several cm of lead
Penetrating power (relative)	1	100	10 000
Ionising power (relative)	10 000 strong	100 moderate	1 weak
Path through matter	straight	bent	straight
Is it deflected by a a magnetic field?	yes	yes	no
Is it deflected by an electric field?	yes	yes	no

Points to note:

- α-particles produce ions by knocking electrons from atoms with which they interact.
- At the end of its path each α-particle absorbs two electrons and becomes a neutral helium atom.
- β-particles cause less ionization than α-particles.
- γ-radiation does not produce ions directly.

28.4 Radioactive decay

Changes occur in the nucleus of atoms when radiations are emitted.

The emission of α-particles

An alpha (α) particle is a composite of two protons and two neutrons. It has an atomic number of 2 and a mass number of 4. When a parent nuclide emits an α-particle:

- atomic number decreases by 2;
- mass number decreases by 4.

The change may be represented as follows:

$$\underset{\text{parent radioactive nuclide}}{^{A}_{Z}X} \rightarrow \underset{\text{daughter nuclide}}{^{A-4}_{Z-2}Y} + \underset{\text{alpha particle}}{^{4}_{2}He^{++}}$$

Examples of α decay

$$^{238}_{92}U \rightarrow \,^{234}_{90}Th + \,^{4}_{2}He^{++}(\alpha) \tag{i}$$

$$^{234}_{90}Th \rightarrow \,_{88}Ra + \,^{4}_{2}He^{++}(\alpha) \tag{ii}$$

$$_{88}Ra \rightarrow \,^{226}Rn + \,^{4}_{2}He^{++}(\alpha) \tag{iii}$$

The emission of β-particles

The beta (β) particle is a high-speed electron. It can be represented as $^{0}_{-1}e$. The β-particle, then, has an atomic number of -1 and a mass number of 0.

When a parent nuclide emits a β-particle:

- atomic number increases by 1;
- mass number stays the same.

$$\underset{\text{parent radioactive nuclide}}{^{A}_{Z}X} \rightarrow \underset{\text{daughter nuclide}}{^{A}_{Z+1}Y} + \underset{\text{beta particle}}{^{0}_{-1}e}$$

Examples of β decay

$$^{32}_{15}P \rightarrow \,^{32}_{16}S + \beta(^{0}_{-1}e) \tag{iv}$$

$$^{73}_{30}Zn \rightarrow \,^{73}Ga + \beta(^{0}_{-1}e) \tag{v}$$

$$_{54}Xe \rightarrow \,^{140}_{55}Cs + \beta(^{0}_{-1}e) \tag{vi}$$

During a beta decay a neutron spontaneously changes into a proton and an electron (a beta particle): the beta particle is ejected from the nucleus. This change from neutron to proton leads to an increase in atomic number. However, the mass number stays the same.

The emission of γ-radiation

Neither atomic number nor mass number is altered by the emission of γ-radiation. However, the emission leaves the nucleus in a less excited (i.e. a more stable) state.

Think of γ-radiation as packets of energy. Gamma radiation may be emitted on its own or may be emitted after the emission of an alpha or beta particle.

Gamma radiation has a higher frequency than X-radiation but most of its other properties are quite similar to those of X-rays. Whereas X-rays are produced when high-speed electrons lose energy on striking a solid target, γ-radiation results from the changes within the nucleus of an unstable atom.

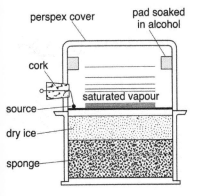

Figure 28.3 *The cloud chamber.*

- *The upper chamber contains vapour.*
- *The lower portion of the upper chamber is cooled (by dry ice) and it becomes supersaturated.*
- *Emissions from the source produce ions.*
- *Supersaturated vapour condenses on the ions, producing visible tracks.*

Thinking it through

Why do the alpha particle tracks in Figure 28.4(a) have different lengths?

28.5 Methods of detecting radiation

Making tracks – the cloud chamber

A cloud chamber is a device which makes the paths of moving charged particles visible (remember that charged particles cannot be seen directly). It works on the principle that ions are produced by radioactive particles and that these ions can act as centres on which vapour condenses. In the cloud chamber of Figure 28.3 vapour condenses on ions, causing tracks to be formed in the chamber (Figure 28.4). Light, shone from the side, enables the tracks to be clearly seen from above. The tracks are similar to the trails made by jet planes.

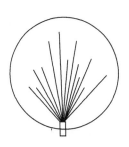
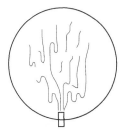
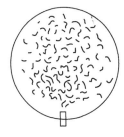

a) Tracks of α-particles are thick and straight

(b) Tracks of β-particles are thin, bent and tortuous

(c) Tracks of γ-radiation produce secondary tracks

Figure 28.4 *Tracks as seen in the cloud chamber.*

The Geiger–Müller tube

The Geiger–Müller, or Geiger, tube detects ions produced when
α-particles, β-particles or γ-radiation interact with a gas at low
pressure.

Each burst of radiation that enters the tube produces a pulse of
current. When radiation hits argon atoms electrons are knocked
off and positive ions are produced. The freed electrons accelerate
towards the positively charged metal wire, colliding with more
argon atoms which, in turn, become positive ions. At the same
time, positive ions move towards the negatively charged metal
cylinder, generating a pulse of current. The current pulse is
amplified and registered on an electronic counter or loudspeaker
(Figure 28.5).

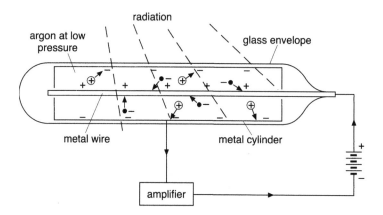

Figure 28.5 *The Geiger–Müller tube at work.*

28.6 Comparing the ranges of α, β and γ-radiation

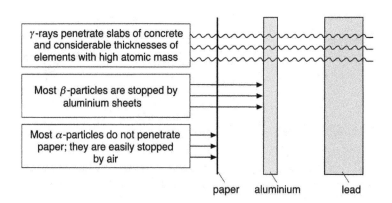

Figure 28.6 *The ranges of ionizing radiations compared.*

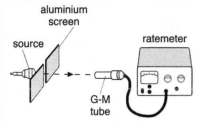

aluminium screen
source
ratemeter
G-M tube

Figure 28.7 *Finding the range. Use an aluminium screen with a small gap in front of the α-source. This prevents overload of the G–M tube. The screen is not needed with β or γ-sources.*

How to compare the ranges of ionizing radiation

1 Determine the background radiation count (the reading on the meter with no source present: see p. 341).
2 Place a pure α-source such as americium-241 less than 5 cm from the G–M tube as shown in Figure 28.7.
3 Record the ratemeter reading.
4 Place a thin sheet of paper between the source and the G–M tube and record the new ratemeter reading.
5 Repeat using thick sheets of paper.
6 Repeat the above steps with a pure β-source, many sheets of paper, sheets of aluminium of different thicknesses.
7 Try again with a γ-source, many sheets of paper, sheets of aluminium and sheets of lead of different thicknesses.

Things to do

1 Design an experiment to determine how many metres of air will stop α-particles.
2 Design an experiment to test the statement: 'the intensity of the radiation from a γ-ray source falls off inversely as the square of the distance from the source'.

Note

In general, nuclides with 2, 8, 20, 28, 50 and 82 protons and 2, 8, 20, 28, 50, 82 and 126 neutrons are unusually stable. These numbers of protons and neutrons are known as 'magic numbers'.

Note

Whenever the nuclide from a radioactive decay is itself radioactive further decay takes place.

From this experiment you should find that:

- α-particles are stopped by thick sheets of paper;
- β-particles are stopped by between 5 mm and 1 cm of aluminium;
- γ-rays are not completely stopped even by 10 cm of lead.

28.7 Unstable nuclei

The nuclei of many atoms are stable and do not undergo nuclear changes readily. However, some nuclei, in particular the heavy ones, are unstable or radioactive. These nuclei disintegrate (decay) spontaneously and become stable by emitting energy in the form of ionizing radiation.

During radioactive changes:

- new elements may be formed;
- particles or packets of radiation may be ejected from the nucleus;
- energy is released.

A parent nucleus continues to emit some form of radiation until stable daughter or granddaughter, etc., nuclei are obtained. For example:

uranium-238 $\rightarrow$ thorium-234 $\rightarrow$ protactinium-234
parent nuclide   daughter nuclide   granddaughter nuclide

Figure 28.8 shows that thorium-232 is involved in a series of radioactive decays which end in the formation of lead-208. Other unstable nuclides, e.g. actinium-228, radium-228 and polonium-216, are formed along the route.

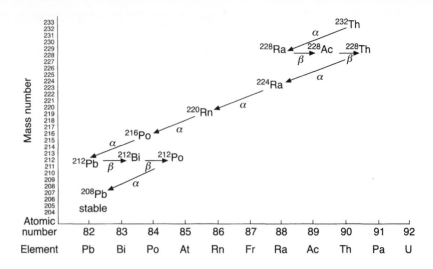

Figure 28.8 *The sequence of decays shown, starting with thorium-232, is an example of a radioactive decay series.*

28.8 Background radiation

Background radiation is unavoidable radiation. It is the natural radioactivity that is always present in our surroundings. Sources of background radiation include:

- radioactive materials in the Earth, e.g. deposits of uranium and radium;
- the air, which contains small amounts of thoron and radon, two radioactive gases;
- the testing of atomic weapons and nuclear wastes, e.g. from nuclear reactors;
- food and drink – radioactive materials in the Earth's surface are taken up by plants and animals or become dissolved in our water supply;
- cosmic rays, high energy radiations which enter the Earth's atmosphere from space. Cosmic rays collide with atoms in the atmosphere, producing radioactive species. The further you are away from the Earth's surface, the more cosmic radiation you receive.

28.9 Radioactive decay is a random process

The decay of radioactive substance is random. Any atom may decay at a given time. Moreover, three atoms may decay in the first second, six in the second, none in the third and so on.

Note

The reading of the G–M or other counter, in the absence of a source, is the 'background count'.

How to show that radioactive decay is a random process

1 Set up the apparatus of Figure 28.7 but excluding source and screen. In this arrangement the G–M tube detects only background radiation.
2 Record the count rate or the number of clicks heard over successive ten-minute periods.

Thinking it through

Use the data in Table 28.2 to calculate the average background count for the given place.

Table 28.2 gives data for ten 1-minute periods for a given place.

Table 28.2										
Number of clicks	19	13	23	17	20	22	22	23	21	26
Time period	1	2	3	4	5	6	7	8	9	10

The lack of any pattern in the clicks indicates that radioactive decay is a random process.

How to illustrate random processes using dice

Do this as a class exercise. You will need 100 dice.
1 Throw all 100 dice and count the number which land 'sixes up'. Consider these to be 'decayed' dice and remove them from succeeding throws.
2 Repeat, throwing the 'undecayed' dice (those that did not land 'sixes up' in the first throw).
3 Once again, remove the 'decayed' dice, and repeat throws with 'undecayed' dice.
4 Record the number of 'undecayed' dice and number of throws.
5 Draw a graph of your results.

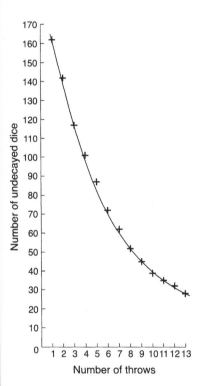

Figure 28.9 *Graph of 'undecayed' dice against number of throws.*

The result for one such dice throwing exercise is given in Table 28.3. A plot of number of undecayed dice against number of tosses is shown in Figure 28.9.

Curves of the kind in Figure 28.9 are known as 'exponential decay curves'. The number of dice which turn up 'sixes' decreases with each successive toss. The number of 'sixes' depends on the number of dice in the throw and on the probability of throwing a 'six' (1/6). Radioactive decay follows the same general pattern.

Table 28.3												
Number of 'undecayed' dice	162	140	117	101	87	72	62	52	45	39	35	32	28
Throw number		1	2	3	4	5	6	7	8	9	10	11	12	13

Half-life

For a given radioactive element, there is a specific time during which half the original number of radioactive nuclei will decay. This time is the **half-life** of the radioactive element.

Imagine that you are provided with a sample of sodium-24, the half-life of which is 15 hours. Let the initial activity of the sample be 2000 Bq. A half-life of 15 hours means that in 15 hours the activity will fall to 1000 Bq; after 30 hours the activity will be 500 Bq and after 45 hours the activity will be 250 Bq . . . and so on . . .

Table 28.4	Half-lives of some radioactive isotopes					
Radioactive isotope	O-13	Na-24	I-131	Sr-90	Ra-226	U-238
Half-life	0.01 s	15 h	8 days	28 years	1620 years	4.5×10^9 years

Each radioactive element has its own half-life. This means that radioactive elements may be identified by their half-lives.

The variation of activity with time for a radioactive source, e.g. radon-220, may be determined experimentally using a G–M tube and a scaler. Table 28.5 shows the data obtained in one such determination, after correcting for background radiation.

Table 28.5	Activity of radon														
Activity/Bq	240	208	184	168	144	128	112	96	88	78	68	58	50	44	39
Time/s	0	10	20	30	40	50	60	70	80	90	100	110	120	130	140

The graph of activity (Bq) versus time (s) is shown in Figure 28.10.

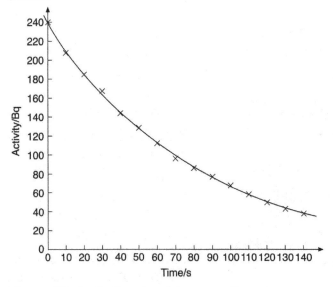

Figure 28.10 Radioactive decay curve for radon.

Thinking it through

- After what time will the activity be 80 Bq?
- What is the activity after 83 seconds?

The graph shows that activity drops from 240 Bq to 120 Bq in about 54 seconds and that activity also falls from 120 Bq to 60 Bq in the same time. The half-life of radon is therefore 54 seconds.

A useful formula

If the initial activity of a source is A_0 and the activity after n half-lives is A_n, then:

$$\frac{A_0}{A_n} = 2^n$$

Example

The initial activity of a source is 16 000 Bq. After how many half-lives will the activity fall to 125 Bq?

Solution

$$\frac{16\,000}{125} = 2^n$$

$$128 = 2^n$$

$$n = 7$$

The activity drops to 125 Bq after 7 half-lives.

28.10 The dangers of radiation
The biological effects of radiation

Radiation damages biological tissue by:

- killing cells (both healthy and diseased);
- causing cells to reproduce in an uncontrolled way, i.e. causing cancers;
- causing mutations, i.e. producing new cells with different characteristics.

Radiation damage may be somatic (showing up in the irradiated person) or hereditary (showing up in the offspring of irradiated persons, e.g. as birth defects).

Both Marie Curie and her daughter Irene died of leukaemia. This disease resulted from their long and repeated exposure to ionizing radiation.

In 1986 a reactor at the Chernobyl nuclear power plant in Russia exploded. Radiation spread into the air and was blown by winds into central Europe thousands of kilometres away.

Note

Large doses of radiation can cause serious injury, cancer or death. The extent of injury depends on both the size of the dose and the period of exposure.

Thinking it through

The actual dose a person receives depends on occupation. A pilot, for example, receives twice as much as the average teenager. Why?

Safety around radiation

- Ensure that radioactive isotopes (**radio-isotopes**) do not enter the body, e.g. through cuts, through the mouth, nose . . . little can be done once a radio-isotope enters the body.
- Handle all radioactive sources with tongs. In some industries, robots are used to handle sources.
- Point all sources away from you and from other people.
- Replace or store sources in their special containers.

For routine handling of radioactive materials:

- keep the amount being handled to a minimum.
- contain radioactive materials properly.
- follow safety rules for washing clothing, disposal, etc.

Figure 28.11 *The damaged Chernobyl reactor encased in concrete.*

Figure 28.12 *Nuclear fuel rods are handled remotely by robots.*

28.11 Uses of radio-isotopes

Gauging materials

The ability of materials to absorb radiation can be used:

- to measure thickness;
- to measure density;
- to control the quantity and consistency of powders or liquids as they are 'filled' into boxes, tins or cartons.

To monitor the thickness of a sheet of material on a production line, a radio-isotope is placed on one side of the moving material, a detector on the other side. Radiation passes through the material as it moves between two rollers (Figure 28.13).

Any significant change in thickness shows up on the detector.

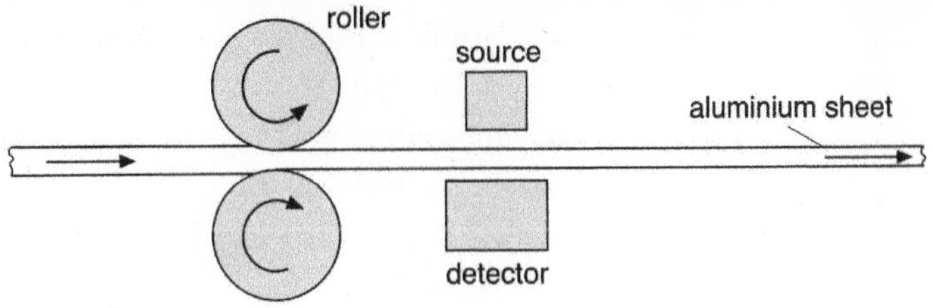

Figure 28.13 *Monitoring the thickness of sheet material.*

Tracer studies

A small amount of a radioactive material is mixed in with the material to be studied. The radioactive material can subsequently be traced or exactly located, because of the radiation it emits, i.e. it acts as a tag or label.

Apart from their use in medical diagnosis, e.g. to locate blood clots and tumours, tracers are used:

- to find which parts of engines or tools are being worn away;
- to trace leaks in pipelines underground; this is especially important in the oil industry;
- to zero in on what is happening in different stages of a chemical process. Here, the radio-isotopes act as 'walkie-talkies'. Tracer analyses are more accurate than ordinary chemical analyses. Why?

Other uses

- To study friction in car engines, piston rings are irradiated in a nuclear reactor and fitted into the engine. As the engine is run, radioactive metal is worn off the rings and gets into the engine oil. The amount of radioactivity in the oil gives an indication of the extent of wear.
- Radiations from radioactive sources can be used to ionize the air near production lines in factories, thus preventing the build-up of dangerous static charges.

Checklist

After studying Chapter 28 you should be able to:

- discuss the contributions of Becquerel and Marie Curie to the field of radioactivity
- recall that radioactivity is associated with nuclear instability
- recall the nature of the three types of emissions from radioactive substances, i.e. α and β-particles, and γ-rays
- compare the properties of these emissions

- predict the effects of magnetic and electric fields on the path of the emissions
- describe experiments to compare the ranges of the emissions in various media
- discuss changes which occur in the nucleus when radiation is emitted
- discuss methods of radiation detection
- represent and interpret nuclear reactions written in nuclide form
- describe an experiment to demonstrate radioactive decay's random nature
- discuss background radiation
- define the term 'half-life'
- use random decay graphs to show that random decay processes have constant half-lives
- recall and use the formula $A_0/A_n = 2^n$
- solve simple half-life problems
- recall that decay is independent of external conditions
- discuss the effects of radiation
- discuss the useful applications of radio-isotopes
- discuss the need for safety around radiation

Questions

1 A radio-isotope of radon (Rn) decays according to the equation:

$$^{222}_{86}\text{Rn} \rightarrow ^{M}_{L}\text{Po} + ^{N}_{2}\text{He}$$

What is the value of L, M and N?

2 You start at the same time with 2 g of radio-isotope X, half-life 8 hours, and 16 g of radio-isotope Y, half-life 4 hours. What mass of each isotope remains after 1 day?

3 Copy and complete the following nuclear reactions:

a $^{28}_{14}\text{Si} + ^{4}_{2}\text{He} \rightarrow ?$

b $^{207}_{82}\text{Pb} + ? \rightarrow ^{208}_{82}\text{Pb}$

c $^{123}_{53}\text{I} + \gamma\text{-ray} \rightarrow ?$

d $^{243}_{95}\text{Am} + \alpha + \beta \rightarrow ?$

e $^{238}_{92}\text{U} - ? \rightarrow ^{234}_{90}\text{Tb}$

4 Explain why the nuclei of some atoms are stable, while the nuclei of other atoms are unstable.

5 Explain what is meant by the following:
a radioactive decay is a random process,
b background radiation,
c the half-life of a radio-isotope.

6 $^{140}_{54}\text{X}$ is a beta-emitter.
a What is meant by the term 'a beta-emitter'?
b What is the atomic number and the mass number of the nuclide formed when $^{140}_{54}\text{X}$ decays?

7 Discuss three major uses of radio-isotopes.

8 Why are gamma rays not affected by either magnetic or electric fields?

9 Using a tabular format, compare α and β-particles under the following headings: range, ionizing power, effect on a photographic plate, and possible damage to biological tissue.

10 List three precautions that should be taken when handling radioactive sources. Why are these precautions necessary?

Figure 29.1 *Albert Einstein (1879–1955, humanist and outstanding theoretical physicist). He was awarded the Nobel Prize in Physics in 1921 for work on the photoelectric effect.*

Note

The **mass–energy equation**, $E = mc^2$, is considered the most 'famous' equation of the twentieth century. It was formulated by Einstein in 1905.

In 1919, Rutherford fired alpha particles at ordinary nitrogen nuclei and obtained oxygen-17 nuclei and protons as product particles.

$$^{14}_{7}\text{N} + ^{4}_{2}\text{He} \rightarrow ^{17}_{8}\text{O} + ^{1}_{1}\text{p}$$

This 'discovery' meant that 'man could get inside the atomic nucleus and play with it if he could find the right projectile'. Since that time many other types of nuclear reaction have been achieved.

Energy – called **nuclear energy** – is released whenever nuclear reactions take place. This energy may be used for both peaceful and destructive purposes.

29.1 The mass–energy balance sheet

The product particles in all nuclear reactions have a total mass which is less than the sum of the masses of the original (starting) nuclei or particles. Nuclear reactions, then, always result in a 'mass deficit'.

The energy (ΔE) released in a nuclear reaction is related to the mass deficit (Δm) by Einstein's equation:

$$\Delta E = \Delta m \times c^2$$

where c = the speed of light = $3 \times 10^8 \, \text{m s}^{-1}$.

29.2 Nuclear fission

During the 1930s many researchers fired neutrons at uranium nuclei. There were conflicting interpretations of the results of these experiments, until Hahn and others set things on the right course. A paper entitled 'A new type of nuclear reaction' appeared in the scientific journal *Nature* on 16 January 1939, in which the authors said '. . . It seems possible that the uranium nucleus has only a small stability of form (*i.e. it is unstable*) and may, after neutron capture, divide itself into two nuclei of roughly equal size . . .'. The authors were talking of **nuclear fission**.

Uranium-235 is the only naturally occurring 'fissionable' nucleus. The fission of uranium-235 can be represented as follows:

$$^{235}_{92}\text{U} + ^{1}_{0}\text{n} \rightarrow ^{236}_{92}\text{U} \rightarrow \text{X} + \text{Y} + 2 \text{ or } 3 \, ^{1}_{0}\text{n} + \text{energy}$$

Note

Energy locked 'up' in the nucleus is released on a large scale during nuclear fission and nuclear fusion reactions.

Note

Fission fragments have very long half-lives. They pose serious threats to health unless carefully stored.

where X and Y are highly radioactive fission fragments. These fragments have mass numbers in the range 85–105 and 130–150 respectively.

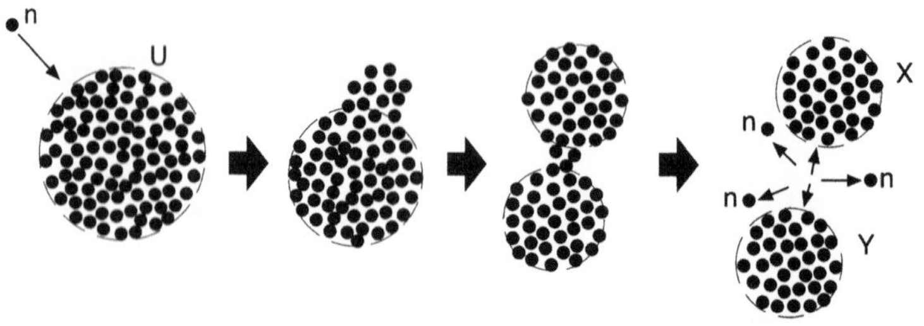

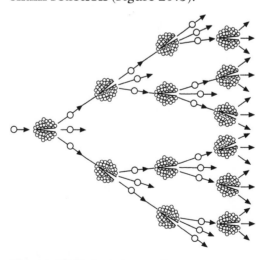

Figure 29.2 *Fission of a uranium nucleus.*

Note

Fission fragments are all β-emitters. Their atomic numbers change with time but their mass numbers remain unchanged.

The steps in the fission of uranium-235 may be summarized as follows (Figure 29.2):

- a slow neutron hits a uranium-235 nucleus and is absorbed (captured);
- the uranium nucleus becomes less stable and loses its shape;
- the unstable intermediate nucleus splits into radioactive fragments and two or three neutrons are released;
- energy is released.

The fission fragments fly apart at high speeds.

Chain reaction

The neutrons ejected in each fission may be captured by other nearby uranium atoms. This leads to more fissions. The number of neutrons builds up in a short time and so too does the energy released. This kind of self-sustaining fission reaction is called a **chain reaction** (Figure 29.3).

Figure 29.3 *A chain reaction.*

The energy released in chain reactions may reach explosive proportions. Chain reactions may be controlled, as in nuclear reactors for the generation of electricity, or uncontrolled, as in the explosion of a fission bomb.

Consider the hypothetical situation in which:

- each fission reaction releases two neutrons;
- sufficient uranium is present;
- there is no loss of neutrons.

Starting with one neutron, 2^4 neutrons will be present after 4 fissions and 2^{80} neutrons will be present after 80 fissions, i.e. sufficient neutrons, theoretically, to bring about the fission of all the atoms present in 1 mole of uranium. The whole process, lasting a mere fraction of a second, will result in a massive explosion.

If the mass of uranium is too small, too many neutrons will escape from the surface and the chain reaction will fizzle out. The smallest mass of uranium that will produce a self-sustaining chain reaction is called the **critical mass**. A fission bomb is produced when two masses of fissionable material smaller than critical mass, are brought together to exceed critical value.

Example

The Sun produces energy as a result of the following fusion reaction:

$$4\,_1^1\text{H} \rightarrow\ _2^4\text{He} + 2\,_1^0\text{e}$$

Calculate:

a the mass deficit in kg and u.

b the energy released in J and MeV.

$$\text{Mass of } _1^1\text{H} = 1.008\,13\,\text{u}$$
$$\text{Mass of } _2^4\text{He} = 4.003\,87\,\text{u}$$
$$\text{Mass of } _1^0\text{e} = \text{negligible}$$

Solution

a Mass deficit Δm = mass of reactants − mass of products
$$= (4 \times 1.008\,13) - (4.003\,87)$$
$$= 4.032\,52 - 4.003\,87$$
$$\Delta m = 0.028\,65\,\text{u}$$
$$1\,\text{u} = 1.66 \times 10^{-27}\,\text{kg}$$
$$\text{so: } \Delta m = 0.028\,65 \times 1.66 \times 10^{-27}$$
$$= 4.7559 \times 10^{-29}\,\text{kg}$$

Note

The unit of atomic mass, u, is one-twelfth the mass of a carbon atom.

$$1\,\text{u} = 1.66 \times 10^{-27}\,\text{kg}$$

The electronvolt (eV) is a unit of energy used in atomic and nuclear physics.

$$1\,\text{eV} = 1.6 \times 10^{-19}\,\text{J}$$

b Energy released $E = \Delta m c^2$

$$= 4.7559 \times 10^{-29} \times (3 \times 10^8)^2$$

$$= 4.28 \times 10^{-12}\,\text{J}$$

$$1\,\text{J} = \frac{1}{1.6 \times 10^{-19}}\,\text{eV}$$

so: $4.28 \times 10^{-12}\,\text{J} = \dfrac{4.28 \times 10^{-12}}{1.6 \times 10^{-19}}\,\text{eV}$

$$= 2.675 \times 10^7\,\text{eV}$$

$$= 26.75\,\text{MeV}$$

29.3 From uranium to electricity

Uranium exists naturally as two isotopes – uranium-235 and uranium-238 in the ratio 1 : 141. The element was used for hundreds of years to colour ceramics and glass but is now an important energy resource for producing electricity.

Whereas 1 kg of uranium can produce 60 000 kilowatt hours (kW h) of electricity, 1 kg of coal produces only 3 kW h and 1 kg of oil only 4 kW h.

The energy released by controlled fission of uranium or plutonium is converted to heat in nuclear reactors.

Nuclear reactors may also be used to produce neutrons for research or radio-isotopes for medical uses, agriculture and industry.

A nuclear reactor produces electricity as follows (Figure 29.4):

- slow neutrons are fired at enriched uranium fuel in the reactor core;
- these neutrons are captured, fission occurs and energy is released;
- coolant circulates through the core extracting heat;
- the coolant flows through heat exchangers and the heat extracted is used to produce steam;
- the steam turns turbines which rotate generators: electricity is generated and fed into the national grid;
- the coolant, having given up its heat, is recycled through the core.

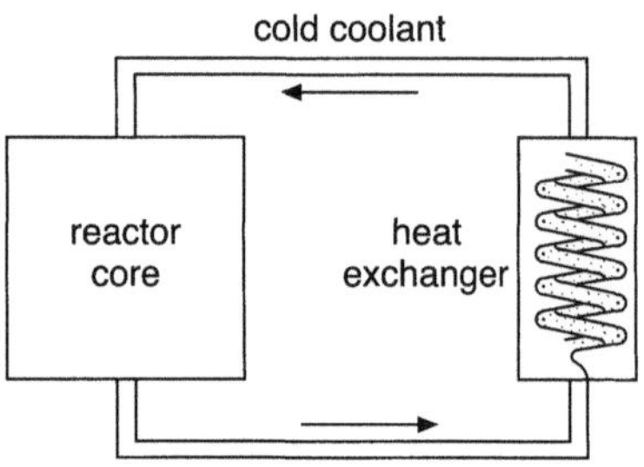

> **Note**
>
> The uranium used as a fuel is **enriched**. The ratio of uranium-235 to uranium-238 is increased to 1 : 140.

Figure 29.4 *The main parts of a nuclear reactor.*

> **Note**
>
> A heat exchanger is any device for transferring heat from one fluid to another.

> **Thinking it through**
>
> - Why are heat exchangers made of metal, with coiled tubes and fins?
> - What physical and chemical properties do you expect the ideal reactor coolant to have?

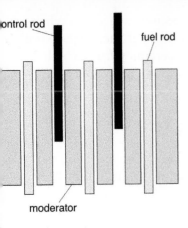

control rod

fuel rod

moderator

Figure 29.5 *The reactor core.*

There are many different designs for reactor cores. The efficiency of the reactor core depends on its operating temperature.

Figure 29.5 shows a close-up of the reactor core.

- The enriched uranium is stored as **fuel rods** inside the core.
- Neutrons are slowed down, but not absorbed, by the graphite **moderator**.
- The slowed-down or thermal neutrons are easily absorbed by the uranium in the fuel rods.
- To avoid the build-up of an uncontrolled chain reaction, boron **control rods** are used to absorb some of the neutrons produced in the fission reaction. These control rods are raised or lowered as required to maintain a steady rate of fission.

The pros and cons of nuclear power

Arguments for

- The reserves of uranium, although not limitless, are many times those of the fossil fuels.
- Uranium produces many times more energy than the same mass of any fossil fuel.
- It does not produce 'greenhouse' gases (p. 169) as burning fossil fuels does.

Arguments against

- High levels of radioactivity are associated with nuclear wastes.
- There are fears that the construction requirement standards for nuclear reactors are not high enough to ensure safety. Runaway fission reactions could have consequences on the scale of the Chernobyl disaster or worse.
- Workers at nuclear plants may be damaged by radiation.
- Materials used to cool nuclear reactors may cause thermal and other pollution.

Radioactive wastes are produced in the mining of uranium and at all stages in the conversion of the uranium to electricity. The big challenge is, how do we keep these wastes isolated for thousands of years?

If radioactive wastes are liquefied and stored in steel containers the following problems arise:

- how to provide continuous cooling for the system, since the fission products generate heat;
- how to get rid of any radioactive gases produced.

There are proposals for scattering the wastes in space. But why should space be littered in this way? The most attractive proposal of all involves 'glassifying' the wastes in ceramics, from which radioactive materials cannot leach. However, this will be very costly.

What happens if a reactor core overheats – meltdown

A meltdown is the most feared of nuclear accidents. It may be caused in the following way:

- A coolant pipe bursts and the emergency cooling system falls. This leads to a rise in temperature of the core.
- The nuclear fuel melts and hot radioactive gases fill the core.
- The pressure vessel and containment dome are burnt through.
- Radioactive materials are thrown kilometres into the air and hot radioactive materials burn into the ground.

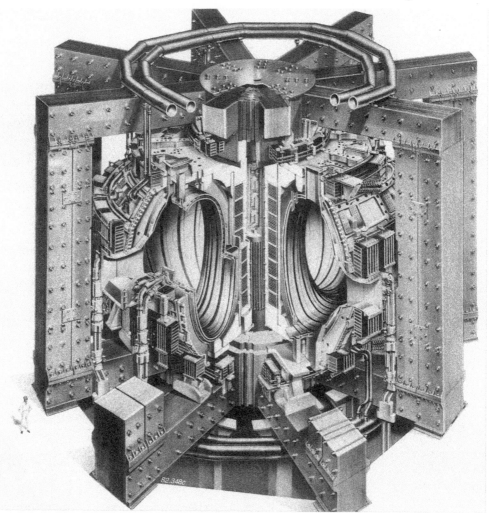

Figure 29.6 *A cutaway of the Tokamak, a doughnut-shaped reactor for producing energy from the fusion of two isotopes of hydrogen.*

Note

Fusion power is today's dream. We are searching for the means to make it tomorrow's reality.

Note

Lasers have been used on a small scale to bring hydrogen nuclei together.

Note

The overall biological hazard from fusion is low. Fusion plants may be built near centres of high population density.

Note

At age 16, Einstein gave up his German citizenship and became a citizen of the world.

29.4 Nuclear fusion

Fusion is the process by which light nuclei are brought together with so much energy that the repulsion between them is overcome. They fuse to become a new single nucleus with the release of vast amounts of energy. The Sun and other stars produce energy in this way – when hydrogen nuclei fuse to form helium nuclei.

People first achieved an uncontrolled fusion reaction in 1952, when they exploded a hydrogen bomb and wiped an island in the Pacific off the map. This bomb had about 500 times the explosive power of the bomb which destroyed Hiroshima.

Unfortunately, we have yet to produce fusion energy in a controlled manner on a large scale and make it work for our benefit. For successful fusion a temperature of about $1 \times 10^8\,°C$ is required. Further, there must be an effective means of confining the hot hydrogen gas. Neither of these problems has yet been solved completely.

One possibility for fusion is the reaction between two isotopes of hydrogen, deuterium (D) and tritium (T):

$$[D]^2_1H + [T]^3_1H \rightarrow {}^1_0n + {}^4_2He + \text{fusion energy}$$

While much energy is needed to start fusion reactions, the energy released during the reaction will be thousands of times greater than the starting energy.

Fusion is attractive because:

- there is no fear of runaway reactors, as is the case with fission;
- deuterium, the major fuel, is both cheap and readily available;
- there is little or no radioactive waste from fusion.

29.5 Einstein

The years 1895 to 1940 were the most exciting in the history of the physical sciences. During this 'golden age' many brilliant and dedicated scientists worked to uncover the laws governing the ultimate structure of matter. Of these, Albert Einstein is the best remembered and the most revered. His revolutionary thinking changed how we view our physical world. He was concerned all his life with providing better explanations for the way things work.

In 1905 Einstein published papers in different areas of physics:

- He satisfactorily explained Brownian motion.
- He used Planck's quantum theory to explain the photoelectric effect – the emission of electrons from metal

Note

Einstein was a 'theoretical' physicist. He left it to others to experiment and validate the predictions of his 'thought experiments'.

surfaces which have been exposed to light or ultraviolet radiation. His explanation of the photoelectric effect provided a platform on which much of later physics was built.

- In a third paper entitles 'On the electrodynamics of moving bodies', he published the first part of his theory of relativity, commonly called the 'Special Theory of Relativity'. The second part of the theory of relativity, the 'General Theory of Relativity', was published in 1915.

Things to do

Using the information provided, together with additional research material, write a paper on 'Einstein and his contributions to the scientific world'.

Checklist

After studying Chapter 29 you should be able to:

- associate the release of energy in a nuclear reaction with a change in mass
- recall and use Einstein's equation $\Delta E = \Delta mc^2$
- define nuclear fission
- discuss the steps involved in nuclear fission
- discuss chain reactions
- discuss the use of uranium as an energy source
- define and discuss nuclear fusion
- discuss the advantages and disadvantages of the use of nuclear energy

Questions

1 What equation relates energy to mass, and who was responsible for its development?

2 Define mass deficit.

3 Compare and contrast nuclear fission and nuclear fusion.

4 Calculate the energy released as a result of the following nuclear reactions (see table).

a $^{14}_{7}\text{N} + ^{4}_{2}\text{He} \rightarrow ^{17}_{8}\text{O} + ^{1}_{1}\text{p}$

b $^{2}_{1}\text{H} + ^{2}_{1}\text{H} \rightarrow ^{4}_{2}\text{He}$

c $^{6}_{3}\text{Li} + ^{2}_{1}\text{H} \rightarrow 2 ^{4}_{2}\text{He}$

Nuclide	Relative molecular mass
$^{1}_{1}\text{p}$	1.007 59 u
$^{2}_{1}\text{H}$	2.014 74 u
$^{4}_{2}\text{He}$	4.003 87 u
$^{6}_{3}\text{Li}$	6.017 00 u
$^{14}_{7}\text{N}$	14.003 07 u
$^{17}_{8}\text{O}$	16.999 13 u
1 u = 1.66×10^{-27} kg	

F   Revision questions for Section F

Multiple-choice questions

Questions 1–3

A Dalton   B Thomson
C Chadwick   D Bohr

Which of the scientists A–D:

1 is credited with the discovery of the neutron?

2 proposed the 'plum pudding' model of the atom?

3 provided evidence that electrons occupy permissible levels (shells) in the atom?

4 Elements are placed in the Periodic Table on the basis of increasing:
A atomic number   B mass number
C atomic mass   D formula mass?

Questions 5–7

	Relative charge	Relative mass
A	0	0
B	0	1
C	−1	0
D	1	1

Which of A–D is:

5 a γ-ray photon?

6 a neutron?

7 an electron?

Questions 8–10

A groups   B periods
C isotopes   D nuclides

Match labels A–D with the descriptions in questions 8–10.

8 Species which have the same number of protons but different number of neutrons.

9 Species which have the same number of electrons in their outer shell.

10 Species characterized by their atomic number, mass number and symbol.

Questions 11 and 12

A number of protons
B number of protons plus number of electrons
C number of neutrons
D number of protons plus number of neutrons

Which of A–D:

11 is always the same as the atomic number?

12 is the nucleon number?

13 Sodium-24 emits an α-particle. What are the atomic number and the mass number of the resulting nucleus?

	A	B	C	D
Atomic number	9	10	11	12
Mass number	20	20	24	24

14 α-decay and β-decay are respectively the emission from an unstable nucleus of:

	α-decay	β-decay
A	an electron	a proton
B	a neutron	a proton
C	a helium nucleus	an electron
D	a proton	a helium nucleus.

15 The initial activity of a sample of a radio-isotope is 96 Bq. After 60 hours the activity falls to 6 Bq. What is the half-life of the radio-isotope?
A 40 h   B 30 h   C 15 h   D 10 h

Questions 16–20

A Curie   B Becquerel
C Einstein   D Rutherford

Which of A–D:

16 discovered radioactivity?

17 received Nobel Prizes in both Physics and Chemistry?

18 did significant work in the field of radioactivity?

19 proposed a planetary model for the atom?

20 related mass deficit in nuclear reactions to the amount of energy released?

Questions 21 and 22

A genetic mutations
B nuclear fission
C nuclear fusion
D radioactive disintegration

Select from the options A–D the one which best fits the descriptions in questions 21 and 22.

21 The emission of ionizing radiations from unstable nuclei.

22 The process in which energy is released when small atomic species join to form larger ones.

Questions 23–25

	Relative atomic mass	Relative penetrating power	Deflected by a magnetic field
I	4	low	yes
II	0	moderate	yes
III	1	high	no
IV	0	high	mo

A I and IV only   B II and III only
C III only   D I only

23 Which of I–IV described above is a neutral particle?

24 Which of I–IV is likely to be an electron?
     A I   B II   C III   D IV

25 Which of I–IV is likely to be an α-particle?
     A I   B II   C III   D IV

26 What energy, in kilojoules, is released in a nuclear reaction in which the mass deficit is 10^{-13} kg?
     A 3   B 9   C 30   D 900

27 Gamma (γ) rays are:
     A high frequency electromagnetic radiation
     B high speed electrons
     C helium atoms which have lost their electrons
     D radiations with wavelengths of about 2 metres.

28 The chemical properties of an element are determined by its:
     A atomic number
     B mass number
     C number of neutrons
     D number of isotopes.

Questions 29–32

The graph shows how the activity of a radioactive source varies with time.

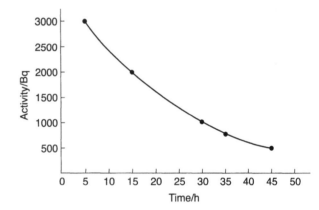

29 What is the most likely value for the initial activity of the source?
     A 2500 Bq   B 3500 Bq
     C 4000 Bq   D 4500 Bq

30 What is the half-life of the source?
     A 10 h   B 15 h   C 30 h   D 45 h

31 What is the activity of the source after 40 hours?
     A 3500 Bq   B 700 Bq
     C 600 Bq   D 500 Bq

32 After what time will the activity be 1500 Bq?
     A 21 h   B 17 h   C 15 h   D 11 h

Structured and free-response questions

1 What is meant by:

a atomic number?

b mass number?

A nuclide has atomic number 15 and mass number 31.

c What is meant by the term nuclide?

d How many of each of the following particles are present in this nuclide?
(i) protons (ii) electrons (iii) neutrons

2 What does $^{138}_{56}$Ba mean?

3 **a** Outline the experiments which led Rutherford to propose the 'planetary model' of the atom.

b In what way(s) is Rutherford's 'planetary model' of the atom better than Thompson's 'plum pudding model'?

4 **a** Give a biographical sketch of Niels Bohr.

b What improvements did Bohr make to Rutherford's 'planetary model' of the atom? Why were such improvements necessary?

5 The distribution of electrons in the shells of krypton, atomic number 36, is 2, 8, 18, 8. Write the electron configuration for each of the following elements.

a calcium (atomic number 20)

b bromine (atomic number 35)

c sulphur (atomic number 16)

6 Copy and complete the table below.

Element	Atomic number	Mass number	Protons	Neutrons	Electrons
Carbon	6	14			6
Oxygen			8	8	8
Bismuth		209	83	126	
Gold	79	197		118	
Caesium		113		78	55

7 Write short notes on the contributions of Henri Becquerel, Marie Curie and Ernest Rutherford to the field of radioactivity.

8 What is the effect of ionizing radiation on a photographic film? What part did photographic plates (film) play in the 'accidental' discovery of radioactivity?

9 **a** *Radioactive decay* is a *random process*.

b Each radio-isotope has its own *half-life*.

c When measuring the *activity* of a radioactive source, make allowance for *background radiation*.

Explain the meaning of each term in italics in sentences **a–c** above.

10 Describe an analogue experiment to illustrate random processes. In your description, include:
- the readings to be taken
- a discussion of the use that you will make of these readings.

11 What are the main sources of background radiation?

12 Why are some nuclides radioactive?

13 Use a table to compare α-particles, β-particles and γ-radiation under the headings: relative charge, relative mass, relative ionizing power, relative penetrating power (range) and behaviour in electrical fields.

14 Draw a composite diagram to show the effect of a strong magnetic field on α, β and γ-radiations. Clearly indicate the direction of the magnetic field.

15 Given a pure β source, outline an experiment that you would carry out to show that β-particles are deflected in a magnetic field. Give a diagram of the experimental set-up.

16 Describe experiment(s) to compare the ranges of α and β-particles in such materials as paper and aluminium.

Appendix: School Based Assessment (SBA)

The SBA skills tested by CXC (through your physics teacher) are:

- Observation/Recording/Reporting (O/R/R);
- Manipulation/Measurement (M/M);
- Planning and Designing (P&D);
- Analysis and Interpretation (A&I).

Whereas O/R/R, M/M and P&D are tested under the CXC profile 'Experimental Skills', A&I is tested under the profile 'Use of Knowledge'.

Your teacher should assign a mark between 0 and 10 for each skill tested. For the purpose of reporting to CXC your teacher should formally assess each skill *at least twice* in each of your fourth and fifth form years.

What is CXC looking for when the skill O/R/R is tested?

Your ability to:

- select the observations most appropriate to the given experiment;
- record the results of all measurements accurately;
- present data in an acceptable format (e.g. diagrams, tables and graphs);
- write a complete, accurate, logically sequenced and concise report;
- report on, and cross-check, discrepant/ unexpected results or observations.

What is CXC looking for when the skill M/M is tested?

Your ability to:

- follow instructions (however, you are not expected to do so slavishly);
- set up standard laboratory or other apparatus to take measurements conveniently and accurately and make appropriate observations;
- use measuring instruments accurately and quickly;
- safely and competently set up electrical circuits (circuit diagrams may or may not be given).

What is CXC looking for when the skill P&D is tested?

Your ability to:

- formulate a hypothesis appropriate to the 'problem/question under study';
- identify the manipulated variable and responding variable as well as the variables that should be controlled;
- clearly indicate the readings to be taken and the observations to be made;
- outline how the readings/observations are to be used to test the hypothesis or solve the problem;
- logically sequence the steps in a feasible experimental procedure to test the hypothesis or solve the problem;
- modify the original procedure if difficulties are encountered or unexpected results are obtained;
- reformulate the original hypothesis, if necessary;
- discuss the limitations of the experimental procedure;
- identify sources of errors and suggest ways to reduce their effects.

What is CXC looking for when the skill A&I is tested?

Your ability to:

- calculate accurately;
- infer from data;
- make reasonable predictions from data (which may be given or obtained experimentally);
- use data to identify patterns;
- use data to identify relationships;

- identify sources of errors and ways of minimizing their effects;
- recognize the limitations of the experimental method used or of the data collected.

Sample SBA activities and possible mark schemes

1 To determine the refractive index of a block of transparent glass

Apparatus and materials

Rectangular block of glass, optical pins (4), sheets of duplicating paper, protractor, ruler.

The mark scheme

Three skills could be tested with this activity. These are O/R/R, M/M and A&I.

Observation/Recording/Reporting

- For results presented in acceptable format, e.g. in a table which is correctly headed
  award 2 marks
- For procedure which is accurately but concisely described (point form may be used) *award 2 marks*
- For the graphical display of data (a possible maximum 6 marks may be awarded):
  (a) Title of graph correctly given
  award 1 mark
  (b) Axes labelled/units stated
  award 1 mark
  (c) Scales chosen so that graph occupies at least 80% of graph space *award 1 mark*
  (e) Accurate plotting of points
  award 2 marks
  (f) For drawing line of best fit *award 1 mark*

  Total of O/R/R = 10 marks

Manipulation/Measurement

- For outline of glass block carefully traced *award 1 mark*
- For careful replacement of block after it has been removed *award 1 mark*

- For use of a pencil with a fine point
  award 1 mark
- For use of search pins/no parallax approach to locating images *award 2 marks*
- For neat ray tracing *award 2 marks*
- For accurate measurement of angles
  award 3 marks

  Total of M/M = 10 marks

Analysis and Interpretation

- For accurate determination of the gradient $\sin i$ vs $\sin r$ graph; the answer being given to the same number of significant figures as the individual readings
  award 2 marks
- For recognizing that refractive index is a dimensionless quantity *award 1 mark*
- For use of maximum triangle in determining gradient *award 1 mark*
- For indicating on the graph the coordinates used to determine the gradient
  award 1 mark
- For recognizing the significance of the gradient of the graph *award 1 mark*
- For identifying the limitations of the experiment *award 1 mark*
- For identifying the sources of error
  award 2 marks
- For stating steps taken to reduce the effect of errors *award 1 mark*

  Total of A&I = 10 marks

2 To determine the specific heat capacity of the material of which a standard 100 g mass is made

Apparatus and materials

100 g mass, beaker with distilled water, plastic cup (calorimeter), Bunsen burner, thermometer, string.

The mark scheme

The skills M/M and A&I could be tested by this activity.

Manipulation/Measurement

- For carefully suspending the mass by a string and allowing it to remain in contact with boiling water long enough for its temperature to reach $100\,°C$ *award 1 mark*

- For rapid transfer of mass to water in calorimeter while not transferring drops of hot water   *award 2 marks*

- For correct use of measuring cylinder:
  - (a) cylinder placed on horizontal surface
  - (b) bottom of meniscus read
  - (c) parallax error avoided   *award 3 marks*

- For correct use of thermometer:
  - (a) bulb of thermometer completely immersed
  - (b) water (mixture) in plastic cup thoroughly stirred to ensure uniform temperature throughout
  - (c) bulb not rubbing against side of cup
  - (d) thermometer left long enough in water   *award 4 marks*

$$\textit{Total of M/M} = 10\ \textit{marks}$$

Analysis and Interpretation

- For identifying the underlying scientific principle – the principle of conservation of energy   *award 1 mark*

- For identifying the relevant formula   *award 1 mark*

- For making correct substitution into the formula   *award 1 mark*

- For showing all steps in the calculation   *award 2 marks*

- For giving the correct units   *award 1 mark*

- For quoting answer to the correct number of significant figures   *award 1 mark*

- For identifying not less than three sources of errors   *award 2 marks*

- For appreciating that the effects of the above errors may be minimized
      award 1 mark

$$\textit{Total of A\&I} = 10\ \textit{marks}$$

Other examples of activities that may be used to test O/R/R, M/M and A&I

3 To investigate the relationship between load and extension for a spiral spring

Apparatus and materials

Stand and clamp, spiral spring, scale pan, weights, optical pin to serve as pointer

Outline of skills

- Obtain values of load corresponding values of extension.
- Plot a graph of load versus extension.

4 To use a simple pendulum to determine a value for the acceleration due to gravity (*g*)

Apparatus and materials

Retort stand and clamp, string, Plasticine (to make a bob), metre rule, stop watch.

Outline of skills

- Time 20 swings for different lengths of the pendulum.
- Obtain corresponding time for one swing (T).
- Plot a graph of time (T) for one swing versus corresponding length of pendulum.
- Use the relationship $T^2 = 4\pi^2\, l/g$ to obtain a value for g.

5 To investigate the dependence of the resistance of electrical wires on (a) diameter (b) length (c) the type of material of which the wire is made

Apparatus and materials

Constantan, manganin and nichrome wires of different gauges, a multimeter with resistance measuring capability (alternatively an ammeter/voltmeter method may be used for determining the resistance of the wires).

Outline of skills

- Control both length and type of material when investigating the dependence of resistance on diameter.
  Control length and diameter when investigating the dependence of resistance on type of material of the wire.

6 To use dynamics trolleys and ticker timer to establish the relationship between force, mass and accleration

- Your teacher will provide the details.

7 To investigate the current–voltage characteristics of a filament lamp

Apparatus

Filament lamp, voltmeter, ammeter, power supply, rheostat.

Outline of skills

- Set up the relevant circuit (see the diagram below) with rheostat at a high setting.

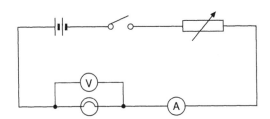

- Vary the rheostat setting and record the corresponding currents and voltages.
- Plot a graph of current (y-axis) versus voltage (x-axis).
- Comment on the graph.

8 To determine the 'melting point' of stearic acid and to investigate the effects of impurities on the melting point of stearic acid

Apparatus and materials

Boiling tube with solid stearic acid, thermometer, stirrer, timer.

Outline of skills

- Heat the stearic acid until it melts completely and insert the thermometer.
- Record temperatures, initially at half-minute intervals and then at 1 minute intervals.
- Plot a temperature/time graph.
- Use the flattened portion of the graph to determine the melting point of the acid.
- Discuss with your teacher how you will investigate the effects of different impurities on the melting point of stearic acid. (Will the change, if any, in the melting point of the acid depend on the amount of impurity added?)

Example of a completed activity

The activity: Using a balancing method to determine the mass of a metre rule

Apparatus and materials

Retort stand and clamp, knife edge, metre rule, masses (5 g–50 g), string.

Procedure

- The metre rule was placed horizontally on the knife edge.
- A 5 g mass was placed at the 5 cm mark.
- The metre rule was adjusted so that it balanced on the knife edge.
- The distance, l, between the point of suspension of the mass and the knife edge was recorded in a table.
- This procedure was repeated using nine other masses, all hung from the 5 cm mark.

Instructions to the students

- Plot a graph of mass (m) in kg against reciprocal length ($1/l$) in m^{-1}.
- Find the gradient, s, of the line.
- Find the mass of the metre rule given that the slope $s = 0.45m$.
- The mass of the rule is also equal to the value of the y-intercept.

The student drew the graph as shown on the next page.

TABLE OF
RESULTS

Mass/g	Mass/kg	L/cm	L/m	$\frac{1}{L}$/m^{-1}
5	0.005	44	0.44	2.273
10	0.010	42	0.42	2.381
15	0.015	41	0.41	2.439
20	0.020	40	0.40	2.500
25	0.025	38	0.38	2.632
30	0.030	37	0.37	2.703
35	0.035	36	0.36	2.778
40	0.040	35	0.35	2.857
45	0.045	34	0.34	2.941
50	0.050	33	0.33	3.030

Comments:

(1) Give the table a Title or heading

(2) Length can be measured with greater precision. The metre rule measures length to the nearest 0.1 centimetre

This lack of precision will affect your final answer.

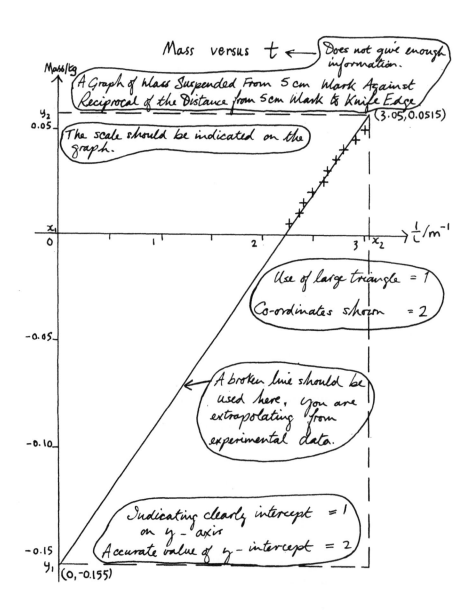

Mass versus $\frac{1}{L}$ ← Does not give enough information.

A Graph of Mass Suspended From 5cm Mark Against Reciprocal of the Distance from 5cm Mark to Knife Edge

The scale should be indicated on the graph.

(3.05, 0.0515)

y_2 0.05

x_1 0 ... 1 ... 2 ... 3 x_2 $\frac{1}{L}$/m^{-1}

Use of large triangle = 1
Co-ordinates shown = 2

A broken line should be used here, you are extrapolating from experimental data.

-0.05

-0.10

Indicating clearly intercept = 1 on y - axis
Accurate value of y - intercept = 2

-0.15
y_1 (0, -0.155)

The teacher assessed this activity for the skill Analysis and Interpretation, using the mark scheme summarized below. The teacher's corrections and comments are shown in the student's report.

The mark scheme for A&I

- For use of maximum triangle in determining the gradient   *award 1 mark*
- For clearly showing on the graph the coordinates $(x_1, y_1$ and $x_2, y_2)$ used in determining the gradient of the line
                                                                        award 2 marks
- For substituting correct values into the formula $s = 0.45m$ *award 1 mark*
- For accurate determination of the mass from the gradient   *award 2 marks*
- For giving answer to correct number of significant figures   *award 2 marks*

- For stating the correct units   *award 1 mark*
- For clearly showing the y-intercept on the graph   *award 2 marks*
- For accurately determining the value of the y-intercept   *award 2 marks*
- For giving the correct units and sig. figs. for the y-intercept   *award 2 marks*
- For correctly identifying relevent sources of errors   *award 3 marks*
- For discussion on action taken to reduce the effects of errors   *award 2 marks*

Total for A&I = 20 marks

The teacher will divide the student's score by 2 for CXC.

The student treated the results as follows.

Scale
    $4\,cm$ to $1\,m^{-1}$ on the x-axis
    $4\,cm$ to $0.05\,kg$ on the y-axis   *(This should be indicated on the graph.)*

Gradient of the line between $(0, -0.155)$ and $(3.05, 0.0515)$ ✓

$$s = \frac{0.0515 + 0.155}{3.05 - 0}$$

(Accurate determination of mass from gradient ① correct units ② significant figures ① Substituting correct values into formula ①.)

$$= 0.068\,Kgm^{-1}$$

→ $s = 0.45\,mass$

$0.068 = 0.45\,mass$ *(Final answer should not be given to a greater number of significant figures than the individual readings.)*
∴ Mass $= 0.150\,kg$

The intercept on y-axis $= 0.155\,kg$

∴ Mass $= 0.155\,kg$ *(Correct sig. fig. (units) ②)*

Discussion
  Sources of errors and action taken to reduce their effects
1. Metre rules with chips and dents will, in effect, be non-uniform. ✓   *(Identifying sources of errors – 2½)*
  (a) A metre rule with as few visible defects as possible was chosen. ✓
  (b) A small bit of plasticine was used to ensure that the metre rule balanced at the 50 cm graduation. ✓   *(For action etc. 1½)*

2. In experiments like this one, draughts and vibrations make it difficult to obtain the point of balance. ✓   *(Total of AEI = $\frac{16}{20}$)*
  (a) The experiment was carried out in a draught-free part of the lab. ✓   *(Mark for CKC purposes = $\frac{8}{10}$)*
  (b) The experiment was done during the lunch hour – at a time when there was minimum movement in the lab. ✓

3. A no parallax method used in determining the distance L above. ✓

Examples of activities that could be used to test the skill Planning and Designing

1 Design an experiment to determine the power rating of a candle.

2 Given a Bunsen burner, a beaker, a measuring cylinder, water, a stopwatch, a thermometer, design an experiment to determine a value for the specific latent heat of vaporization (L_v) of water.

3 Make careful observations of the way a golf ball (or ping pong ball) when released from different heights, bounces off the surface of the floor of the laboratory. Based on your observations (i) formulate a testable hypothesis (theory), (ii) design an experiment to test the hypothesis.

4 'Orange juice cools more rapidly than juices made from other citrus fruits', says the proprietor of the Village Snackette. As stated, this is not a testable hypothesis. (i) Reformulate the hypothesis. (ii) Design experiment(s) to test your reformulated hypothesis.

5 A model enthusiast wants to determine which surface is most suitable for racing his model cars. (i) Formulate an appropriate hypothesis. (ii) Design experiments to test the hypothesis.

6 A form-one student lives near a housing development. He notices that the masons always add steel rods when casting concrete. He asks why and is told that the steel rods reinforce the concrete, i.e. make it stronger. She wonders whether other metals could be used to reinforce concrete.

Design an experiment that will help the student to solve her problem. In your design pay attention to the meaning of strength and how the strength of concrete may be determined.

7 An 8-year-old boy observed how water emerged from a hole near the base of a plastic bottle. By filling the bottle to different heights with water he concluded that the rate at which water emerged from the hole depended on the height of the water in the bottle.

Design an experiment to test the boy's hypothesis.

Try it yourself!

On the following two pages is a student's attempt to test a hypothesis (or theory) concerning the behaviour of rubber bands. A teacher's mark scheme for assessing the student's work is given below. Carefully read the student's report and apply the teacher's mark scheme.

Teacher's mark scheme for Planning and Designing

- For identifying the underlying scientific principles (Hookes' law, etc.) *award 3 marks*
- For correctly identifying the variables involved:
  - (i) Manipulated – masses *award 1 marks*
  - (ii) Responding – the extension produced *award 1 marks*
  - (iii) Controlled – brand, batch, thickness, etc. of the rubber bands. *award 1 marks*
- For a logically sequenced procedure to test the hypothesis *award 3 marks*
- For outlining the readings/observations to test the hypothesis *award 2 marks*
- For clearly indicating how the readings/observations are to be used to test the hypothesis *award 3 marks*
- For a reasoned discussion on the limitations of the experimental procedure: *award 2 marks*
  - (a) sources of errors should be identified *award 2 marks*
  - (b) means of reducing the effects of the errors should also be identified *award 2 marks*

Total = 20 marks

OMARI FELIX
SBA 11
26/09/96

AIM
To design an experiment to test the hypothesis
''Rubber Bands Obey Hooke's Law''.

THE UNDERLINING PRINCIPLE
Hooke's Law states that provided the
stretching force does not extend a material
beyond its elastic limit, the extension of the
material is directly proportional to the
stretching force.
  In this experiment one has to decide if
equal increases in load lead to the same
change in the extension of the rubber bands
under test.

VARIABLES
Manipulated – Masses

Responding – Extension produced when force is
              applied

Controlled – Rubber bands of identical brand,
              type, size, thickness, weight,
              batch, etc.

APPARATUS
• Retort stand and clamp to hold up system.
• Rule held vertically in a clamp.
• Rubber bands as described above.
• A scale pan to hold the masses.
• A pin attached to the base of the scale pan
  to serve as a pointer.

DIAGRAM OF APPARATUS

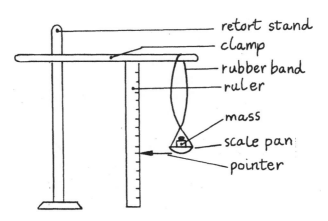

PROCEDURE
The apparatus should be set up as shown above
excluding the mass.
  The initial pointer reading should be
recorded in the table below.

Increasing masses should then be added to the scale pan and the final pointer readings should be recorded.

The differences in length should be recorded.

TABLE OF RESULTS

Weight/N	Final length/mm	Initial length/mm	Extension/mm

TREATMENT OF RESULTS

A graph of extension against force will be plotted [force ≡ weight ≡ mass × gravitational force]

If the graph is not linear in any way, then Hooke's Law is not obeyed.

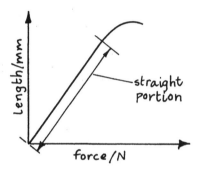

Results of the graph would be stated in the conclusion.

DISCUSSION AND LIMITATION

Parallax errors may have contributed to inaccuracies. The use of a straight pointer helped to eliminate this error.

If rubber bands were not identical, the lengths stretched for same forces may differ. See controlled variables for elimination of this error.

Rubber bands tend to be kinky, this may also vary the extension lengths of similar rubber bands. This was avoided by discarding kinky rubber bands.

Answers to numerical problems

Chapter 1

Page 6
Problem 1 a 1.5789×10^4
b 1.495×10^2 **c** 5.9×10^{-3}
d 7.02×10^{-2}
Problem 2 a 7.6×10
b 4.52×10^2

Page 7
Problem 1 a 2350 **b** 5.502
c 4802.1
d 466.82
Problem 2 a 2.35×10^3
b 5.502 **c** 4.8021×10^3
d 4.6682×10^2

Page 9
Margin question $\frac{5}{18}$
Problem 2 a $50\,\text{m}$ **b** $0.050\,\text{g}$
c $2500\,\text{nm}$

Pages 10–11
5 $7.0 \times 10^{-7}\,\text{m}$
6 a $1.7 \times 10^1\,\text{nm}$, conversion factor 10^9
b $1.7 \times 10^{-5}\,\text{mm}$, conversion factor 10^3
c $1.7 \times 10^{-11}\,\text{km}$, conversion factor 10^{-3}
9 $299\,792\,458\,\text{m s}^{-1}$
12 $3.0 \times 10^{-7}\,\text{m km}^{-1}$

Chapter 2

Page 12
Things to do 1 $223.5\,\text{mm}$ $(22.35\,\text{cm})$

Page 13
Things to do
1 $8.5\,\text{mm}$
Things to do
2 $72.5\,\text{mm}$ $(7.25\,\text{cm})$

Page 14
Things to do
1 $38.37\,\text{mm}$
Things to do
2 $29\,580\,\text{mm}^3$ (4 sig. fig.)

Page 17
Things to do 8.0–$9.0\,\text{cm}^2$

Page 18
Things to do i $19.8\,\text{mm}$
ii $4.70\,\text{mm}$ **a** $325.7\,\text{mm}^2$
b $341.8\,\text{mm}^3$

Page 19
Thinking it through $9\,\text{cm}^3$

Page 21
Things to do Helium $0.000\,17$, air 0.0012, gasoline 0.8, aluminium 2.7, steel 7.92, copper 8.9, glass 2.5, soft wood 0.5–0.6, gold 19.3
Problem i $0.800\,\text{g cm}^{-3}$
ii $800\,\text{kg m}^{-3}$

Pages 23–24
1 Top reading $1.56\,\text{mm}$, bottom reading $4.70\,\text{mm}$; Diameter $3.14\,\text{mm}$, area $31.0\,\text{mm}^2$, volume $16.2\,\text{mm}^3$
3 $1.64\,\text{N kg}^{-1}$
4 390–$430\,\text{km}^2$
6 Mass $6.5 \times 10^{-3}\,\text{kg}$
8 Reading (a) $203.5\,\text{mm}$ $(20.35\,\text{cm})$, reading (b) $23.5\,\text{mm}$ $(2.35\,\text{cm})$
a $88\,300\,\text{mm}^3$ (3 sig. fig.)
b $15\,900\,\text{mm}^2$ (3 sig. fig.)
c $4.83\,\text{g cm}^{-3}$
10 0.29
12 a $0.02\,\text{m}$ **b** $0.1\,\text{V}$
c $1.24\,\text{V m}^{-1}$
13 a i $1.96 \times 10^{-3}\,\text{m}^3$
ii $7860\,\text{kg m}^{-3}$ **b** $11.2\,\text{kg}$

Multiple-choice questions for Section A

Pages 34–35
1 C **2** A **3** B **4** D **5** D **6** D **7** B **8** D
9 C **10** D **11** D **12** C **13** B **14** C
15 A **16** D **17** B **18** B **19** D
20 A **21** D **22** B **23** B **24** B
25 A **26** D **27** C **28** C **29** A
30 C **31** A **32** C **33** D **34** C
35 B **36** B **37** A

Structured and free-response questions for Section A

Pages 36–37
1 a i 3 (in this case) **ii** 3
c $10.1\,\text{m s}^{-1}$
2 a 4.3×10^{-2} **b** 3.26×10^2
c 3.0×10^4
3 a 7.26 (7.26×10^0)
b 4.28×10^3
c 57.8 (5.78×10^1)
8 a $0.5\,\text{s}$ **b** 0–$60\,\text{s}$ **c** $36.5\,\text{s}$
13 $1.7 \times 10^{-3}\,\text{m}^3$ (2 sig. fig.)
19 a $0.454\,\text{kg}$ **b** $6.5 \times 10^2\,\text{cm}^3$, $6.5 \times 10^{-4}\,\text{m}^3$
c $7.0 \times 10^2\,\text{kg m}^{-3}$

Chapter 4

Page 41
Thinking it through
1 Force increases 16-fold

Page 42
Things to do
5 Weight $4\,\text{N}$, mass $0.41\,\text{kg}$

Page 46
Problem 2 $2.5\,\text{m}$

Page 53
Problem 1 a $312.5\,\text{J}$ **b** $281.25\,\text{J}$

Pages 54–55
1 b $510\,\text{N}$ at $4\,\text{m}$ mark, $260\,\text{N}$ at $1\,\text{m}$ mark
3 b $110\,\text{N}$ (2 sig. fig.)
4 $410\,\text{N}$ (2 sig. fig.)
5 d i 6.5 **ii** $11.7\,\text{cm}$
6 b $400\,\text{N}$
7 b i 4 **ii** 5.33 **iii** 75%

Chapter 5

Page 58
Problem 1 $7500\,\text{N}$ (2 sig. fig.)

Page 60
Problem a $80\,\text{km h}^{-1}$
b $22.2\,\text{m s}^{-1}$

Page 61
Problem $8\,\text{m s}^{-2}$

Page 64
Things to do a $0.12\,\text{s}$
c $8.75\,\text{m}\,\text{s}^{-1}$

Page 66
Problem a $21\,\text{m}\,\text{s}^{-1}$ **b** $6\,\text{m}\,\text{s}^{-2}$
c $63\,\text{m}$

Page 67
Problem 1 $1452.6\,\text{m}$
Problem 2 $28\,\text{m}$

Page 75
2 $30\,\text{m}\,\text{s}^{-1}$
3 $9\,\text{s}$, $202.5\,\text{m}$
5 $200\,\text{m}\,\text{s}^{-2}$
6 $5\,\text{m}\,\text{s}^{-2}$
7 a $0.25\,\text{m}\,\text{s}^{-1}$
b $0.50\,\text{m}\,\text{s}^{-1}$ **c** $0.05\,\text{m}\,\text{s}^{-1}$
8 $2.5\,\text{m}\,\text{s}^{-2}$, $1437.5\,\text{m}$
9 $8\,\text{m}\,\text{s}^{-2}$, $40\,\text{m}$
10 $12.5\,\text{m}\,\text{s}^{-1}$, $3.33\,\text{m}\,\text{s}^{-2}$, $41.7\,\text{m}$

Chapter 6

Page 81
Problem
a $4\,\text{m}\,\text{s}^{-2}$
b $12\,\text{m}\,\text{s}^{-1}$
d $42\,\text{m}$

Page 83
Problem Assign the left to right direction as positive
a $30\,\text{kg}\,\text{m}\,\text{s}^{-1}$ **b** $-10\,\text{kg}\,\text{m}\,\text{s}^{-1}$
c $20\,\text{kg}\,\text{m}\,\text{s}^{-1}$ **d** $8\,\text{kg}\,\text{m}\,\text{s}^{-1}$
e $12\,\text{kg}\,\text{m}\,\text{s}^{-1}$
f $2.4\,\text{m}\,\text{s}^{-1}$, left to right

Page 90
1 a $32.5\,\text{m}$ **b** $2.5\,\text{s}$
2 $2.5\,\text{m}\,\text{s}^{-2}$
3 a $50\,\text{m}\,\text{s}^{-2}$ **b** $45\,000\,\text{N}$
4 $5\,\text{m}\,\text{s}^{-1}$
5 $2\,\text{m}\,\text{s}^{-1}$
6 $0.05\,\text{m}\,\text{s}^{-1}$
7 a Y **b** both same **c** both same
8 Momentum increases four-fold, $3 \times$ initial momentum

Chapter 7

Page 97
Problem $2\,\text{m}\,\text{s}^{-1}$ (1 sig. fig.)

Page 99
1 b $80\,\text{m}\,\text{s}^{-1}$ **4** $400\,\text{W}$

5 a $78\,125\,\text{W}$ **b** $2500\,\text{kg}\,\text{s}^{-1}$
6 c 16.7%

Chapter 9

Page 116
Thinking it through
Part 4 $2600\,\text{m}^2$ (2 sig. fig.)

Pages 120–121
1 $10^5\,\text{N}$ downwards
3 $103\,920\,\text{Pa}$, $42\,\text{N}$
4 a $245\,\text{N}$
5 a $150\,\text{N}$ **b** $15.3\,\text{kg}$
c $0.0153\,\text{m}^3$
6 a $6.0 \times 10^{-4}\,\text{m}^3$
b $5.9\,\text{N}$ **c** $110\,\text{kg}\,\text{m}^{-3}$ (2 sig. fig.)
7 $10.2\,\text{m}$
8 $6120\,\text{kg}$
10 Before $14.7\,\text{N}$, after $11.0\,\text{N}$

Multiple-choice questions for Section B

Pages 122–125
1 D **2** B **3** B **4** C **5** D **6** C **7** B **8** B
9 A **10** A **11** A **12** D **13** B **14** D
15 C **16** D **17** A **18** C **19** B
20 B **21** B **22** D **23** B **24** C
25 A **26** C **27** A **28** B **29** C
30 C **31** A **32** B **33** C **34** D
35 A **36** B

Structured and free-response questions for Section B

Pages 125–130
3 c i $42\,\text{N}$ **ii** $8\,\text{N}$
4 a $1.5 \times 10^5\,\text{N}$ **b** $6 \times 10^4\,\text{kg}$
c $2.4 \times 10^5\,\text{N}$
5 a $362.25\,\text{kJ}$
c i $28\,350\,\text{kg}\,\text{m}\,\text{s}^{-1}$
ii $28\,350\,\text{kg}\,\text{m}\,\text{s}^{-1}$ **iii** $28\,350\,\text{N}$
iv $18\,\text{m}\,\text{s}^{-2}$
d i $12\,150\,\text{J}$ **ii** $60\,750\,\text{N}$
7 b i Momentum doubles
e $3.75\,\text{m}\,\text{s}^{-1}$
8 c ii $500\,\text{N}$
14 a $2 \times 10^5\,\text{J}$
b $1.5625 \times 10^4\,\text{J}$
17 a $1.2 \times 10^5\,\text{Pa}$
c ii $30\,\text{N}\,\text{cm}^{-2}$ **iii** $165\,\text{N}$
19 a i $0\,\text{m}\,\text{s}^{-1}$ **ii** $15\,\text{m}\,\text{s}^{-1}$

iii $15\,\text{m}\,\text{s}^{-1}$
iv $0\,\text{m}\,\text{s}^{-1}$ **c i** $3.75\,\text{m}\,\text{s}^{-1}$
ii $13.75\,\text{m}\,\text{s}^{-1}$ **iii** $11.4\,\text{m}\,\text{s}^{-1}$
d $1237.5\,\text{m}$
20 b ii initial speed/3
c i $40\,\text{m}\,\text{s}^{-1}$, $0.21\,\text{kg}$
ii $8.4\,\text{kg}\,\text{m}\,\text{s}^{-1}$
iii $-4.2\,\text{kg}\,\text{m}\,\text{s}^{-1}$
iv $12.6\,\text{kg}\,\text{m}\,\text{s}^{-1}$, $504\,\text{N}$
21 b i 25% **ii** $625\,\text{N}$

Chapter 10

Page 138
2 a i $50°\text{C}$ **ii** $60°\text{C}$ **b** $13.6\,\text{cm}$
3 $1064°\text{C}$
4 a $-6°\text{C}$ **b** $118°\text{C}$

Chapter 11

Page 144
1 a $45\,\text{cm}$ **b** $48°\text{C}$

Chapter 12

Page 154
1 $30\,\text{m}^3$

Chapter 13

Page 157
Problem $4400\,\text{J}\,\text{kg}\,\text{K}^{-1}$ (2 sig. fig.)

Page 158
Problem $50°\text{C}$

Page 164
1 a $4.32 \times 10^5\,\text{J}$ **b** $60\,\text{W}$
2 $33°\text{C}$
3 $3.36 \times 10^3\,\text{J}$
4 $4000\,\text{J}\,\text{kg}^{-1}\,\text{K}^{-1}$
5 a $327°\text{C}$ **c ii** $2.7 \times 10^5\,\text{J}\,\text{kg}^{-1}$
e $1300\,\text{J}$ **f** $80.7°\text{C}$

Multiple-choice questions for Section C

Pages 172–174
1 C **2** B **3** C **4** B **5** B **6** D **7** B
8 A **9** B **10** C **11** C **12** D **13** B
14 A **15** D **16** C **17** A **18** B
19 A **20** D **21** C **22** D **23** D
24 A **25** D **26** B **27** B **28** C

Structured and free-response questions for Section C

Pages 174–178
5 e 2940 J **f i** 2940 J **ii** 49 J
6 c 1.8 × 10⁵ J
d 5.6 × 10⁵ J kg⁻¹
7 e i 156 s **ii** 390 J kg⁻¹ K⁻¹
12 b i 4.8 MJ **c** 960 s
15 c 133 020 J

Chapter 15

Page 183
Thinking it through
Part 4 a T quadruples
b increases **c** decreases
Pages 188–189
1 0.20 m
2 a 2.0 cm **b** 1.0 cm s⁻¹ **c** 0.5 Hz
4 No sound
5 c 2.3 × 10⁷ Hz
6 a 1.6 × 10⁻⁵ s **b** 14.7 s

Chapter 16

Page 195
2 a 1 × 10¹⁰ Hz

Chapter 17

Page 202
1 225 m

Chapter 18

Page 209
1 5 m

Chapter 19

Page 213
Problem 1.33
Page 223
1 1.5 **2** 1.67 × 10⁸ m s⁻¹
3 a 20 cm **b** 80 cm **c** 16 cm
4 b 56.4°

Multiple-choice questions for Section D

Pages 224–227
1 B **2** D **3** C **4** D **5** A **6** B **7** A **8** B

9 B **10** D **11** A **12** B **13** A **14** D
15 A **16** B **17** D **18** B **19** B
20 C **21** A **22** A **23** B **24** D
25 A **26** B **27** C **28** B **29** B **30** C
31 B **32** A **33** C **34** B **35** B
36 A **37** B **38** A **39** D **40** A
41 A **42** D **43** C **44** D **45** B
46 C **47** B **48** A **49** B **50** C **51** B
52 D **53** A **54** A **55** C **56** B
57 B **58** A **59** B **60** A **61** C

Structured and free-response questions for Section D

Pages 228–231
2 a 300 m s⁻¹
3 b 10 cm
c 0.2 s **d** 5 Hz **e** 2 m
4 a 0.25 m **b** 0.375 m **c** 2 m s⁻¹
6 b 4 Hz
7 b 0.24 m **c** 1375 Hz **d** 1.38 m
11 a 3 m **b** 2.5 × 10⁹ Hz
18 1.33

Chapter 20

Page 236
Thinking it through F/4

Chapter 21

Page 247
Problem 500 s
Page 248
Problem 5.04 × 10⁷ J
Page 252
1 2.5 × 10¹⁹ electrons/s
2a 15.8 V **b** 44.7 V
3 0.46 A
4 90 A

Chapter 22

Page 259
Thinking it through
1 21 MΩ
2 6.5 kΩ
3 40 kΩ

Chapter 24

Page 285
Top problem 1 15.2 A, 14 kWh (units)
Top problem 2 4.5 kWh (units)
Margin problem 2 a 7.5 units
b 5 units **c** 6 units **d** 1.5 units
e 1.25 units

Page 289
1 20 A
2 a i 7623 units **ii** 5728 units
b 1895 units
c 6.8 × 10⁹ J
d $284.25
4 a 26.25 A
5 a 295 W **b** 3.54 kWh
c 0.296 kg

Chapter 26

Page 309
Problem a 50 V **b** 5 **c** 20 A
Page 310
Problem a Voltage 100 V, power 10 000 W **b** Voltage 40 V, power 400 W
Page 311
1 a 50 **b** 24 W **c** 2.5 A
3 b 0.75 W **c** 0.075 W (assuming 100% efficiency)
d 0.006 25 A **e** 12.5

Multiple-choice questions for Section E

Pages 312–319
1 D **2** A **3** A **4** C **5** A **6** C **7** C
8 D **9** C **10** B **11** B **12** C **13** B
14 D **15** C **16** B **17** B **18** A
19 B **20** C **21** B **22** B **23** A
24 D **25** D **26** C **27** D **28** C
29 A **30** C **31** D **32** C **33** B
34 C **35** C **36** A **37** B **38** A
39 B **40** D **41** A **42** B **43** D
44 D **45** D **46** B **47** A **48** D
49 D **50** B **51** C **52** A **53** D
54 D **55** B **56** C **57** C **58** C **59** D
60 B **61** A **62** D **63** B **64** A
65 C **66** B **67** B **68** B **69** B **70** C
71 D **72** B **73** D **74** A **75** C
76 A **77** D **78** B **79** A **80** D
81 C **82** C **83** D **84** B **85** B **86** C

Multiple-choice questions for Section E (*cont.*)

Pages 312–319 (*cont.*)
87 A **88** B **89** B **90** C **91** B **92** B
93 B **94** C **95** C **96** C

Structured and free-response questions for Section E

Pages 320–325
1 a coulomb **b** newton **c i** $F \times 2$
ii $F/\sqrt{2}$
2 a i $300\,\Omega$ **ii** $0.04\,A$ **iii** $4\,V$
b i $200\,\Omega$ **ii** $0.06\,A$
9 d $2.47\,M\Omega$ **e** $68\,\Omega$
f $360\,\Omega$, **7 g** $490\,\Omega$
10 a i $10\,\Omega$ **ii** $2\,A$ **iii** $1.33\,A$
iv $4\,V$ **v** $10.67\,W$ **b i** $14\,\Omega$
ii $1.43\,A$
11 a $8660\,kg\,m^{-3}$
b i $7.2\,g\,cm^{-3}$
14 a i $1.667\,A$ **ii** $5.4\,\Omega$ **b** $1.11\,A$
c $3.6\,\Omega$
16 a $10\,\Omega$ **b** $0.15\,A$, $16.67\,\Omega$
c $0.40\,A$ **d** $6.25\,\Omega$
18 c $720\,\Omega$ **d** $0.5\,A$ **e** $0.167\,A$
39 c $120{:}22$
40 a i $2.0 \times 10^{8}\,W$
ii $3.3 \times 10^{4}\,V$ **f** $8.64 \times 10^{11}\,J$
g $2.3 \times 10^{4}\,m^{3}$

Chapter 27

Page 328
Thinking it through $2,8,3$

Page 330
Thinking it through
1 83 neutrons

Page 332
Things to do
1 $Z = 54$
2 Xe
3 group 8, period 5

Page 332
1 a 32 **b** 15 **c** P
2 126

Chapter 28

Page 336
Thinking it through
Part 1 Ra mass number
$A = 230$
Part 2 $Z = 86$
Thinking it through
Part 1 Ga atomic number
$Z = 31$
Part 2 $A = 140$

Page 341
Thinking it through 20.6
counts/min

Page 343
Thinking it through
Part 1 $90\,s$

Part 2 $85\,Bq$

Page 346
1 $L = 84$, $M = 218$, $N = 4$
2 X mass $0.25\,g$, Y mass $0.25\,g$
6 b $Z = 55$, $A = 140$

Chapter 29

Page 354
4 a $0.205\,MeV$ **b** $23.98\,MeV$
c $22.42\,MeV$

Multiple-choice questions for Section F

Pages 355–356 **1** C **2** B **3** D **4** A
5 A **6** B **7** C **8** C **9** A **10** D **11** A
12 D **13** D **14** C **15** C **16** B
17 A **18** A **19** D **20** C **21** D
22 C **23** C **24** B **25** D **26** B
27 A **28** A **29** C **30** B **31** B
32 A

Structured and free-response questions for Section F

Page 357 **1 d i** 15 **ii** 15 **iii** 16
5 a $2, 8, 8, 2$ **b** $2, 8, 18, 7$
c $2, 8, 6$

Index